LIVING WITH CLIMATE CHANGE

How Communities Are Surviving and Thriving in a Changing Climate

LIVING WITH CLIMATE CHANGE

How Communities Are Surviving and Thriving in a Changing Climate

Jane A. Bullock • George D. Haddow
Kim S. Haddow • Damon P. Coppola

CRC Press
Taylor & Francis Group
Boca Raton London New York

CRC Press is an imprint of the
Taylor & Francis Group, an **informa** business
AN AUERBACH BOOK

Cover photo, center row, center photo:

Napa River Dry Bypass—Napa River Flood Protection. US EPA. http://www.epa.gov/region9/mediacenter/napariver/gallery/tour4.jpg

CRC Press
Taylor & Francis Group
6000 Broken Sound Parkway NW, Suite 300
Boca Raton, FL 33487-2742

Version Date: 20150805

International Standard Book Number-13: 978-1-4987-2536-1 (Paperback)

Visit the Taylor & Francis Web site at
http://www.taylorandfrancis.com

and the CRC Press Web site at
http://www.crcpress.com

We dedicate this book to Dave Dickson whose untimely passing has robbed us all of a great leader in disaster risk reduction.

Contents

Acknowledgments

The authors acknowledge the following individuals whose contributions have helped to make this book possible: Mayor Tom Bates, Ron D. Brunner, Karen Carney, Arrietta Chakos, Brian Cowan, Mayor Frank Cownie, Josh Foster, Mayor George Heartwell, Eric Holdeman, Crystal Kline, Cooper Martin, John R. Nordgren, Megan O'Grady, Satya Rhodes-Conway, Alexis St. Juliana, James Schwab, Missy Stults, and Jason M. Vogel. Their stories are the heart of this book and their dedication to protecting their communities and fellow citizens serves as a lesson to us all.

We also thank Jennifer Abbott at Taylor & Francis for the assistance she provided in the development of this book.

Authors

Jane A. Bullock is a former research scientist and adjunct professor at the Institute for Crisis, Disaster and Risk Management at The George Washington University in Washington, DC. She is a principal in Bullock & Haddow LLC, a homeland security and disaster management consulting firm. She worked for 22 years at the Federal Emergency Management Agency (FEMA) where she served as chief of staff to FEMA Director James Lee Witt. She is a coauthor of numerous university-level textbooks, including *Introduction to Emergency Management* (Butterworth-Heinemann, 2013) and *Introduction to Homeland* Security (Butterworth-Heinemann, 2015).

Damon P. Coppola is an emergency management systems engineer and partner at Bullock & Haddow LLC. Coppola provides technical assistance to governments at the federal, state, and local levels, international organizations, foreign governments, nongovernmental organizations, and the private sector. He is the author of *Introduction to International Disaster Management* (Butterworth-Heinemann, 2015), and coauthor of several professional and academic texts, including *Introduction to Emergency Management* (Butterworth-Heinemann, 2013), *Introduction to Homeland Security* (Butterworth-Heinemann, 2015), and *Communicating Emergency Preparedness: Strategies for Creating a Disaster Resilient Public* (CRC Press, 2009).

George D. Haddow is a principal in Bullock & Haddow LLC, a homeland security and disaster management consulting firm. Haddow currently serves on the adjunct faculty in the Homeland Security Studies Program and the Disaster Resilience Leadership Academy (DRLA) at Tulane University in New Orleans, Louisiana. He is the coauthor of several university and professional textbooks, including *Introduction to Emergency Management*, Fifth Edition (Butterworth-Heinemann, 2013), *Introduction to Homeland Security*, Fourth Edition (Butterworth-Heinemann 2015), and *Disaster Communications in a Changing Media World*, Second Edition (Butterworth-Heinemann, 2014). Haddow is the former deputy chief of staff to James Lee Witt during his tenure as director of the US Federal Emergency Management Agency (FEMA). At FEMA, Haddow was responsible for policy formulation in the areas of disaster response and recovery, public/private partnerships, public information, environmental protection, and disaster mitigation.

Kim S. Haddow is the president of Haddow Communications, in New Orleans, Louisiana—a company specializing in strategic media planning, messaging, and developing research-driven media content, branding, and advertising materials for nonprofits. Clients have included the Rockefeller Family Fund, Sierra Club, Center for Media and Democracy, Make It Right Foundation, US State Department, Public Campaign, and the Trust for America's Health. Haddow worked as a media consultant at Greer, Margolis, Mitchell, Burns (GMMB), a Washington, DC–based firm, advising political campaigns and nonprofits. Haddow began her career at WWL–AM in New Orleans where she managed the news department.

Contributors

Ronald D. Brunner is a policy scientist and professor emeritus at the University of Colorado, Boulder, where he joined the faculty in 1981 and retired in 2007. A native of Colorado, he earned a BA and a PhD at Yale University. He specializes in the integration of theory and practice in the policy sciences. Working with policymakers at various levels of government, he has applied central theory to problems in energy, social welfare, space, education, natural resource, and climate change policy. His most recent books, coauthored, are *Finding Common Ground* (Yale University Press, 2002) and *Adaptive Governance* (Columbia University Press, 2005) on natural resource policy, and *Adaptive Governance and Climate Change* (American Meteorological Society, 2010).

Karen Carney, managing scientist at Stratus Consulting, is a terrestrial ecologist with more than 20 years of experience in various aspects of environmental science, including global climate change analysis, terrestrial ecosystem ecology, and biodiversity conservation. She currently leads a wide range of analytical efforts at Stratus Consulting, many of which are focused on understanding, mitigating, and adapting to the impacts of climate change on ecosystems. Dr. Carney's recent work includes analyses of (1) the impacts of climate change on ecosystem services in the United States, (2) the impacts of climate change on critical ecosystem threshold dynamics, and (3) the performance of forest management offset protocols developed for voluntary carbon markets. She has also been extensively involved in Stratus Consulting's evaluation practice, serving as both a team leader in real-time evaluations and a lead evaluator. Dr. Carney has also conducted academic research related to carbon cycling in natural ecosystems, including a study that demonstrated for the first time that elevated atmospheric carbon dioxide causes soils to lose carbon over time by enhancing microbial degradation of soil organic matter. Her research has also addressed how land-use change and plant diversity in tropical forests affect the amount of carbon stored by ecosystems, a factor that is critical to predicting and mitigating the contributions of land-use to climate change. Dr. Carney earned a PhD in geological and environmental sciences from Stanford University and a BA in biology from Kalamazoo College.

Arrietta Chakos is a public policy advisor specializing in community resilience and risk reduction. She confers with public sector institutions on practical ways to improve disaster and climate resilience, reliability of energy and public utility services, and financial resources for safer communities. She was assistant city manager in Berkeley, California, serving in the public sector for 18 years. Her position as research director at the Harvard Kennedy School's "Acting in Time" Advance Recovery Project focused on disaster recovery in American communities. The City and County of San Francisco, the Association of Bay Area Governments, City of Palo Alto, the California Energy Commission, and District of Columbia's Homeland Security and Emergency Management Agency are recent clients. As a member of the Resilient America Roundtable at the National Academy of Sciences, she advises on national-level resilience and disaster recovery policy and implementation. She is an appointee to the National Research Council's Committee to Advise the US Global Change Research Program mandated by Congress to "... to understand, assess, predict, and respond to human-induced and natural processes of global change." She has served as a technical advisor on panels for FEMA, GeoHazards International, the Organization for Economic Cooperation and Development (OECD), the World Bank, California's Office of Emergency Services, and the University of Pittsburgh Medical Center (UPMC) Center for Biosecurity.

Brian Cowan worked on programs for natural hazards management, principally in the Federal Emergency Management Agency, over a period of 25 years. He also directed for 10 years the implementation of a financial assistance program for firefighters that provided several billion dollars, through tens of thousands of competitive grants, directly to local fire departments.

David Dickson authored this case study for *Global Warming, Natural Hazards, and Emergency Management* (G.D. Haddow, J.A. Bullock, and K.S. Haddow, CRC Press, 2008) prior to his untimely death in June 2011. The *Napa Valley Register* described Dickson as "a major force behind community passage of the Napa flood control project in 1998." Prior to his death, Dickson served as a senior consultant to MIG, Inc., a California-based planning and design firm. Dickson consulted with the Federal Emergency Management Agency (FEMA), Environmental Protection Agency (EPA), Army Corps of Engineers, the University of California, and George Washington University in the areas of watershed management, restoration, disaster management, and financial planning. His public agency work included positions as chief financial officer of the Napa County Flood Control and Water Conservation District and community development director for the Napa County Administrator's Office. He was project manager for the Napa Valley's Living River Flood Management Plan—a comprehensive watershed-wide plan for flood damage reduction, river and watershed restoration, and economic revitalization in the City of Napa. He was the architect and manager of the Community Coalition planning process and the financing plan of this countywide effort that now totals over $500 million in public investment. He earned a BA in political science from San Diego State University and had completed master's-level coursework at the Public Finance Institute, University of California, Davis.

Josh Foster has more than 25 years of experience working on climate change science and policy issues in the federal and nonprofit sectors, including more than 19 years working on climate adaptation. He works with the Oregon Climate Change Research Institute (OCCRI) as a program manager for both the NOAA-funded Climate Impacts Research Consortium (CIRC) and the Department of Interior–funded regional Climate Science Center. Foster earned a BA in international relations and environmental policy from the University of Massachusetts at Amherst and MA/MES degrees from Yale University in international relations and in environmental management.

Cory Hamilton is a doctoral student at Eastern Michigan University, where he focuses on educational leadership and experiential education.

Eric Holdeman is a nationally known emergency manager. He has worked in emergency management with federal, state, and local governments. Today he serves as director, Center for Regional Disaster Resilience, which is part of the Pacific Northwest Economic Region (PNWER). In 2007, he was recognized by *Government Technology* magazine as one of the Top 25 people in the nation who "challenge convention, confront entrenched bureaucracy and promote innovation." He is also a consultant and a prolific writer, authoring numerous articles for professional journals and opinion pieces for local, regional, and national newspapers including the *Washington Post*. He is a writer for *Emergency Management* magazine where he contributes feature articles and also has a regular column, "Eric's Corner." An experienced and accomplished public speaker, he is sought after to present at national and regional conferences. Holdeman has the United States' most popular blog on the topics of emergency management and homeland security at www.disaster-zone.com.

Rachel Jacobson is a recent graduate focused on urban climate adaptation from the University of Michigan Ford School of Public Policy and the School of Natural Resources and Environment at the University of Michigan.

Crystal Kline is a charter member of Tulsa Project Impact and Citizen Corps, a member of the Disaster Resistant Business Council, and a member of the Board of Directors of Tulsa Partners, Inc. She is the owner and operator of CK Consulting LLC, a professional consulting firm specializing in disaster preparedness and planning. She is Oklahoma's first female federally certified Master Exercise Practitioner, and is an adjunct instructor at FEMA's Emergency Management Institute in Emmitsburg, Maryland. She has responded to multiple disasters throughout the years, and was a part of the team that designed and facilitated Community Emergency Response Team training in

Haiti as part of the Clinton Global Initiative. She is coauthor of the book *Disaster Preparedness: A Living Free Guide* published in 2014 by Penguin Publishing.

Cooper Martin is the program director for the Sustainable Cities Institute at the National League of Cities (NLC). The SCI program provides information, tools, and guidance to strengthen communities and enable them to thrive while facing the challenges presented by a changing climate and uncertain global economy. His areas of expertise include climate resilience, community development, transportation, and emergency management policy. Martin serves on the board of directors for The Infrastructure Security Partnership and the steering committee of the STAR Communities Index. Prior to joining the NLC, he held multiple positions in technical assistance, research, and advocacy at the American Institute of Architects and worked one session in the Kansas House of Representatives. Martin's educational background includes a bachelor's degree in political science and international studies from the University of Kansas and a master's degree in public policy from American University.

Sara Meerow is a PhD candidate at the University of Michigan's School of Natural Resources and Environment, where she focuses on urban resilience to climate change.

John R. Nordgren is a senior advisor for the Michigan-based Kresge Foundation, where his work is focused on advancing a national strategy for supporting community-based climate change adaptation. From 2009–2014 Nordgren was a senior program officer for Kresge's Environment Program, leading development and implementation of Kresge's strategies for advancing the field of climate change adaptation through place-based and field-building investments, which approached $50 million over that period. Nordgren serves on the National Academy of Sciences Board on Atmospheric Sciences and Climate (BASC) and has worked closely with the White House Council on Environmental Quality in support of the administration's Climate Adaptation Task Force. He is cochair of the Consultative Group on Biological Diversity's Conservation Science Program and a founding member of the board of directors for EcoAdapt, Inc. Nordgren served on the committee to advise the City of Chicago on the implementation of its Climate Adaptation Plan (2011), and has served for 5 years as a member of the advisory committee for the Wildlife Conservation Society's Climate Adaptation Fund. Prior to joining Kresge, John was a program officer for the Henry P. Kendall Foundation in Boston, where he led the foundation's Northeastern Landscape Conservation Program and developed the foundation's Climate Adaptation Program. He was previously director of programs for the Massachusetts-based Manomet Center and worked in the Legislative Program of the Natural Resources Defense Council in Washington, DC. Nordgren earned degrees in public policy from The American University in Washington, DC, and Tufts University in Medford, Massachusetts.

Megan O'Grady, senior associate at Stratus Consulting, specializes in resource management, policy, and climate change adaptation issues, working especially on scientist–stakeholder partnerships. Her work includes creation of decision-making and risk assessment tools, and project and program evaluation. O'Grady provides framing and analytical support to several adaptation projects, drawing on her extensive domestic and international experience working directly with a wide variety of experts and stakeholders to provide scientifically sound and user-friendly guidance around adaptation issues. Prior to joining Stratus Consulting, O'Grady was at Columbia University where she managed the New York City Panel on Climate Change, the technical advisory body for Mayor Bloomberg's Climate Change Adaptation Task Force. She coauthored, coordinated, and edited climate change adaptation publications of record for New York City; designed a three-dimensional risk assessment framework to address multiple layers of uncertainty; served as the NPCC representative to the 40 task force stakeholders; and organized a New York City knowledge network meeting with over 100 attendees. O'Grady earned her MPA in environmental policy from

Columbia University's School of International and Public Affairs. She earned a BA in environmental studies from Saint Olaf College.

Melisa Ongun is a recent graduate from the Masters Program in Sustainable Systems at the University of Michigan, where she focused on urban climate adaptation and environmental policy.

Ann Patton is a Tulsa-based writer and consultant specializing in disaster management, urban affairs, and grassroots partnership building. She has 45 years of experience in program management, journalism, consulting, and book publishing. She heads Ann Patton Company LLC, a professional writing and consulting firm, and APLcorps Books LLC. A former newspaper reporter, she has published three recent books: *The Tulsa River*, a coffee-table-style community book about the Arkansas River at Tulsa; *Dan's War on Poverty, A Grassroots Crusade for Social Justice*, focused on local poverty and social justice issues; and *Fifty Years Remembered*, Tulsa District, US Army Corps of Engineers, 1989, a coffee-table history of water resources in the Tulsa region (a 50-year history for the Tulsa District, US Army Corps of Engineers). Patton has also published hundreds of articles, technical reports, book chapters, and presentations, special reports on flood mitigation and tornado safety, and materials about building disaster-resistant communities.

For the City of Tulsa, she was founding director of Tulsa's Project Impact program, the local Citizens Corps, and Tulsa Partners Inc. As a consultant, she has worked for clients such as FEMA, the Department of Homeland Security, the Medical Reserve Corps, the National Wildlife Federation, and the National Mitigation Association's Resilient Neighbors Network.

Satya Rhodes-Conway is the managing director of the Mayors Innovation Project and a senior associate at the Center on Wisconsin Strategy (COWS). She works with cities across the country to implement innovative policy that promotes environmental and economic sustainability and builds strong, democratically accountable communities. She has researched and written extensively about local policy that promotes sustainability, equity, and democracy. Rhodes-Conway served three terms on Madison, Wisconsin's City Council, giving her a practical perspective on local government and policy. Before coming to COWS, she analyzed state endangered species programs for Defenders of Wildlife, researched and wrote about progressive environmental policy at the State Environmental Resource Center, and taught undergraduate biology and ecology. She earned degrees from Smith College and the University of California–Irvine.

James Schwab, AICP, is the manager of the Hazards Planning Center for the American Planning Association (APA). He has led numerous hazards-related training and research projects for APA over the past two decades, including both the original "Green Book," *Planning for Post-Disaster Recovery and Reconstruction* (PAS Report No. 483/484, 1998) and its second version, *Planning for Post-Disaster Recovery: Next Generation* (PAS Report No. 576, 2014), both underwritten by the Federal Emergency Management Agency (FEMA). He also managed APA's pioneering exploration of the integration of hazard mitigation into the local planning process, *Hazard Mitigation: Integrating Best Practices into Planning* (PAS Report No. 560, 2010), also with FEMA support. In addition, he has led APA's involvement in the NOAA Digital Coast Partnership since 2010, organized recovery training workshops in New York and New Jersey after Hurricane Sandy, and has worked on collaborative projects with other federal agencies including the Forest Service and US Geological Survey. In 2008, he served as a visiting fellow for the Centre for Advanced Engineering in New Zealand, consulting on national hazards policy. Schwab earned MA degrees in both journalism and urban and regional planning at the University of Iowa and a BA in political science at Cleveland State University.

Alexis St. Juliana is an environmental policy analyst with extensive experience assessing stratospheric ozone protection programs, and climate change adaptation, mitigation, and impacts. Her recent project work includes the assessment of options for mitigating black carbon emissions, evaluation of potential human health impacts from climate change, and appraisal of climate change adaptation actions at national and international levels. St. Juliana's projects have been conducted for a range of clients including state, national, and international public institutions, as well as foundations and private sector clients. Her work is incorporated in a number of government reports and peer-reviewed journal articles. St. Juliana earned an MPP from the University of Michigan and a BA in political science and Spanish from Indiana University.

Missy Stults is a science research fellow at the University of Michigan where she is simultaneously pursuing her doctoral degree on urban adaptation/resilience to climate change and providing consulting services to assist local governments around the United States in enhancing their resilience to existing and projected future changes in climate. As part of this work, Stults served as supporting author for the Adaptation Chapter and the Urban Technical Input of the 2014 US National Climate Assessment. In addition, Stults is currently working on a project to profile the state of community-based adaptation initiatives unfolding across the United States as well as another project to assess the types of services and resources available to support local and regional climate adaptation activities. Before attending the University of Michigan, Stults was the national climate director for ICLEI–Local Governments for Sustainability where she worked with more than 600 local governments to advance their climate mitigation, climate adaptation, and sustainability efforts. Stults earned her bachelor's degree in marine biology and environmental science from the University of New England and her master's degree in climate and society from Columbia University.

Jason M. Vogel, managing analyst at Stratus Consulting, has worked extensively on climate change adaptation issues and has experience in policy analysis, climate change impact and adaptation assessments, project and program evaluation, decision analysis, risk assessment, translating climate science for decision-making audiences, qualitative and quantitative methods of information elicitation, and more. He has used these skills for municipalities and water utilities by conducting science-based analyses, translating climate change science, evaluating and conducting vulnerability assessments, and providing strategy and policy support. Dr. Vogel has also used these skills to bring climate resilience to bear on development planning. This includes producing guidance, vulnerability assessments, policy support, and capacity building to support international financing institutions and development banks. He has also worked with institutions such as the World Bank, the Global Environment Facility, and the Green Climate Fund to provide high-level advice and support for their climate adaptation efforts. Dr. Vogel has also worked extensively on evaluating the performance of nonprofit organizations and foundations. He was trained as an atmospheric scientist and chemist before turning his attention to public policy. Dr. Vogel earned a PhD in environmental studies and an MS in astrophysical, planetary, and atmospheric sciences from the University of Colorado, Boulder, as well as a BS in chemistry and a BA in humanities from the University of Texas at Austin.

Introduction

Less than a decade ago the idea of adapting to the inevitable effects of a changing climate was controversial. Today it is considered common sense by a growing number of cities and communities around the world that are taking concrete steps to reduce their vulnerability and increase their resilience.

Just as cities took the lead on cutting carbon dioxide and other greenhouse gases that are disrupting the climate, they are driving resilience planning and implementation. Cities are not waiting for the scientific or political debate to be settled. They cannot afford to.

"For us, it's serious. It's right now. Drought. Sea level rise. Storms. They aren't abstract to us. We see them. It's our job to deal with them," according to Berkeley, California, Mayor Tom Bates (Bates 2014). As former New York City Mayor Michael Bloomberg explained, "Around the world, city leaders are not wasting time debating the science of climate change or waiting around for international treaties to be signed; we are taking action. There's simply too much to do and too much at stake...Mayors are pragmatists, not partisans; innovators, not ideologues. We are responsible for delivering results, not debating politics" (Bloomberg 2014). Grand Rapids, Michigan, Mayor George Heartwell noted, "Things get done at the city level. All over the world, cities are getting the work of mitigation and adaption done" (Heartwell 2014).

Communities are beginning to recognize that they must adapt to the new normal of more-frequent hot days, more flooding, and more impacts on people and property. There are cities working now to reduce their risk, to prepare for a new and changing environment and to adapt to "potential damages, take advantage of opportunities and cope with the consequences" of climate change (US Global 2014). At the state, local, tribal, and territorial levels, leaders are deciding on ways to "invest in more resilient infrastructure, revise land use, update building codes and adjust natural resource management and other practices to improve the resilience of their communities to climate impacts" (White House 2014).

But not enough cities are moving fast—or far—enough. The most recent National Climate Assessment warns "the pace and extent of adaptation activities are not proportional to the risks to people, property, infrastructure, and ecosystems from climate change" (US Global 2014). Actions to date "appear to be incremental changes, not the transformational changes" that [will] be needed to protect us from the climate changes that are coming (Bierbaum 2012). According to UN Secretary-General Ban Ki-moon, "The world is largely ill-prepared for the risk from a changing climate. The good news is that if we act now, we have the means to build a more sustainable world" (UN 2014).

Building a more sustainable world requires "preparing for, and adjusting pro-actively to climate change—both negative impacts as well as potential opportunities" (World Bank 2011). Many of the city leaders who are actively developing innovative strategies to reduce the risks from extreme weather "recognize that increasing their cities' resilience to climate change will keep people and businesses out of harm's way, and—if done right—drive economic growth and improve the quality of life in metro areas" (Kelley 2013). Resilience, according to Judith Rodin, president of the Rockefeller Foundation, is "the capacity of individuals, communities, organizations and systems to survive, adapt and grow in the face of shocks and stresses, and even transform when conditions require it." It is "a lever for unlocking greater economic development, as well as improved social services and more broadly shared prosperity" (Rodin 2014).

This book examines obstacles cities and other entities face in their efforts to plan and take action to reduce the effects of climate change. It also details how leaders have overcome those obstacles and implemented plans that will help their communities not just survive, but thrive—to genuinely become resilient to climate change. These essays and case studies have been authored by individuals who were directly involved in the successful design and implementation of community-based resilience programs. Collectively, these essays and case studies provide a clear and procedural roadmap for building safer, stronger communities (see box "Mainstreaming Climate Adaptation Comes of Age …").

Chapter 1 provides a snapshot of the state of resilience planning, obstacles, needs, and the common best practices of the cities that have moved from planning to implementation. The ability to move into the action phase of resilience-building begins with the recognition and acceptance of three basic concepts: Humans cannot tame nature and we need to work with and learn how to survive from nature; there will be no returning to normal after a natural disaster—simply restoring what has been lost won't protect us from what's coming; and preparing for the future helps provide us with benefits now in the realms of disaster risk reduction, community livability, and improved health conditions. The chapter also identifies eight common traits and experiences shared by cities that have succeeded in implementing resilience.

Chapter 2 presents an extended essay concerning the risks poised by climate change to communities. The ongoing changes to the climate now and in the future will force communities to consistently review and update their assessment of community hazard risks. This chapter examines the current and future risks poised by a variety of hazards aggravated by climate change and discusses how these changing risks may impact the ability of communities to plan and adapt.

Chapter 3 presents two essays concerning the role urban and regional planners and local emergency managers can play in designing and implementing climate change adaptation programs and measures. The first essay examines how communities can include climate change projections and risk assessments into community comprehensive plans and community hazard mitigation plans. The second essay discusses the role that local emergency managers can play in helping their community deal with climate change.

Chapter 4 presents three essays, the first examines ongoing federal, state, and local government programs designed to address climate change and provides background information concerning what government agencies across the country are doing relative to climate change adaptation. A second essay examines federal strategies for risk management that provide a potential roadmap for how communities can address the challenges of climate change and design adaptation programs and actions that will reduce the impacts of climate change. The third essay examines the results of a Kresge Foundation–funded effort by the University of Michigan to assess available resources to communities seeking to conduct climate change adaptation actions, how these resources are meeting the needs of climate change practitioners, and finally, to identify potential new opportunities.

Chapter 5 presents two essays by two individuals who have been involved in community adaptation practices for many years. The first essay examines the results of a February 2012 workshop that brought together 80 practitioners from across the country involved in climate change adaptation at the community level. The second essay presents a critique of the adaptation chapter in the Third National Climate Assessment published in May 2014. Also included throughout this chapter are four sidebars that briefly examine the work done by the American Society of Adaptation Professionals (ASAP), profile the scope of the 100 Resilient Cities that was pioneered by the Rockefeller Foundation (100RC) program, summarize the findings of the 2014 Global Cities report developed by CDP, and describe the goals and objectives of the ongoing National Disaster Resilience Competition sponsored by the US Department of Housing and Urban Development (HUD) in collaboration with the Rockefeller Foundation.

Chapter 6 presents the first of a series of seven case studies of communities from around the country that have taken an aggressive, community-based approach to reducing the impacts of their chronic hazard risks and climate-related risks. The first set of three case studies presented in this chapter concerns long-term disaster risk reduction actions taken by Tulsa, Oklahoma; Berkeley, California; and Napa, California.

Chapter 7 presents a second set of case studies that examine climate change adaptation programs in Flagstaff, Arizona; Grand Rapids, Michigan; and Norfolk, Virginia. In addition, a case study of the adaptation measures designed and implemented in New Jersey and New York as part of their recovery efforts from 2012's Hurricane Sandy is included in this chapter.

MAINSTREAMING CLIMATE ADAPTATION COMES OF AGE ...

Josh Foster

The implementation of climate change adaptation, planning, and preparedness to improve local resilience to climate-related disasters has accelerated over the last decade. In 2005, when I attended my first local government climate conference, elected officials rarely if at all considered climate change adaptation as a priority, preferring instead to focus on emissions reductions via energy efficiency and renewables. Leaders still believed that adaption was giving up on preventing global warming and impacts were in the future.

Between then and now, increasing scientific understanding and local awareness of the inevitability of global warming and related impacts (via global and US national climate impacts assessments), direct experience with current extreme climate events tied to climate change, such as Hurricane Sandy, and the practical advantages of increasing resilience of water, health, transportation, energy, and emergency management systems to weather disasters have resulted in a rapid renaissance in thinking and acting at local and state governments about climate adaptation and disaster preparedness. Recommendations from the White House Task Force on Climate Preparedness and Resilience mainstreaming climate adaptation in federal agencies supporting state, local, and tribal government policies have added national policy momentum to complement local activities.

This top-down policy approach couples well with recent bottom-up experimentation with peer learning among local government policies and practices. These issues show that the climate adaptation policy process is maturing from "is it happening" to "what do we do about it?" Climate policies are now as much about floods and droughts, public health responses, water supply reliability, sewer system robustness, and urban tree planting as installing solar panels, building windmills, or increased energy efficiency.

The next steps will be further mainstreaming of climate resilience in practice at local levels, particularly for disaster management. A democratic debate now is needed about where, when, and how to adapt similar to other local public policy issues of the past. The public and decision makers will all have ideas about the appropriate level and mix of government and private sector action, where to spend funding on adaptation and preparedness, and who should pay, especially as needs metastasize in the face of climate changes while the available funding remains limited.

Inevitably, there will be—and already are—local legal and regulatory conflicts over climate-related disaster mitigation and preparedness similar to environmental laws of the 1970s and 1980s. Emerging needs for property buyouts, flood insurance rate increases, zoning and building code changes, and questions about whether to approve and finance real estate development in risky locations provide a further impetus for mainstreaming climate governance.

These actually are good signs that adaptation has arrived on the scene and by necessity has achieved a measure of legitimacy. An evolving goal will be to provide constructive ways to resolve anticipated conflicts that inevitably will arise, while also building integrated, practicable, actionable, implementable, measurable, and beneficial local solutions.

Chapter 8 examines several case studies of community climate change adaptation efforts from around the globe in order to understand how climate change adaptation practices around the world occur and what guidance can be gleaned from these practices that can help chart the course of climate change adaptation in the United States and other parts of the world. In addition, a brief discussion concerning integration of climate change adaptation and disaster risk reduction is included at the end of this chapter to highlight what we believe is an opportunity to leverage the efforts in both practices to achieve more resilient communities across the globe.

Chapter 9 offers conclusions and recommendations based on the experiences and ideas presented in the essays and case studies included in the book. Those common features that can be found in each essay and case study are highlighted and what impact or role they had in the successful design and implementation of hazard reduction programs. Based on these conclusions, a series of recommendations are presented concerning how officials and agencies in the federal, state, and local governments, and the private and nonprofit sectors, can support and promote programs in their communities that will reduce the impacts of future disasters influenced by global warming.

Our hope is that the information presented in this book will make it clear to community leaders that there are successful models for building the types of community partnerships that will be needed to reduce the impacts of future floods, droughts, wildfires, and other disasters influenced by climate change and make it possible for all of us to live and thrive in a changing climate. It is to that end that this book is dedicated.

1 Learning to Survive and Thrive in a Changed Climate

Kim S. Haddow

INTRODUCTION

The climate has changed. Sea levels are rising, glaciers are shrinking, snow and ice cover are disappearing, the oceans are warmer and extreme weather events—including heat waves in Europe, Asia, and Africa, heavier rain and snowstorms in North America and Europe, and droughts worldwide—are happening with increasing frequency and force.

"Simply put, we are rapidly remaking the planet and beginning to suffer the consequences," says Michael Oppenheimer, professor of geosciences and international affairs at Princeton University (Borenstein 2014).

Climate change is an indisputable and growing threat, and more cities and communities around the world are taking concrete steps to reduce their vulnerability and to adapt and make themselves more resilient. But not enough cities are moving fast—or far—enough. Most adaptive actions to date "appear to be incremental changes, not the transformational changes" that (will) be needed to protect us from the climate changes that are coming (Bierbaum et al. 2012). We have 10–15 years, not even one generation, to get it right.

A STEPPED-UP TIMETABLE HAS INCREASED THE NEED FOR URGENT ACTION

Climate change is now upon us. According to the latest (May 2014) US National Climate Assessment conducted every four years to inform the nation "about observed changes, the current status of the climate, and anticipated trends for the future":

> Americans are noticing changes all around them. Summers are longer and hotter, and extended periods of unusual heat last longer than any living American has experienced. Winters are generally shorter and warmer. Rain comes in heavier downpours. People are seeing changes in the length and severity of seasonal allergies, the plant varieties that thrive in their gardens, and the kinds of birds they see in any particular month in their neighborhoods.
>
> Other changes are even more dramatic. Residents of some coastal cities see their streets flood more regularly during storms and high tides. Inland cities near large rivers also experience more flooding, especially in the Midwest and Northeast. Insurance rates are rising in some vulnerable locations, and insurance is no longer available in others. Hotter and drier weather and earlier snow melt mean that wildfires in the West start earlier in the spring, last later into the fall, and burn more acreage. In Arctic Alaska, the summer sea ice that once protected the coasts has receded, and autumn storms now cause more erosion, threatening many communities with relocation. (US Global Change 2014)

Hot weather now kills more Americans than all other natural disasters combined and experts predict that summer heat waves will worsen and lead to more deaths (Brody 2014).

Worldwide, extreme weather has also increased. According to the Centre for Research on the Epidemiology of Disasters in Belgium, which tracks the world's catastrophes, there have been more than 6600 major climate, weather, and water disasters worldwide since 1992 causing more than \$1.6 trillion in damage and killing more than 600,000 people. From 1983 to 1992 the world averaged

FIGURE 1.1 Completed in 1986 in the Netherlands, the Eastern Scheldt Storm Surge barrier has movable gates that can close when floodwaters threaten. (Credit: Rijkswaterstaat. https://www.dropbox.com/sh/02hcu7g56via56r/AACxcp-YYLs2QPE76X_0XTLAa/Images#lh:null-Oosterschelde.jpg.)

147 climate, water, and weather disasters each year. From 2004 to 2014, that number jumped to an average 306 a year (Borenstein 2014) (see Figure 1.1).

In a speech to the United Nations General Assembly in September 2014, President Obama noted the ubiquitous evidence of the changing climate and urged immediate action. "In America, the past decade has been the hottest on record. Along our eastern coast, the city of Miami now floods at high tides. In our west, wildfire season now stretches for most of the year. In our heartland, farms have been parched by the worst drought in generations, and drenched by the worst spring in our history. A hurricane left parts of this great city dark [New York] and underwater, and some nations already live with far worse" (Fritz 2014).

The President concluded: "We have to adapt" (Fritz 2014).

"TIME IS NOT ON OUR SIDE"

The Fifth Assessment Report of the Intergovernmental Panel on Climate Change (IPCC) released in November 2014 contained the panel's strongest statement of scientific certainty (95 percent) that climate change was happening and caused by humans since its first assessment in 1990. For the first time, the panel created by the United Nations (UN) and the World Meteorological Organization (WMO) used the word "irreversible," warning that "Warming by the end of the 21st century will lead to high to very high risk of severe, widespread and irreversible impacts globally" (UN 2014).

The opportunity to determine whether we will surpass the level of warming that is considered dangerous by world leaders "will slip away within the next decade," according to UN Secretary-General Ban Ki-moon. "Time is not on our side…leaders must act" (UN 2014).

"The world is largely ill-prepared for the risk from a changing climate," according to the Secretary-General. "The good news is that if we act now, we have the means to build a more sustainable world" (UN 2014).

LIMITING AND LIVING WITH CLIMATE CHANGE

There are two ways to minimize the damage from climate change and increase resilience: reduce the greenhouse gas emissions that are disrupting the climate *and* adapt effectively to existing and future impacts.

THE ANTICIPATED EFFECTS OF CLIMATE CHANGE

Hotter temperatures: If emissions keep rising unchecked, then global average surface temperatures will likely rise at least 2°C (3.6°F) above preindustrial levels by 2100, and possibly 3°C or 4°C or more.

Higher sea-level rise: The expert consensus is that global sea levels will rise somewhere between 0.7 and 1.2 meters by the end of the century if global warming continues unchecked (that's between 2 and 4 feet). And that's only the average. In regions like the eastern United States, sea-level rise could be even higher.

Heat waves: A hotter planet will mean more frequent and severe heat waves.

Droughts and floods: Across the globe, wet seasons are likely to become wetter and dry seasons drier. As the IPCC puts it, the world will see "more intense downpours, leading to more floods, yet longer dry periods between rain events, leading to more drought."

Hurricanes: It's not yet clear what impact global warming will have on tropical cyclones. The IPCC said it was likely that tropical cyclones would get stronger as the oceans heat up, with faster winds and heavier rainfall. But the overall number of hurricanes in many regions was likely to "either decrease or remain essentially unchanged."

Heavier storm surges: Higher sea levels will increase the risk of storm surges and flooding when storms do hit.

Agriculture: In many parts of the world, the mix of increased heat and drought is expected to make food production more difficult. The IPCC concluded that global warming of 1°C or more could start hurting crop yields for wheat, corn, and rice by the 2030s, especially in the tropics. (This isn't uniform, however: some crops may benefit from mild warming, such as winter wheat in the United States.)

Extinctions: As the world warms, many plant and animal species will need to shift habitats at a rapid rate to maintain their current conditions. Some species will be able to keep up, others likely won't. Coral reefs, for instance, will have difficulty adapting if the oceans continue warming and become more acidic. The National Research Council has estimated that a mass extinction event "could conceivably occur before the year 2100."

Long-term changes: Most of the projected changes above will occur in the twenty-first century. But temperatures will keep rising after that if greenhouse gas levels aren't stabilized. That increases the risk of more drastic longer-term shifts. One example: if West Antarctica's ice sheet started crumbling, that could push sea levels up significantly. The National Research Council deemed many of these rapid climate surprises unlikely this century, but a real possibility further into the future.

Source: Plummer, B. 2014b. *Everything you need to know about global warming.* Vox. Available at http://www.vox.com/cards/global-warming/how-has-global-warming-affected-the-world-so-far#E5499185.

Many cities have been working for years to reduce their carbon emissions, and they are making progress. "There's lots of cities now making a major impact on their carbon emissions—some of them reducing their emissions between 10–15%—some as much as 25%," explained Cooper Martin, Program Director of the National League of Cities' Sustainable Cities Institute (Martin 2014).

The most common ways cities are reducing their emissions are by increasing energy efficiency and switching to renewable energy. Cities are also experiencing what Mayor George Heartwell of Grand Rapids, Michigan, calls, "triple bottom line benefits" (Heartwell 2014). Mitigation tactics—including upgrading city facilities, switching to light-emitting diode (LED) streetlights, powering city-owned fleets with renewable and less polluting fuels, and converting to solar and wind power and facilitating their use in residential and commercial buildings—are generating economic, environmental, and social bottom-line benefits. Cities using renewable energy and exercising energy efficiency are saving money—and in tight, tough economic times, that kind of cost-savings program wins supporters and in some cities has opened the door to other climate impact-abating measures.

Mitigation efforts to cut carbon emissions are essential, but not enough. Even if greenhouse gas emissions were reduced to zero today, the earth's atmosphere is already loaded with enough carbon dioxide to raise global temperatures by at least 1°C (or 1.8°F) over preindustrial levels, and even a mild temperature increase will likely mean more heat waves, higher sea-level rise, and so on. That means adapting *and* building resilience are imperative.

"Recent events like Hurricane Sandy in the Northeast, flooding throughout the Midwest and severe drought in the West have highlighted the vulnerability of many communities to the impacts of climate change. In 2012 alone, the cost of weather disasters exceeded $110 billion in the United States, and climate change will only increase the frequency and intensity of these events. That is why, even as efforts to reduce greenhouse gas emissions continue, communities must be prepared for the impacts of climate change that can no longer be avoided" (White House 2014) (see Figure 1.2).

The realities of this changed climate are "jolting many city leaders into action to protect public health, safety and the local economy" (Kelley and Bhattacharyya 2013). Communities are beginning to recognize that they must adapt to this new normal of more frequent hot days, more flooding, and more impacts to people and property.

Now some—but not enough—cities are working to reduce their risk, to prepare for a new and changing environment, to adapt to "moderate potential damages, take advantage of opportunities and cope with the consequences" of climate change (US Global Change 2014). At the state and local, tribal, and territorial levels, leaders are deciding on ways to "invest in more resilient infrastructure, revise land use, update building codes and adjust natural resource management and other practices to improve the resilience of their communities to climate impacts" (White House 2014).

These recent adaptation actions and ongoing mitigation efforts to cut greenhouse gas emissions are linked: effective mitigation reduces the need for adaptation in the future. Disaster risk reduction (DRR) and adaptation planning are also closely connected. The actions a community takes to reduce its vulnerability to a natural disaster are also what a community must do to limit the impacts of climate change. The steps taken to reduce the risk of coastal flooding are very similar to the actions needed to adapt to sea-level rise. The main difference is that sea-level rise adaptation assumes a longer time frame for impact, and therefore a longer need and implementation time

FIGURE 1.2 Remains of a home in Union Beach, New Jersey, on November 8, 2012, stand as a stark reminder of the storm's power after Hurricane Sandy swept through the area. FEMA worked with state and local officials to assist residents who were affected by Hurricane Sandy. (Photo by Patsy Lynch/FEMA. https://www.fema.gov/media-library/assets/images/66443.)

FIGURE 1.3 Federal resources are supporting states as they rebuild infrastructure.

frame. Sea-level rise planning also assumes an increase in the vulnerability of areas already subject to coastal flooding and adaptation projects must take their increased vulnerability into account.

Resilience provides a critical point of integration for adaptation, disaster risk reduction, and mitigation strategies. All three are essential parts of a comprehensive climate change response strategy to increase a community's resilience. "Resilience" is defined in the 2014 National Climate Assessment as the "capability to anticipate, prepare for, respond to, and recover from significant multi-hazard threats with minimum damage to social well-being, the economy, and the environment" (US Global Change 2014) (see Figure 1.3).

According to the World Bank, "A resilient city is one that is prepared for existing and future climate impacts [and capable of] limiting their magnitude and severity...a resilient city is able to respond quickly and effectively in an equitable and efficient way" (World Bank 2011).

But Judith Rodin, president of The Rockefeller Foundation, believes the concept of resilience encompasses more than survival and recovery: that it includes the notions of growth and new opportunities. Resilience, according to Rodin, is "the capacity of individuals, communities, organizations and systems to survive, adapt and grow in the face of shocks and stresses, and even transform when conditions require it." It is "a lever for unlocking greater economic development, as well as improved social services and more broadly shared prosperity" (Rodin 2014a).

Rodin believes investing in resilience creates a dividend—the package of cobenefits investing in building resilience, in creating new systems and technologies that can yield to a city's near-term benefits—"job creation, economic opportunity, social cohesions and equity," while ensuring the community is better prepared for an extreme weather event or other shock (Rodin 2014a).

CITIES LEAD THE WAY

Why should cities be the tip of the spear? Cities drive economic growth and innovation. They are also home to the majority of people around the globe. In the United States, more than 80 percent of the population now lives in a metropolitan area and that percentage will climb to near 90 percent by 2050 (Kelley and Bhattacharyya 2013) "The ability of cities to build resilience and achieve sustainable growth will have a profound impact on the quality of life in America, today and into the future" (Kelley and Bhattacharyya 2013) (see Figure 1.4).

Cities are often the first responders to climate impacts. As the State, Local and Tribal Leaders Task Force on Climate Preparedness and Resilience noted, "From repeated low level flooding and extreme storms to increasing temperatures and drought, climate change hits every community differently. State, local, tribal, and territorial leaders are at the forefront of dealing with these impacts and preparing their communities for future changes. These leaders recognize the need to act now to protect their communities, and are doing so with their own authorities and resources while working with diverse partners including business, community organizations, various levels of government, and citizen groups" (White House 2014).

FIGURE 1.4 Planning paves the way to community resilience. (From http://www.fema.gov/media-library/assets/images/68025.)

"Climate change is happening much differently and much more quickly than anyone imagined," explained Frank Cownie, the mayor of Des Moines, Iowa, and a member of the White House's State, Local and Tribal Leaders Task Force on Climate Preparedness and Resilience. "We are already dealing with the consequences of inaction in the face of an ever-changing climate." Given the gridlock on Capitol Hill, "It would be foolish for us to sit back and assume that the federal government will step up if it's bad enough" (Beamish 2014).

"Today the change is to adaptation, because we see climate change coming so quickly," explained the mayor of Carmel, Indiana, Jim Brainard, who also served on the White House Task Force. "Local communities have been working it very hard," but limited resources and capacity are constraining efforts (Beamish 2014).

Cities are not waiting for the scientific or political debate to be settled. They cannot afford to. "Around the world, city leaders are not wasting time debating the science of climate change or waiting around for international treaties to be signed; we are taking action. There's simply too much to do and too much at stake…Mayors are pragmatists, not partisans; innovators, not ideologues. We are responsible for delivering results, not debating politics. And as the world becomes increasingly more urban, the importance of bold local action—particularly on climate change—will continue to grow" explained former New York Mayor Michael Bloomberg (Bloomberg 2013).

According to a joint 2014 survey by the Massachusetts Institute of Technology (MIT) and ICLEI–Local Governments for Sustainability, 74 percent of US cities have perceived changes in the climate, including increased storm intensity (31%), higher temperatures (30%), and more precipitation (28%) (Aylett 2014). Martin at the National League of Cities noted there has been a "really dramatic shift in the number of cities addressing climate change in the last 5–10 years." According to Martin, many city leaders are "done arguing if climate change is happening. They don't care why it's happening. They are living with it" (Martin 2014).

Cities are preparing for climate change in a variety of ways, but particularly in the planning, installation, maintenance, and improvement of infrastructure. The exact nature of adaptation planning depends on the geographic location of the city and the projected climate effects in that region. Coastal regions, for example, must adapt to rising seas and increased flooding; mountainous regions are likely to see less snow pack and more ice; and dry regions are likely to experience increasingly severe droughts and more fires.

Because cities are dynamic systems that face unique impacts, their adaptation must be location-specific and tailored to local circumstances, according to the World Bank. "The starting point in managing risks and building long-term resilience is for a city to understand its exposure and sensitivity to a given set of impacts, and develop responsive policies and investments that address these vulnerabilities" (World Bank 2011).

CITIES TAKE THE LEAD ON CLIMATE

Satya Rhodes-Conway
Managing Director of the Mayors Innovation Project

Unlike the federal and state governments in the United States, cities have widely acknowledged that the global climate is changing, that human action is contributing to that change, and that we need to act now to both slow the changes and prepare for them. Cities are not waiting to act—in fact, some cities have been working to reduce their greenhouse gas (GHG) emissions for years, and many more are now taking action to both mitigate and prepare for climate change.

Cities are the front line for climate impacts—they control the majority of our transportation, water, and waste infrastructure, which is under increasing pressure as a result of climate disruption and the changing temperature and precipitation patterns it brings. Cities need to maintain, upgrade, and replace our aging, outmoded infrastructure with greener, more efficient, and more effective infrastructure, and they are trying to do so, despite limited resources.

WHAT CITIES CAN DO

Local governments have jurisdiction or significant influence over a number of GHG emissions sources, including building energy use, transportation, water infrastructure, sewer treatment, and waste management. Buildings, for example, account for almost 40 percent of US GHG emissions, and their energy use is subject to municipal regulation in the form of building codes, benchmarking requirements, zoning, and more. Transportation produces 30 percent of US GHG emissions and is heavily influenced by local planning and investment. Most transit systems are run by local or regional authorities. The waste sector is responsible for 2.3 percent of US GHG emissions and is almost entirely under local control.

Cities have jurisdiction over adaptation and resilience as well. In the three big areas where resilience is needed—physical infrastructure, social connections, and financial resources—cities have a good degree of control. Many aspects of resilience planning overlap heavily with things city governments already do, such as emergency management and infrastructure improvements.

Cities must prepare for climate change in numerous ways, but particularly in the planning, installation, maintenance, and improvement of infrastructure. The exact nature of adaptation planning depends on the geographic location of the city and the projected climate effects in that region. Coastal regions, for example, must adapt to rising seas and increased flooding; mountainous regions are likely to see less snow pack and more ice; and dry regions are likely to experience increasingly severe droughts and more fires. Cities will, in general, experience more extremes in temperature and more intense storm events.

SETTING GOALS

Many cities have climate action plans or have signed climate protection agreements, but these plans must have some muscle behind them. Far too often they remain entirely aspirational. Bellingham, Washington, set a goal in 2007 of reducing GHG emissions 70 percent by 2020, relative to consumption in 2000. The City currently buys 100 percent renewable power for all city facilities and has installed rooftop generation on some city buildings. Local university students agreed to a slight raise in tuition so that the Western Washington University campus could be powered entirely by renewable energy. The city government also sponsored a community green power challenge, resulting in the purchase of enough green power to meet

11 percent of the community's total energy use. The additional cost of this generation is partially offset by the comprehensive energy retrofit and energy-efficient new construction programs the city runs. A preferential loan program for solar-energy installation has also been created for local businesses. The city supports energy efficiency in a variety of ways, including energy-efficiency challenges, energy-efficient land-use planning, and technical assistance to the public on how to construct green buildings.

REDUCING GHGS

The most common ways cites are reducing their emissions are through energy efficiency and renewable energy. Many do so by upgrading their own facilities. The City of Los Angeles, for instance, installed more than 141,000 LED streetlights, cutting energy use by 63 percent and saving the city $7 million a year in electricity costs. The City of Reno, Nevada, combined energy efficiency with solar and wind generation into a $20 million project that will save them $1.3 million a year, and has retained or created 279 jobs.

Others work with the private sector to reduce community-wide emissions. To promote energy efficiency, Arlington County, Virginia, permits larger or taller buildings than zoning would otherwise allow for developers who achieve Leadership in Energy & Environmental Design (LEED) certification. Seattle, New York City, Washington, DC, San Francisco, Austin, Texas, Boston, Chicago, and Minneapolis have all adopted commercial benchmarking and disclosure laws. These are relatively simple, low-cost ways to bring market forces to bear to spur investment in building energy efficiency. A city passes a law requiring that on a fixed yearly schedule or when a building is put on the market, its energy consumption is disclosed either to prospective buyers or the entire public.

Many communities are promoting distributed renewable energy generation. In Sonoma County, the California Energy Independence Program offers PACE financing for commercial and residential properties. Since it launched in 2009, it has financed improvements, both energy efficiency and renewable energy, on 1700 residential properties and 57 commercial ones, worth $58 million, creating or retaining 714 jobs. Portland, Oregon, runs the Solarize Portland program, where the city assists neighborhood associations and other organizations with program design, finding contractors, and marketing to participants. Since 2009, the program has purchased 600 solar installations.

Lancaster, California, passed one of the strongest mandates for residential renewable energy by requiring all builders of new dwelling units to provide solar energy systems for new homes, with a minimum generation of 1 kW per unit. This policy helps to reduce costs to homeowners by using a bulk purchase of solar panels to lower the per unit cost of the solar panel installation, as well as wrapping the cost of the solar energy into the price of the home, thereby effectively allowing homeowners to finance solar photovoltaic systems through a home mortgage, all while saving money on utility costs.

Of course, there are other ways to reduce GHGs as well. By using smart growth policies, local governments can promote a future vision of mixed-use development with housing and transportation choices near jobs, shops, and schools, reducing reliance on single-occupancy vehicles. These policies can achieve roughly 20 to 60 percent reductions in transportation-related GHG emissions alone, according to many studies.

INCREASING RESILIENCE

While we tend to think about them as mitigating climate change, many of the actions cities take to reduce their GHG footprints also promote resilience. Distributed renewable energy production, for example, has the potential to keep providing power to a city if more conventional sources go off-line, and a strong, multimodal transportation system is more likely to provide options to get people where they need to go in a severe weather event than one that relies only on cars. Still, there are many additional things cities are doing to adapt to climate change.

Adaptation may seem daunting, but cities already respond to many of the challenges climate change will bring, albeit at a less severe level. What is important is to integrate climate change into planning and to think about the resilience of the community as a whole, with a particular emphasis on vulnerable populations. For example, probably the single best way to increase the ability of low-income families to respond to climate change is to increase their overall economic stability.

Land-use planning is a particularly important tool in adaptation planning since the resiliency of a community is highly dependent on its land-use policy. Local governments can increase resiliency by discouraging development in areas vulnerable to hazards such as flooding, wildfire, and land erosion, and by protecting natural capital such as wetlands and forests. Boston is completing a review of its water and sewer system to make sure it's ready for projected sea levels and increased precipitation, and has assessed the climate vulnerabilities of city-owned buildings. Boston also added consideration of climate resilience and flooding into its review process for new construction in the private sector. Punta Gorda, Florida's extensive plan to deal with rising seas and increased flooding includes trying to become energy-independent, adjusting building codes to require more flood-resistant buildings, and buying and protecting coastal mangrove forests. Chula Vista, California, is adapting to rising temperatures and deadly heat waves with cool roofs, cool pavements, and by expanding the urban forest.

Energy infrastructure is another important focus for adaptation. Denver, Colorado's Climate Adaptation Plan includes actions that increase energy efficiency, especially measures that reduce usage in high-demand times such as during heat waves. Denver has also developed the Denver Energy Assurance Plan, which helps the city prepare for disruptions in the energy supply during a crisis like a winter storm.

It is undeniable that action at the federal and state level, or with large private-sector actors like investor-owned utilities, can have a larger impact than working city by city. However, in recent years it has been politically much more difficult to work at these levels than we would like, and we have seen little progress and even some backsliding. Local governments provide an opportunity for success—perhaps the most likely opportunity we have at the moment—and success at the local level can resonate upwards. Furthermore, because local government is the level of government closest to people's day-to-day lives, it is well suited to create a policy environment that makes living a low-carbon lifestyle possible, or even preferable, and to prepare for the wide-ranging impacts of our changing climate.

Berkeley, California, Mayor Tom Bates explains: "We know our vulnerabilities. We know what we need to do. It hasn't rained here in three years. Odds are we will have to make severe water cuts just to get through next year. We have a 60-day water supply. For us, it's serious. It's right now. Drought. Sea level rise. Storms. They aren't abstract to us. We see them. It's our job to deal with them" (Bates 2014).

Grand Rapids Mayor Heartwell explained that "Climate change is an abstract for Congress, an ideological debate." At the Michigan statehouse, "the Democrats and Republicans fight all day long about investing in protecting the natural assets of the state. But at the local level—at the level of mayors and city councils we don't have that luxury—we have the floods and the tornados." According to Mayor Heartwell, "Things get done at the city level. All over the world cities are getting the work of mitigation and adaptation done" (Heartwell 2014).

The National League of Cities' Martin noted that it's logical for cities to take the point on resilience planning and implementation. "Cities are the closest to the issue. States have risks spread over different corners. It's easier to assess and act on the vulnerabilities of one, smaller place. It's easier to identify and focus on specific threats when you are closer to the ground" (Martin 2014).

Martin also believes mayors have a political advantage: "If you are governor—you are term limited—and that forces you to focus on short-term political strategies. Mayors are often elected term after term. Mayors are closer to real governing and providing services and further away from partisan politics. Mayors are often not focused on next political job opportunity. They tend to be focused on practical solutions as opposed to ideological debates" (Martin 2014).

According to the World Bank, cities are well-positioned to lead because "adaptation in cities requires collaborate problem-solving and coordination across sectors. Cities are well-positioned to act as conveners of a wide range of partners. Climate change [may impact] many sectors: land use, housing, transportation, public health, water supply and sanitation, solid waste, food security and energy. Adaptive efforts in any of these sectors will often involve multiple government agencies, as well as broad partnerships that include other governments, local communities, nonprofit organizations, academic institutions and the private sector" (World Bank 2011).

The Center for American Progress (CAP) agrees that a city's ability to coordinate across sectors is crucial: "Increasing resilience requires looking across transportation, energy, storm and drinking water management, and other infrastructure systems and sectors to understand where cities are the most vulnerable and to identify the best and most cost-effective strategies to manage climate change risks. Without a cross-system approach to resilience, metro areas will likely continue to face the cascading disruptions of power and transit outages caused by extreme weather events that leave millions of people stranded and unable to get to work or school, seek medical care, or even leave their high-rise apartments" (Kelley and Bhattacharyya 2013).

The World Bank explains, "Adaptation is the process of preparing for, and adjusting proactively to climate change—both negative impacts as well as potential opportunities" (World Bank 2011). Many of the city leaders who are actively developing innovative strategies to reduce the risks from extreme weather "recognize that increasing their cities' resilience to climate change will keep people and businesses out of harm's way, and—if done right—drive economic growth and improve the quality of life in metro areas" (Kelley and Bhattacharyya 2013). These leaders know that upgrading public transit and other infrastructure, providing cleaner and more reliable energy, and expanding parks and green spaces will create jobs, attract new businesses, improve air quality, and yield other quality-of-life benefits.

An example from CAP helps make the case: "Strengthening grid resilience, for example, improves energy reliability, which reduces economic losses during storms and can lower emissions by increasing the availability of clean distributed power. Planting more trees and creating more parks reduces stormwater runoff while increasing property values, carbon storage, and quality of life. Creating diverse transportation and energy systems provide backups and alternatives that allow people to get to work, businesses to stay open, and the lights to stay on, and communities to stay safe, even when parts of these systems are damaged by a storm" (Kelley and Bhattacharyya 2013).

ADAPTATION PLANNING IS PICKING UP, BUT IMPLEMENTATION LAGS

According to a May 2014 survey of 468 cities worldwide—including 298 in the United States—conducted by MIT in partnership with ICLEI–Local Governments for Sustainability, climate adaptation planning is happening, but it is less of a priority for US local governments than their non-US counterparts:

- Fifty-nine percent of US cities are pursuing adaptation planning compared to 68 percent worldwide. The United States had the lowest percentage for any region, with Latin American and Canadian cities having the highest (95 percent and 92 percent, respectively).

RISING SEA LEVELS: A SNAPSHOT OF ONE CLIMATE CHANGE THREAT

Rising seas are one of the major consequences of climate change and are closely linked to stormwater management and urban flooding problems, the climate change impact that ranked number one (72 percent) in a survey of close to 300 US cities conducted by the Massachusetts Institute of Technology and ICLEI–Local Governments for Sustainability (Aylett 2014).

Clearly rising sea level is "no longer an abstract challenge for a future day" (Harvard Kennedy School 2014):

- US cities along the Atlantic Ocean now have routine "nuisance flooding" or "sunny-day flooding"—1–2 feet of flooding above high tides that rarely occurred in the past. "Today, when the tide is extra high, people find themselves splashing through down-town Miami, Norfolk and Annapolis on sunny days and dealing with flooded roads in Atlantic City, Savannah and the coast of New Hampshire. In parts of New York City and elsewhere, homeowners are dealing with flooded basements, salt-poisoned yards and falling property values, not only because of catastrophic storms, but because tides, aided by sea level rise, now cause flooding where they live," explained Melanie Fitzpatrick, a climate scientist at Union of Concerned Scientists (Union of Concerned Scientists 2014).
- In London, closures of the Thames Barrier built to protect central London from flooding have increased every decade since it became operational in 1982. In 2014, the barrier was closed more than 50 times—a record (de Castella 2014).
- Global ocean levels have risen about 19 centimeters in the last century and the rate of rise has accelerated. Washington, DC, Wilmington, North Carolina, and Annapolis, Maryland, have already reached a tipping point, which is defined as more than 30 days of nuisance flooding a year. Other cities on the East and Gulf Coasts are following close behind and will see 30 or more days of flooding each year by 2050 because of sea-level rise, according to the National Oceanic and Atmospheric Administration (Sheppard 2014).

Communities that are subject to sea-level rise may experience an increase in coastal vulnerability. Impacts to communities may include

- Increased flooding and drainage problems
- Destruction of natural resource habitats
- Higher storm surges and increased evacuation areas and evacuation time frames
- Increased shoreline erosion
- Saltwater intrusion
- Loss of infrastructure and existing development (Florida Department of Economic Opportunity)

The State of Florida's primer on adaptation outlines the three main strategies a community may use to adapt to sea-level rise:

- *Protection.* Protection strategies involve "hard" and "soft" structurally defensive measures to curb the impacts of rising seas, such as shoreline armoring or beach renourishment, in order to decrease vulnerability yet allow structures and infrastructure in the area to remain unaltered. Protection strategies may be targeted for areas of a community that are location-dependent and cannot be significantly changed structurally (i.e., downtown centers, areas of historical significance, water-dependent uses, etc.).

- *Accommodation.* Accommodation strategies do not act as a barrier, but rather alter the design through measures such as elevation or stormwater improvements, to allow the structure or infrastructure system to stay in place. Adaptation measures do not prevent flooding or inundation of the property but do protect the structure. Accommodation strategies may be suitable for location-dependent structures that could be changed to accommodate water, without compromising the use (i.e., bridge elevation, residential home elevation, downtown stormwater improvements, etc.).
- *Retreat.* Retreat strategies involve the actual removal of existing development and possible relocation to other areas and the prevention of future development in these high-risk areas. Retreat options usually involve the acquisition of vulnerable land for public ownership, but may also include other strategies such as transfer of development rights, purchase of development rights, rolling easements, conservation easements, etc.

Source: Florida Department of Economic Opportunity. Florida adaptation planning. Available at http://www.floridajobs.org/community-planning-and-development/programs/technical-assistance/community-resiliency/adaptation-planning.

A number of cities and agencies already have had to take steps to protect their communities and property from rising seas. For example, Tybee Island, Georgia, and Miami Beach have started to upgrade their stormwater sewer systems to prevent seawater from backing up into pipes. Norfolk, Virginia, is returning some of its coastal parks to wetlands. The US Naval Academy in Annapolis, Maryland, is using door dams to protect building entrances from flooding (Union of Concerned Scientists 2014). Post-Hurricane Sandy, New York City is planning to invest in a system of floodwalls, levees, and dikes and to restore swampland and sand dunes. The city of Rotterdam in the Netherlands has built three floating pavilions that are anchored off the city's waterfront and are able to rise with changing sea levels (Gallucci 2013).

- Thirteen percent of US cities have completed an assessment of their vulnerabilities and risks compared to nineteen percent globally. Again, US cities were the least active in this area, while Africa, Australia, New Zealand, and Canada were the most active (Aylett 2014).

Tracking done by the Georgetown Climate Center at the Georgetown University Law Center shows that fewer than half the states have adopted adaptation plans or even have similar work in progress: 14 states have finalized state-led adaptation plans (Alaska, California, Colorado, Connecticut, Florida, Maine, Maryland, Massachusetts, New Hampshire, New York, Oregon, Pennsylvania, Virginia, Washington). Another nine have some level of planning underway (DC, Delaware, Hawaii, Michigan, Minnesota, New Jersey, Rhode Island, Vermont, Wisconsin) (Georgetown Climate Center 2015).

According to the Georgetown Climate Center, "Most state and local governments are at the early phases of adapting. The dominant focus has been on 'soft' activities like planning, vulnerability assessments, and capacity building." Few states "are implementing their plans by making 'hard' changes in law or policy that alter regulatory and management decisions in light of projected climate change" (Grannis et al. 2014).

Even though a growing number of states and cities are taking steps that will make their communities more resilient, the Climate Center says their "research also shows that many states are still not treating this issue with the urgency that is called for" (Grannis et al. 2014). The National Climate Assessment warns "the pace and extent of adaptation activities are not proportional to the risks to people, property, infrastructure, and ecosystems from climate change" (US Global Change 2014).

Most of the plans are considered inadequate. With the exception of New York City's plan "which detailed 250 climate adaptation strategies and put a price tag on most of them—few other cities have outlined specific actions or provided concrete details on how government agencies should implement initiatives or pay for them" (Gallucci 2013).

"A lot of [these adaptation plans] tend to be an overarching, big vision document," or focus on a single, massive project, like a floodwall, said JoAnn Carmin, a professor in urban studies and planning at MIT. "In some cases, there's no clear work plan in place" (Gallucci 2013).

This planning is occurring in an ad hoc matter, which is insufficient, according to the Climate Center: "While planning is occurring at all levels of government and plans are becoming more sophisticated in their analysis of potential impacts and consideration of policy responses, planning is occurring in an ad hoc manner. This is primarily because there are no incentives, mandates or dedicated sources of funding for adaptation. As a result, adaptation planning is taking many make-shift forms that reflect different local champions (for example, senior policymakers, agency staff, or community activists), threats (such as sea-level rise and urban heat-island effect), and vulnerable sectors (such as transportation and ecosystems)" (Grannis et al. 2014).

In addition, implementation of adaptation plans lag, often because of a lack of resources. According to the National Climate Assessment, "Substantial adaptation planning is occurring in the public and private sectors and at all levels of government; however, few measures have been implemented and those that have appear to be incremental changes" (US Global Change 2014). According to the 2012 report, "A comprehensive review of climate adaptation in the United States: more than before, but less than needed," few of the plans "have been implemented and even fewer have been evaluated" (Bierbaum et al. 2012).

But Aaron Ray, a policy analyst at the Georgetown Climate Center, noted that the trend is clear and that a growing number of cities and states are starting to take action. "There's certainly a long way to go, but we think states are recognizing the need to make those changes and are doing that" (Beamish 2014).

The New York City post-Hurricane Sandy adaptation plan, "A Stronger, More Resilient New York," raised the bar according to Jessica Grannis, a staff attorney for the Georgetown Climate Center. "It's a huge step forward in terms of the quality of adaptation plans that are coming up" and could serve as a model for other cities (Gallucci 2013).

Steve Adams, climate adaptation director at the Institute for Sustainable Communities, believes "We have made rapid progress in a short amount of time. It's astonishing how clear it is in 2014 that climate change adaptation, or resilience, is really an idea in good standing, where the conversation in 2008 or 2009 was just beginning to emerge" (Gallucci 2013).

CHALLENGES CITIES FACE

Progress preparing for the next storm has been made, including improved drainage and stormwater systems, stronger building codes, and elevated infrastructure. Investments have been made in more resilient infrastructure, in building smarter and stronger so communities can withstand and rebound more quickly from the next storm. But, "despite the positive momentum in recent years, barriers to implementation still impact action in all sectors and across scales" (Bierbaum et al. 2012).

Cities face a common set of obstacles: access to climate risk information, dealing with the uncertainty of climate change forecasts, addressing skepticism about climate change, achieving collaborative decision-making across sectors and jurisdictions, and funding both the planning and implementation stages of adaptation strategies (Chastain et al. 2012).

The MIT–ICLEI–Local Governments for Sustainability survey of 468 cities worldwide identified challenges in climate adaptation planning. The highest-ranked challenges were

- Approximately 95 percent of US cities report that securing funding for adaptation is a challenge

- Thirty-six percent of US cities, the highest rate globally, reported that the federal government does not understand the realities that they face with respect to adaptation
- Communicating the need for adaptation to and gaining support for adaptation from elected officials was highly rated globally and in the United States (Aylett 2014)

In addition to the United States having the lowest percentage of cities engaged in adaptation planning of every region surveyed, US cities had fewer staff dedicated to the task. "Globally, 63 percent of cities say they have between one and five employees dedicated to climate-change planning; North American cities are most likely to have just one staff member focused on the topic" (Dizikes 2014). According to the report's executive summary notes, "A lack of funding to hire sufficient staff to work on climate change is a significant challenge for 67 percent of cities" (Aylett 2014).

Brian Holland, director of ICLEI USA Climate Programs, noted that "US cities continue to show leadership on climate change." But these surveys make it clear that "local governments need more federal and state support to accelerate their efforts" (Knapp 2014).

According to Martin of the National League of Cities, the demand for funding for adaptation solutions far outstrips the supply. "Just look at the Department of Transportation's TIGER Grants—they had 15 times more applications than could be funded" (Martin 2014). US Department of Transportation's (DOT) Transportation Investment Generating Economic Recovery (TIGER) program allows local governments to compete for funding for road, rail, transit, and port projects that promise to achieve critical national objectives, including solving environmental problems and reducing the nation's dependence on oil. Martin stressed that there are more cities willing to act than there is money to fund implementation.

In a separate analysis, MIT urban studies professor JoAnn Carmin identified obstacles to adaptation as "a lack of funding to pay for comprehensive analysis, a focus on other municipal priorities and a shortage of qualified staff." She also noted that "local governments rarely have access to data on the specific risks that global warming poses to their particular city" (Gallucci 2013).

The 2014 National Climate Assessment identified several of the same barriers to implementation, including "limited funding, policy and legal impediments, and difficulty in anticipating climate-related changes at local scales....Although decision-makers routinely make complex decisions under uncertain conditions, decision-making in the context of climate change can be especially challenging due to a number of factors. These include the rapid pace of changes in some physical and human systems, long time lags between human activities and response of the climate system, the high economic and political stakes, the number and diversity of potentially affected stakeholders, the need to incorporate uncertain scientific information of varying confidence levels, and the values of stakeholders and decision-makers" (US Global Change 2014).

A 2014 University of Michigan report, *Assessing the Climate Adaptation Resource and Service Landscape,* concluded that a significant number of resources exist to support the early phases of the adaptation process, including undertaking risk and vulnerability assessments (1068 resources) and creating a climate adaptation plan (988 resources). However, resources focused on the later phases of the adaptation process, such as implementing strategies, making legal and policy changes, financing adaptation activities, and monitoring and evaluating progress, are the least common. The analysis also identified one notable example: "the dearth of resources focused on financing climate adaptation activities, despite the knowledge that financial constraints are frequently cited as the largest barrier to moving forward with adaptation activities." The report concluded that "as local and regional practitioners begin moving from assessing and planning into action, more specific and targeted resources focused on helping them implement key adaptation actions will be needed" (Stults et al. 2014).

Bottom line: To do more, cities need more funding, technical, and political support.

HOW CITIES HAVE SUCCEEDED IN MOVING FROM RESILIENCE PLANNING TO ACTION

There is no one-size-fits-all adaptation—every city must prepare for the unique threats and opportunities created by its own geography and climate. But there are commonalities in assets and approaches that have helped move adaptation from planning to action.

These commonalities begin with the recognition and acceptance of three basic concepts:

No more "people versus nature." To survive and thrive in a changing climate we must acknowledge that we cannot control or tame nature and that we must learn to live with nature and look to nature for what are termed soft solutions.

According to the Presidential Task Force on Climate Preparedness and Resilience, "Natural systems are important features within the built environment, providing buffers against flood impacts and storm surge, storing water and recharging aquifers, helping to manage stormwater and moderate local temperatures, and providing vital habitat for native and migratory wildlife. Green infrastructure, also called natural infrastructure or natural defenses, for example wetlands, healthy reefs, living shorelines, dunes, floodplains, and forests, can mitigate risks to life and property while providing other social, economic, and environmental benefits, including carbon sequestration. Utilizing green infrastructure alongside traditional infrastructure can help communities, public agencies, and private industry prepare for and respond to climate change in a cost-effective manner, and enhance natural and social systems" (White House 2014).

Cities are using more nature-based solutions: "Our approach is not to simply build a wall and think the problem is solved, but to explore all options, including looking to nature for solutions, like wetlands," explained Debbie Raphael, director of San Francisco's Department of the Environment (Beamish 2014).

Cities are indeed learning to live with the changes in the climate. "In many cities beset with floods…the mind-set has been to keep the water out, rather than to plan for the water to come in on occasion…But increasingly engineers and planners—as well as building owners and insurance companies, realize that walls and embankments and levees have their limitations, especially when it's clear the sea will continue to intrude farther into our built environment. The paradigm is now changing to design buildings, particularly the ground floors, so they can withstand some defined level of flooding…It's the notion

FIGURE 1.5 Built in 1982 to protect London from flooding caused by tidal surges, the Thames Barrier is made up of 10 steel gates that span more than a quarter mile across England's largest river. (Credit: ©mirrormere–Fotolia.com. https://www.dropbox.com/sh/02hcu7g56via56r/AACxcp-YYLs2QPE76X_0XTLAa/Images?lst#lh:null-Thames%20Barrier.jpg.)

that there are reserve capacities and multiple lines of defense built into any resilient urban design that can deal with the possible failure of the first line of defense…This is the new thinking in the flood world," reports Rodin of The Rockefeller Foundation (Rodin 2014b).

Hard solutions, such as dikes, levees, and seawalls, will continue to have a role in building resilience but it is the combination of hard and soft infrastructure solutions that is proving most effective in protecting lives and property and building resilience (see Figure 1.5).

There will be no returning to "normal." Simply restoring what has been lost won't protect us from what's coming. We will not be prepared for the future if the goal is to return to what was. The scale and frequency of ongoing and future events mean we need to calculate for the broad array of changes that are now evident and increasing and to build in flexibility to continue adapting to ever-changing conditions. "The climate of the past is not a good basis for future planning" (US Global Change 2014).

According to the National Climate Assessment, "After a long period of relative stability in the climate system, climate conditions are changing and are projected to continue to change. As a result, historically successful strategies for managing climate-sensitive resources and infrastructure will become less effective over time" (US Global Change 2014).

"We can't continue to delude ourselves that things will get back to 'normal' someday. They won't. It's a losing game…" (Rodin 2014b).

Building for the future helps us with the now. Preparing for climate change and building resilience for the future has near-term cobenefits. "Actions to reduce emissions, increase carbon uptake, adapt to a changing climate, and increase resilience to impacts that are unavoidable can improve public health, economic development, ecosystem protection, and quality of life…Climate change adaptation actions often fulfill other societal goals, such as sustainable development, disaster risk reduction, or improvements in quality of life, and can therefore be incorporated into existing decision-making processes" (US Global Change 2014). Former President Bill Clinton believes "We can prepare for the unexpected—and by doing so make our communities stronger, more prosperous and more connected in the process" (Rodin 2014a). Cities surveyed by MIT and ICLEI also noted that adapting to climate change "will advance community livability, reduce impacts from natural hazards, and improve health conditions" (Knapp 2014).

COMMON ELEMENTS AND PRACTICES SHARED BY RESILIENT COMMUNITIES

Leadership

"It's not really a scientific problem anymore. Essentially everything that needs to be done to move the needle is political," according to David Victor, one of the authors of the IPCC working group report and a professor of international relations at the University of California San Diego (Shogren 2014).

"We see across the world—and not only in the first-world countries but also in the developing and emerging countries—that if you have a very engaged mayor, not only in terms of climate change but environmental issues in general, a lot can be done…" according to Diana Reckien of Columbia University's Center for Research on Environmental Decisions (Dattaro 2013).

According to the Georgetown Climate Center, "Strong leadership is an essential factor in communities that pursue adaptive action. Implementing adaptive measures requires a significant investment of time, staff, and resources. To devise workable solutions, we must provide support to those actors who show the political will to be bold and experimental, to commit resources, and to build the capacity to implement measures on the ground" (Grannis et al. 2014).

In the analysis of its 2012 survey of cities around the world, MIT concluded that "Cities are facing the dual challenges of limited financial resources and limited commitment from local and national governments to adaptation…results of the survey suggest that without the commitment of local political officials and the acknowledgment of this agenda by national governments, it will be difficult to make rapid advances in planning and to move from planning to implementation. Financial and informational resources are critical, but the commitment of local officials is essential to advancing adaptation planning and implementation" (Carmin et al. 2012).

The Presidential Task Force on Climate Preparedness and Resilience identified examples of innovative and forward-thinking leadership that can be found across the country: "For example, the City of Houston has created a network of mobile solar-powered community support and disaster response stations that can operate off the electric grid and provide basic needs to the community in the aftermath of major disasters; communities from Vermont to Des Moines and Fort Collins to Fond du Lac Reservation have recovered from severe floods and storm damages by rebuilding roads and other infrastructure with specific designs for better withstanding future hazards; four counties in Southeast Florida joined together to establish a coordinated planning effort to adapt to sea level rise; and low-lying states like Delaware and Maryland have established requirements for state-funded construction projects to be designed to accommodate future sea level rise and increased flooding" (White House 2014).

According to the Center for American Progress report *Storm-Ready Cities*, "Many city leaders—such as those in New York City, Washington, DC, Houston and Miami—are developing innovative strategies to reduce the risks from extreme weather...These leaders are working to meet priorities such as upgrading public transit and other infrastructure, providing cleaner and more reliable energy, creating jobs, attracting new businesses, improving air quality and expanding parks and green spaces. To adequately prepare our nation for the impacts of climate change, more metro areas must follow their lead" (Kelley and Bhattacharyya 2013).

The World Bank noted that "Adaptation is not a one-time effort but an ongoing cycle of preparation, response, and revision. It is a dynamic process and one that should be revised over time based on new information. Underpinning the strongest adaptation processes will be leadership and commitment to measuring progress and assessing effectiveness" (World Bank 2011).

A Crisis Sparked Change

Adaptation is disaster-driven. Disasters give communities the rare opportunity to take advantage of heightened public concern and political will to integrate climate protection and resilience into their rebuilding plans. After major disasters, people are most likely to commit to taking action. According to risk communications expert Peter Sandman, "Feeling runs high. The need is obvious. The urgency is there. That's why it's important to move as swiftly as possible in the revitalization process. Because the disruption and its effects are highly available, it is more likely that people will be able to focus on them and be willing to devote energy looking for solutions" (Rodin 2014a) (see Figure 1.6).

Cities need to be prepared after extreme weather events "to seize the moment to steer political consensus to adaptation activities," advises ICLEI–Europe in their report out of their Resilient Cities

FIGURE 1.6 New York building back better. (From https://www.fema.gov/media-library/assets/videos/83072.)

2014 meeting (ICLEI–Europe 2014). The Georgetown Climate Center has found that, "Although adapting to climate change impacts should be a proactive endeavor, the reality is often the opposite, as communities consider changes in the wake of catastrophic events or near misses. Experiences by communities affected by Hurricanes Sandy, Irene, and Katrina, however, show that planners are better prepared to seize the opportunity to rebuild resiliently after a disaster when they have well-developed plans and laws on the books in advance of the disaster. Otherwise, opportunities for improvement are often missed in the rush to restore normalcy" (Grannis et al. 2014).

For New York City, the catalyst was Hurricane Sandy; for Vermont it was Hurricane Irene. A devastating tornado in 2007 led Greensburg, Kansas, to become a model green community and the world's leading community in LEED-certified buildings per capita (Quinn 2013). In Des Moines, Iowa, it was chronic flooding that led Mayor Cownie to ask: "Where should we build? How should we build? Should we be buying out property that's flooded again and again?" (Cownie 2014).

Tom Bates, mayor of Berkeley, California, noted that it took "real natural disasters to provide the sparks" in his community: the Loma Prieta Earthquake in 1989 and a wildfire in 1991. "Once we have an event—we make great progress," Mayor Bates explained (Bates 2014).

Mayor Bates believes that events also teach communities how to better protect lives and property. "They force communities to raise standards. Whenever new housing is needed, new codes are developed. Damage to the transportation system provides us with the opportunity to improve [the] public's ability to get around without a car, lower the number of cars. Schools are modernized. More light and heating come from renewable. We are better prepared for disasters" (Bates 2014).

A 2014 survey by the US Conference of Mayors makes the link between disasters and changes and improvements in emergency response plans for the next event very clear. Cities were asked if they had been affected by a power outage(s) within the past 5 years due to a severe climate event (e.g., major hurricane, tornado, significant snow or ice storm, major flooding). More than 4 in 10 (41 percent) cities responded affirmatively. Today, "three-quarters of cities have developed an energy emergency response plan that maintains key municipal services during power outages. Three in four cities (76%) have plans in place to sustain key municipal functions during power outages" (US Conference of Mayors 2014).

One challenge for cities is to keep the public engaged and supportive between disasters. Berkeley, California, has been named one of The Rockefeller Foundation's 100 Resilient Cities, and Mayor Bates says it's one of the challenges for the new Chief Resilience Officer (CRO) funded as part of the designation: "We continue to have public meetings and citywide exercises to improve our response…One of the reasons we need to have a Chief Resilience Officer is to help us think through how best to keep folks informed and keep their guard up" (Bates 2014).

Climate Preparedness and Resilience Building Were Folded into Existing Plans and Planning Processes

"Decisions made today about where and how communities grow, the infrastructure they build, and the codes and standards they adopt will affect them long into the future, so decision makers must take climate change into account as they plan," according to the White House Task Force on Climate Preparedness and Resilience (White House 2014). In fact, the World Bank believes that incorporating planning for climate impacts into existing efforts yields the most success. "Those cities that are best able to integrate adaptation well within a broad spectrum of existing planning processes and goals—including priorities in disaster risk reduction, sustainable development, and poverty reduction—will be best positioned to thrive in this new era of climate change" (World Bank 2011).

According to the Georgetown Climate Center, "Effective adaptation planning must lead to tangible actions and align with existing priorities in a community if it is to be successful" (Grannis et al. 2014).

The World Bank recommends using DRR "as a platform from which to develop climate change adaptation plans. In practical terms, disaster risk reduction and climate adaptation can be integrated in many instances…Entry points other than DRR, such as development planning, infrastructure upgrades, can also be used for adaptive efforts in cities" (World Bank 2011).

Fred Buckenmeyer, public works director of Anacortes, Washington, notes "Folks like me who are building facilities to last 100 years can't afford not to weigh all the factors." Buckenmeyer "incorporated protective features into a $65 million redesign of his city's aging water treatment plant after learning it will be increasingly vulnerable to flooding and sediment from the adjacent Skajit River" (Beamish 2014).

Grand Rapids Mayor Heartwell was able to mainstream climate change and risk reduction into his city's traditional planning process. "We started with a sustainability plan in 2005 and we kept improving the plan. It required us setting and making progress toward triple bottom line goals—economic, environmental, and social equity. We track some 100 indicators" (Heartwell 2014). Heartwell was able to include resilience planning in the last two cycles, "embedding initiatives into the plan and integrating sustainability and resilience rather than separate them" (Heartwell 2014).

"These are things we ought to be investing in anyway," according to Mayor Heartwell. Stormwater management, for example. "We can project what we will be facing with climate change—we will see more precipitation in the fall/winter and spring months and hotter/dryer summer months. What will those rainfalls look like in 20/50/100 years? And when we move from models—what will that require us to do? Upsize pipes, increase basin capacity for holding water? What other solutions should we be implementing?"

"Much more diffuse on-site treatment of stormwaters. Policies about where roads are and should be built. Pervious parking lots, bio-retention islands, swales in the middle of the street, and rain gardens are all built into the engineering now," Mayor Heartwell explained (Heartwell 2014).

Cities Are Preparing for Climate Change, but Not Calling It That

Some cities "don't advertize that some of the planning changes being made are resilience strategies. Instead they emphasize the economic benefits, making them more resilient to natural disasters," explained Martin of the National League of Cities. In some cities, "You get no extra points for saying this is going to help us prepare for climate change" (Martin 2014).

According to MIT professor JoAnne Carmin, "Many cities in the US are exploring ways they can pursue adaptation in light of the changes they are experiencing or the data that is readily available from regional studies. However, given the politicized views of climate change in this country, it seems that some cities are emphasizing risk management—that way they can get on with the important tasks of reducing risk and safeguarding local residents and municipal assets" (Knapp 2014).

In Taos, New Mexico, climate adaptation recommendations were added to the city's comprehensive land-use plan, "without mention of the word 'climate,' instead working them in under the 'Natural Resources Conservation' element of the plan." The lesson learned, according to the Headwater Economics report, is that "Sometimes the best way to proceed locally is to avoid the words 'climate change'" (Headwaters Economics 2012).

"The City of Boulder has included pieces of what might be considered climate adaptation in the city master plan for years without explicitly calling them that," according to Headwater Economics. For example, the Integrated Pest Management program has begun considering the emerging health issues around new diseases like West Nile virus appearing because of changes in climate and water resources (Headwaters Economics 2012).

Early on, Grand Rapids Mayor Heartwell explained "we talked about 'adaptation planning.' Now we have modified the vocabulary to talk about resilient communities—it doesn't trigger the same response from the climate deniers. It's like the difference between the 'climate change' and 'global warming'—that's a red flag in front of a bull." Climate change is much less politically loaded, according to Mayor Heartwell (Heartwell 2014).

Mayor Heartwell also believes the term "resilience" has advantages over "adaptation." "Resilience is a much bigger basket to put things in—we can include law enforcement and public safety challenges we face in a warmer environment, like seniors without air conditioning in extreme heat" (Heartwell 2011).

DEFINING RESILIENCE AND SUSTAINABILITY FOR DECISION MAKERS

Cooper Martin
Program Director of the National League of Cities' Sustainable Cities Institute

What is the difference between sustainability and resilience? As the movement in support of resilient communities has accelerated, this is the most common question from city leaders and other professionals in architecture, planning, and government. Even people who see value or urgency in pursuing sustainability and resilience have a difficult time explaining the terms to others.

Definitions are easy enough to find for both of these terms in isolation. Yet the definitions alone ignore the heart of the question, which is how these two concepts relate to, interact with, or differ from one another.

Casually, some explain that the terms are "two sides of the same coin."* Often, groups organized around sustainability focus on interconnectedness while those organized around resilience will more often focus on interdependencies. For many decision makers, these are semantic distinctions without a difference.

Crudely, others will confess their belief that sustainability has become too polarizing and resilience is a better brand with wider appeal. The implication again is that the two ideas are fundamentally interchangeable. However, the term "sustainability" has become less effective for building consensus and resilience provides a convenient pivot.

Most worryingly, some claim that the two are different, but that our civilization has missed its chance to be sustainable. Climate change, resource constraints, and other disruptive perils are coming and the pragmatic course of approach is to "forget sustainability, it's all about resilience."† In this framework, the two are at least understood to be different, but resilience is viewed as being more complete.

At best, each of these comparisons is incomplete because they fail to consider implementation. How can leaders with authority over building projects, community master plans, or entire cities actually incorporate the concepts of resilience or sustainability into their decision-making process? And are there important differences between the two?

After discussing this question with colleagues and practitioners from a variety of fields, it seems the best answer is this: Sustainability is the process of limiting the impact a given project has on the environment. Resilience, however, starts with the premise that we should limit the impact that deleterious environmental effects, such as flooding or wind damage, will have on a project. The difference may seem slight, but contrasting the two terms this way enables decision makers to set different goals and apply different metrics to pursue sustainability and resilience in tandem. It is a framework that is both actionable and distinct. Most importantly, this demonstrates why it is so important to embrace both sustainability and resilience, because applying the framework will yield different results when applied at different scales.

Let's consider a simple example: retrofitting an existing house somewhere along the East Coast. Within a given budget, there are a wide variety of upgrades any homeowner could make.

* See, for example, http://www.naturalnews.com/039017_sustainability_resilience_growth.html and http://www.mistraurbanfutures.org/sites/default/files/urbes_factsheet_05_web.pdf.

† Andrew Zolli. Forget Sustainability, It's All About Resilience: http://tumblr.poptech.org/post/34911186572/forget-sustainability-its-about-resilience. The article also appeared in the *New York Times*: http://www.nytimes.com/2012/11/03/opinion/forget-sustainability-its-about-resilience.html?pagewanted=all&_r=1&. November 2012.

A resilience-minded homeowner wants to know what he or she can do to reduce the risk of specific threats such as coastal flooding or wind damage from hurricanes. The metrics that are important to this homeowner will include risk of storms with various wind speeds or storm surges. There are established standards for impact resistant windows, or for flood risk, that can be referenced to improve the durability of the home itself.

The homeowner might choose to elevate his or her house to protect it from coastal flooding. Insulating the home or supplying it with renewable energy can improve habitability if the power grid goes out. Structurally, garage doors can be reinforced, shutters can cover windows, and roof straps can be installed to protect the areas of the house most vulnerable to wind damage.

Throughout the process the stated goal is to save on the financial and emotional cost of a future disaster. Known expenses are balanced against anticipated benefits.

Conversely, a homeowner wishing to improve the sustainability of his or her home will want to understand how to reduce its environmental footprint. With this stated goal, relevant metrics to consider will be energy consumption, water use, or carbon impact.

This homeowner might install renewable energy sources or improve insulation to reduce energy consumption. He or she knows runoff is a significant source of point-pollution and installs rain barrels or porous surfaces. A compost system could be installed to fertilize home-grown food and limit waste.

Known financial costs are balanced against both environmental and economic benefits and the whole process reduces future environmental impact by making thoughtful investments.

By visualizing these improvements from the perspective of a homeowner, it is easy to see the distinction between sustainability and resilience in both process and outcome. Notice that in both cases the homeowner has good reason to reduce energy consumption and upgrade insulation. This strongly suggests the improvements are financially viable and might be prioritized above other potential upgrades, and indeed, it is one of the most implemented upgrades at the start of the green building movement.

But the results of this same exercise can be very different when it is conducted at the whole-community scale.

It is not immediately clear to our resilience-minded homeowner that raising the home above the base flood elevation can also reduce his or her environmental footprint. The most sustainable communities, however, are those that weather the storm with less need to rebuild on the backs of insurers and taxpayers. It is estimated that between 25–40 percent of US solid waste is building-related waste and debris. Disasters at a large scale can compound environmental and economic cost and even jeopardize community social structures.

Similarly, a homeowner who is simply attempting to live sustainably will not notice his or her flood risk change by using an impervious concrete driveway. However, when this decision is magnified hundreds or thousands of times throughout a community, water that cannot be absorbed into the environment will create flash flood hazards where none previously existed. A risk that is miniscule at first is amplified at a larger scale.

By defining resilience and sustainability in relation to one another and embracing both, leaders are empowered to view goals, challenges, and opportunities more holistically. Architects, homeowners, and other professionals working at a building scale can account for costs and benefits. Planners or city managers can identify collective action problems or structure incentives for development that better serves the whole community. With this framework, the similarities and differences between sustainability and resilience can not only be understood, they can be built in to a more comprehensive decision-making process.

Cities Took Actions That Have Co-Benefits

Climate-change adaptation actions often have implicit and beneficial effects that improve the local economy and quality of life, such as improving the cost-effectiveness and productivity of infrastructure, increasing access to clean and reliable electricity, expanding public transit and green spaces, and improving air quality while reducing the risk of disasters, thus limiting their impact and helping cities, institutions, and businesses recovery and rebound more quickly.

According to the Presidential Task Force on Climate Preparedness and Resilience, "Developing resilient communities means more than just building in a way that allows for rapid recovery from and avoidance of future disasters…Characteristics of these communities include clean and abundant water supplies protected for future generations, and energy systems powered by fuels that do not exacerbate climate change or damage public health and are reliable even when disaster strikes" (White House 2014).

The steps taken to reduce disaster risk now—including banning building or buying out property in flood-prone areas, protecting water supplies, improving the reliability of electric and communications systems, restoring natural wind and water barriers, and decentralizing supply distribution centers—all have long-term benefits in preparing communities for future climate change. Siting and designing buildings and infrastructure for long-term climate resilience "can improve cost-effectiveness by helping ensure continuity of operations and minimizing recovery costs after a disaster" (White House 2014).

"Strengthening grid resilience, for example, improves energy reliability, which reduces economic losses during storms and can lower emissions by increasing the availability of clean distributed power. Planting more trees and creating more parks reduces stormwater runoff while increasing property values, carbon storage, and quality of life. Creating diverse transportation and energy systems provide backups and alternatives that allow people to get to work, businesses to stay open, and the lights to stay on, and communities to stay safe, even when parts of these systems are damaged by a storm" (Kelley and Bhattacharyya 2013).

Some cobenefits can be realized immediately. For example, when various departments from around San Francisco Bay showed up to fight a major fire in Berkeley, California, Mayor Bates explained, "They couldn't connect their nozzles to our hydrants. They also couldn't communicate with one another" (Bates 2014). Berkeley has addressed both those issues—and is recognizing the near-term and long-term benefits of a better-coordinated fire-fighting force in the region.

In addition, Berkeley has invested in material and training that will also yield near- and long-term benefits. "We have acknowledged and acted on the fact that it could be days before city services can get help to you and your neighborhood—it could be three days without assistance. That's why they have set up 87 caches of essential materials throughout neighborhoods to help people to take care of their own family and neighborhood including first aid equipment, tools, generators, jaws of life—even dumpsters—so they can start to recover faster. We have trained 5000 people to help the injured, put out fires and shut off gas valves. We have also identified senior citizens and the disabled in our neighborhoods" (Bates 2014). "As part of its Climate Preparedness and Engagement Program, the city has also been training underserved young adults (ages 18–25) to be disaster ambassadors to deliver preparedness education and coordinate preparedness trainings and activities in their neighborhoods" (City of Berkeley 2014).

The Presidential Task Force report cited the example of Houston's decentralized supply distribution centers: "a network of mobile community support and disaster response energy stations that can operate off the grid and provide basic needs to the community. The solar generators/mobile offices, with battery back-up, are designed for emergency relief efforts after hurricanes or cooling centers during times of extreme heat. Support provided by these units includes water and food, charging stations for phones and medical equipment, and case work assistance. When not being used in an emergency, they are used year-round for services, outdoor classrooms or to educate the public and bring awareness to solar projects" (White House 2014).

Post-Hurricane Sandy, new pilings are being produced to replace the old ones that now support New York City's docks and piers in Lower Manhattan. The new piers are made of "digital concrete"—a stone/concrete-like material that's produced with a 3-D printer. "Flexible, adaptive and strong, they bend in response to wave action rather than trying to withstand it, making them far more resilient and cost effective in the long run," according to Rodin of The Rockefeller Foundation (Rodin 2014a).

With help of Mercy Corps and supported by the Rockefeller Foundation, Lamnga villagers in Indonesia replanted the mangrove forest destroyed by the 2004 tsunami with seedlings. "The venture has been so successful that the villagers have developed a new enterprise for themselves: selling seedlings to other vulnerable communities along the coast to generate additional income. As a result the villagers are gaining greater protection for their geography, ensuring their traditional livelihoods (mangroves supported fish, shrimp and crab populations), and creating a new enterprise—another example of the resilience dividend" (Rodin 2014a).

A Broad Array of Local Stakeholders, Including the Public and the Business Community, Have Been Involved in Planning

Locally led and inclusive planning efforts have been essential in the development and implementation of adaptation plans. "Any viable response to climate change adaptation must recognize the primacy of local governments but also work within multiple jurisdictional levels (states, counties, municipalities, regional planning organizations) and recognize their legal limitations," according to the Georgetown Climate Center (Grannis et al. 2014).

"Resilience requires partnerships. Governments, the private sector, academia and civil society all have roles to play and can contribute efforts—and expertise—to a resilience agenda. Resilience cannot be built by one sector alone" (Boltz and Granlund 2014).

"Pulling these resources together to build resilience strengthens a community's capacity to work together and problem-solve before, during and after a disruption arrive," explained The Rockefeller Foundation president Rodin. "These community partnerships are vital not only because they generate creative and collaborative solutions to vulnerabilities but also because they unite disparate groups in the community around a common cause of readiness" (see Figure 1.7).

The City of Boston, for example, has been recognized by the White House as a "Climate Action Champion" for creating a "Green Ribbon Commission," which represents businesses, nonprofits, and community leaders across a variety of sectors working to develop shared strategies for fighting climate change in coordination with the city's Climate Action Plan (Emergency Management 2014).

London Mayor Boris Johnson has developed an adaptation strategy in response to rising concerns about persistent flooding, drought, and extreme heat waves in the city. "The mayor's office kicked off the initiative in early 2010 by calling on London's 8.1 million residents to share ideas

FIGURE 1.7 Locals discuss ideas at Sea Bright 2020 Community Workshop. (From https://www.fema.gov/media-library/assets/images/72414.)

and expertise in an online forum. 'Either we can grow gills, or we'll need to think of other ways of adapting,' Mayor Johnson said in a promotional video. The website still maintains an active discussion" (Gallucci 2013).

Greensburg, Kansas, Mayor Bob Dixson explained that citizen involvement was essential to building his tornado-devastated community back better and stronger: "In the planning and rebuilding process after the May 2007 tornado, Greensburg citizens met at community meetings to plan the future. This process allowed us to address systematic problems that could be corrected in the rebuild. Sustainability and rebuilding 'green' were the focus of being a resilient community" (White House 2014).

The World Bank has concluded that "Building resilience requires not only robust decision making by those in positions of formal authority, but also a strong web of institutional and social relationships that can provide a safety net for vulnerable populations" (World Bank 2011).

Business buy-in is essential, too, and they have responded to two key motivators: the need to minimize downtime caused by a disaster and the opportunity to capitalize on the opportunities inherent in resilience-building, which include new technologies, new markets, and new jobs. "Business is a powerful engine for change in the San Francisco area," according to Berkeley's Mayor Bates. "The business community here is one of the most powerful and influential groups in the world—and they are thinking about climate change because they want to stay in business" (Bates 2014) (see Figure 1.8).

Des Moines Mayor Cownie notes that adapting to climate change has a "tremendous upside for business. Preparing for what's coming—the greening of the infrastructure—leads to a wide variety of new green jobs. There's a huge opportunity for a new economy built upon a new way of thinking" (Cownie 2014).

In addition to convening stakeholders within a community, successful adaptation strategies have been developed and implemented as a result of multicity, county, and regional collaborations. In California, for example, San Diego, Los Angeles, Sacramento, and the San Francisco Bay Area have formed a regional collaborative to coordinate planning across counties and municipalities in each region.

Participation in regional compacts, like the Southeast Florida Regional Climate Change Compact, which is made up of Palm Beach, Broward, Miami-Dade, and Monroe counties, helps local governments set a unified agenda for adaptation, develop consistent sea-level rise scenarios and consistent methods for mapping impacts in the region, and allows state and federal agencies to efficiently

FIGURE 1.8 Converted from an early 1920s auto garage, this popular Seattle coffeehouse was seismically retrofitted with cross-bracing to stabilize and reinforce the original masonry walls. (Credit: Photo by John M. Feit, 2010. https://www.dropbox.com/sh/02hcu7g56via56r/AACxcp-YYLs2QPE76X_0XTLAa/Images?lst#lh:null-X%20Bracing.jpg.)

provide technical assistance and support, reaching several counties at once. The regional approach, according to the Georgetown Climate Center, "also helps the region avoid less effective piecemeal efforts to implementing adaptation strategies. This coordination has helped these Florida counties plan protective measures for power plants and identify nature-based strategies, such as sand dune restoration, to protect against flood impacts" (Georgetown Climate Center 2014a).

Found a Way to Pay for Programs and Activities That Build Resilience

Cities are in a tough spot. Their populations are growing, demand for public transit and other services are increasing, and their streets, bridges, water, sewage systems, and other infrastructure are in decay. Many are cash-strapped and the steps needed to strengthen their resilience to extreme weather will be expensive. But, as the CAP report *Storm-Ready Cities* makes clear, "failing to prepare communities and shore up critical infrastructure to withstand future superstorms and other climate impacts will cost even more in the long run. Finding cost-effective ways to build resilience in cities will save on disaster aid and reduce damages and economic losses from future extreme weather" (Kelley and Bhattacharyya 2013).

According to IPCC chairman Rajendra Pachauri, "The costs will go up enormously if we keep delaying things. The cost of inaction will be horrendously higher than the cost of action" (Morris 2014). Rodin at The Rockefeller Foundation noted: "The collective cost of climate change disasters is estimated at $200 billion every year...Investing in resilience now can help avoid the devastating financial costs of natural disasters—while making day-to-day life better for everyone. Make no mistake: there may be upfront costs to get this done, but you will save money later: It costs 50 percent more to rebuild in the wake of a disaster than to build in a way that can withstand the shock" (Rodin 2014a).

A study by the Multihazard Mitigation Council found that every $1 that the Federal Emergency Management Agency (FEMA) invests in resilience saves the nation $4 in disaster-recovery costs. "And that's just in federal dollars," according to Martin of the National League of Cities. "That doesn't include insured or private costs. Then the number's more like seven or eight dollars saved. The fact is it's much more productive to think about climate change upfront" (Martin 2014).

Martin is confident the "common sense financial aspects of the need to build resiliency will cut through" (Martin 2014). For example, how the insurance industry calculates acceptable risks and what they are willing to cover and pay for is changing. Some of that is tied to the National Flood Insurance Program's Community Rating System (CRS), "a federal voluntary incentive program that adjusts local flood insurance premiums to reflect the level of flood protection and damage reduction a county requires in its flood management program" (FEMA 2015). That means insurance premium rates can be discounted when a community goes beyond the basic flood protection program standards. Those cost savings could create public demand for community adaptation.

Even though the wisdom of up-front investment is spreading and the federal government and philanthropic foundations like The Rockefeller Foundation and Kresge Foundation are investing more money in creating climate-resilient communities, it is local jurisdictions that are being forced to find funding for their adaptation programs. Some examples include

Berkeley, California

The city allows one-third of the 1.5 percent of the transfer fee that's paid to the city when buildings are sold to be deferred if the property owner takes steps to earthquake-proof his or her homes. Today, close to 75 percent of the city's buildings have been retrofitted for earthquakes. The city also passed a ballot measure that included $600 million in taxes and bond measures to retrofit all public schools, libraries, and fire stations and purchase equipment that would allow Berkeley to pump water from the bay or river east of the city in the event of a disaster (Bates 2014).

Chicago

A key feature of Chicago's adaptation approach is the inclusion of supporting non-city entities in the overall approach to creating and implementing a climate change plan. "Chicago has a long-standing tradition of private sector support for city initiatives…the nonprofit sector invested heavily in supporting the city's planning efforts. This included over $1.5 million in direct funding to support the city's process as well as the recruitment of extensive pro-bono professional services and expertise" (Headwaters Economics 2012).

Dane County, Wisconsin

Dane County, Wisconsin, budgeted for climate adaptation as part of capital improvement projects, anticipated changes in temperature, and precipitation from climate change. "Funding was included to purchase 'blizzard-busting' trucks and trimming equipment to address storm-downed trees, and to install larger culverts to reduce flood impacts from higher stream flows" (Georgetown Climate Center 2014a).

Charlotte and Mecklenburg County, North Carolina

Charlotte and Mecklenburg County, North Carolina, use stormwater fees to implement a buyout program that aims to reduce flood damage by buying high-risk properties. "The stormwater fees are levied on all water customers as determined by the location of the property, its square footage of impervious surfaces (e.g., rooftops or driveways), and the cost of providing stormwater services" (Georgetown Climate Center 2014a). The fees are used to fund three different types of floodplain buyouts: "annual buyouts are selected for acquisition based on overall flood risk to the property and other benefits that can be provided to the community through acquisition; 'quick-buy' properties are structures that the county purchases in the immediate aftermath of damage from destructive flooding; and 'orphan' property acquisitions use stormwater fees to buy properties that did not meet the criteria for a federal grant buyout, yet are bought out because they are adjacent to other properties that have been bought out" (Georgetown Climate Center 2014a).

Grand Rapids, Michigan

Grand Rapids took advantage of an ongoing infrastructure problem—poor, potholed streets—to make the community more resilient. As the city's Climate Resiliency Report made clear in 2013, bad streets were only going to get worse as the climate changed and the weather got worse: "Road infrastructure was designed to withstand local weather and climate based on historical weather data and records of extreme storms. However, the EPA warns that 'historical climate is no longer a reliable predictor of future impacts.' Increased precipitation softens and erodes underlying base layers of roads, which results in damaged, warped, and cracked road surfaces" (WMEAC 2013).

Among the climate report's recommendations were to "study and implement new methods of street maintenance and construction, particularly relating to materials and the construction of physical infrastructure resilient to climate change impacts…But, resources are needed to implement the plans" (WMEAC 2013). The report also noted that the lack of funding also prevents other roadway projects "that could improve Grand Rapids' climate-resiliency infrastructure," including pedestrian and bicycle pathways, increased green space in road medians, and rain gardens to collect stormwater (WMEAC 2013).

In March 2014, Grand Rapids voters approved a street tax to implement repairs and resiliency-related infrastructure by "the largest margin on a tax question in 20 years," according to Mayor Heartwell (Vande Bunte 2014).

State of Massachusetts

"Massachusetts communities are receiving grants from a $50 million fund drawn from the state's capital budget and the fees utilities must pay to the state when they fail to meet renewable energy portfolio standards. The money will be used to protect energy systems, fortify sea walls and dams and fight seawater incursion and flooding" (Beamish 2014).

LESSONS LEARNED: PRACTICAL ADVICE ABOUT HOW TO ADVANCE CLIMATE RESILIENCE

Adapted from the Headwaters Economics' report, *Implementing Climate Change Adaptation: Lessons Learned from Ten Examples* (Headwaters Economics 2012)

FOCUS ON AN IMMEDIATE, RECOGNIZABLE THREAT

Start with an immediate, relevant threat—something the community has experienced before. Drought in Taos, flooding in Grand Rapids, forest fires in Boulder, and a hurricane in New York spurred planning and action.

RECOGNIZE WHAT THE COMMUNITY VALUES ARE AND USE THAT AS A PORTAL INTO THE LARGER ADAPTATION DISCUSSION

If the community cares about its beach front, then a constructive entry point may be the effect of climate change on coastal erosion and flooding.

START WITH AN EXISTING PROCESS

Integrating climate change adaptation into existing disaster risk reduction, public health, water management, and other plans and planning opportunities can help change the culture and institutionalize the inclusion of climate impacts into local planning, budget, and implementation.

LOOK FOR LEADERSHIP IN UNEXPECTED PLACES

Leadership can come from public safety workers like firefighters or water managers—folks on the front lines of the climate fight who know well what needs to be done to live with the impacts of a changing climate.

REACH OUT TO THE COMMUNITY

Involve civic, business, academic, and other local communities in climate discussions and planning. It builds participation and trust.

RECOGNIZE LIMITED CAPACITY

Don't get too complicated too soon. Detailed climate models aren't required to communicate what local impacts are likely to be. You don't need to debate the science; focus on community risks and solutions.

DON'T GET TRAPPED BY THE CLIMATE DEBATE

For practical and political reasons, it may be better not to use the word "climate." Talk instead about what's important to the community.

USE ECONOMIC AND FISCAL ARGUMENTS

Economic and fiscal arguments can be important motivators. Keene, New Hampshire's first move on climate adaptation was built on the cost-saving economics of energy efficiency.

ENGAGE IN REGIONAL ALLIANCES

Regional alliances are good ways for local governments to engage with state and federal authorities and benefit from expanded technical assistance and resources.

> Advice for mayors who have come late to the party and don't know how to dress? Do something. Do anything. Replace the light fixtures in City Hall. Buy electric police cars and garbage trucks. Do something and build on that small success.
>
> You will have made progress on climate changes and had a positive impact on your city. You will have saved taxpayer money, reduced emissions, created opportunities to do more. If you start small and leverage that success into more and more progress—you will be building a resilient city.
>
> **George Heartwell**
> *Mayor, Grand Rapids, Michigan*

Washington, DC

Washington, DC, charges a 5-cent fee on all plastic and paper bags used by customers in grocery, convenience, and liquor stores to generate funds to clean up the Anacostia River and support adaptation programs. Programs include a subsidy for green roofs, stream restoration, and community education and outreach (Georgetown Climate Center 2014a).

Adaptation in Action

There are cities and states actively engaged in building climate resilience. The primary mechanisms that local governments are using include requiring climate change impacts to be factored into land-use and other planning, provisions to protect infrastructure and ecosystems, regulations related to the location, design, and construction of buildings, roads, and bridges, and improving emergency preparation, response, and recovery (US Global Change 2014) (see Figure 1.8).

Examples of current adaptation actions are included below.

Requiring That Climate Change Impacts Be Factored into Plans, Programs, and Budgets

- In New York, the Community Risk and Resiliency Act requires state agencies to consider climate impacts in many permitting and funding decisions, including the siting of bulk storage and hazardous waste facilities and the development of oil and gas drilling permits. State agencies are also preparing model laws to help communities incorporate climate risks into local ordinances (Georgetown Climate Center 2014a).
- California has instructed all of its state agencies to prepare for a rise in sea level and other warming effects. The state has developed guidelines for transportation planners and coastal communities and is creating a Sea Level Rise Database as a clearinghouse for adaptation strategies from across the state. In San Francisco, all future construction projects, city and county agencies, including low-lying San Francisco International Airport, must come up with plans to adapt to rising sea level (Beamish 2014).
- The Washington State Department of Transportation (WSDOT) requires the consideration of climate change impacts in environmental review documents. As a result, large state-funded infrastructure projects now consider climate change impacts over the project's lifespan. According to WSDOT's directive: "Project teams are expected to ask and answer the question, 'How will my project be affected by climate change?' That means that decisions about enlarging culverts, for instance, should factor in rapidly melting glaciers and severe rains" (Beamish 2014).

Enacting Provisions to Protect Infrastructure and Ecosystems

- The post-Hurricane Sandy plan, *A Stronger, More Resilient New York*, proposes more than 250 initiatives to reduce the city's vulnerability to coastal flooding and storm surge. "About 80 percent of the $19.5 billion plan will go to repairing homes and streets damaged by

Sandy, retrofitting hospitals and nursing homes, elevating electrical infrastructure, improving ferry and subway systems and fixing leaky drinking water systems. The rest will go to building and researching floodwalls, restoring swamplands and sand dunes, and other coastal flood protections" (Beamish 2014).

- In Houston, massive flooding in July 2012 prompted the launch of Rebuild Houston, "a city-wide effort to build better streets and drainage systems with the goal of reducing the impact of floods on homes and businesses throughout the city" (Johnson 2013).
- Miami Beach already is "reaping flood-control benefits from new stormwater pumps and upgrades that are part of a $400 million plan to fight encroaching tidal waters and storms that flood city streets. The city also requires a minimum elevation of 5.7 feet for new sea walls, up from the previous 3.2 feet, and is elevating some of its roads" (Beamish 2014).
- In Philadelphia, several partners came together to conduct a vulnerability and risk assessment of the Southeastern Pennsylvania Transportation Authority's (SEPTA) Manayunk/Norristown regional rail line. "This line closely parallels the Schuylkill River, which has experienced 10 of its highest 18 crests in recorded history since 2003, resulting in numerous delays and damage. The Pilot Program built capacity and facilitated the beginning of SEPTA's climate planning work, ultimately positioning SEPTA for a competitive award of $86 million in subsequent federal funding through FTA's Emergency Relief Program. SEPTA will use the funds to improve disaster preparedness by building an alternate system control center, stabilizing embankments over commuter railroads, and improving flood protection of tracks" (White House 2014).
- Tybee Island and Miami Beach have started to upgrade their stormwater sewer systems to prevent seawater from backing up into pipes (Sheppard 2014).
- The US Naval Academy in Annapolis is using door dams to protect building entrances from flooding (Sheppard 2014).
- Maryland and several other states are promoting "living shorelines," wetlands and natural landscape features that can provide a buffer against seawater incursion (Beamish 2014).
- Chicago and other localities are using greenscapes and permeable paving to reduce storm runoff flooding (Beamish 2014).
- Norfolk, Virginia, is returning some of its coastal parks to wetlands (Sheppard 2014).
- Massachusetts is investing in natural landscape defenses such as marshes and dune restoration. "A dune rather than a stone revetment will have a lot more ecosystem benefits," as well as climate resilience, said Bruce K. Carlisle, director of the state Office of Coastal Zone Management (Beamish 2014).

Regulating the Location, Design, and Construction of Buildings, Roads, and Bridges

- Maine limits development and redevelopment in high hazard areas. The state's Sand Dune Rules require that structures located in coastal high hazard zones be moved if they are substantially damaged more than one time in a storm event (Georgetown Climate Center 2014a).
- In low-lying Delaware and Maryland, state-funded construction projects now must be built with extreme weather and rising seas in mind. In Maryland, the new standard is 2 feet above the 100-year flood level. In Delaware, all state agencies have been directed to incorporate climate adaptation measures in the siting and design of state-funded projects (Beamish 2014).
- In Washington State, a recently approved project, the Mukilteo Multimodal Ferry Terminal, considered climate change impacts during the environmental review process. "This project will ensure a resilient transportation connection between Whidbey Island and the mainland by improving access, transit connections, and ferry operations. The project was designed to be resilient to future sea-level rise by using fill to elevate the site for the ferry terminal and locating related facilities, such as access roads, in upland areas" (Georgetown Climate Center 2014a).
- Los Angeles plans to reduce extreme heat risks by updating its building codes to promote cool roofs and pavements that reflect heat rather than absorbing it (Beamish 2014).

NATIONAL CLIMATE ASSESSMENT: TERMS

The National Climate Assessment summarizes the impacts of climate change on the United States now and in the future.

Adapt, adaptation: Adjustment in natural or human systems to a new or changing environment that exploits beneficial opportunities or moderates negative effects.

Adaptive capacity: The potential of a system to adjust to climate change (including climate variability and extremes) to moderate potential damages, take advantage of opportunities, and cope with the consequences.

Mitigation: Technological change and substitutions that reduce resource inputs and emissions per unit of output. Although several social, economic, and technological actions would reduce emissions, with respect to climate change, mitigation means implementing actions to reduce greenhouse gas emissions or increase the amount of carbon dioxide absorbed and stored by natural and man-made carbon sinks.

Multiple stressors: Stress that originates from different sources that affect natural, managed, and socioeconomic systems and can cause impacts that are compounded and sometimes unexpected. An example would be when economic or market stress combines with drought to negatively impact farmers.

Resilience: A capability to anticipate, prepare for, respond to, and recover from significant multi-hazard threats with minimum damage to social well-being, the economy, and the environment.

Risk: A combination of the magnitude of the potential consequence(s) of climate change impact(s) and the likelihood that the consequence(s) will occur.

Vulnerability: The degree to which a system is susceptible to, or unable to cope with, adverse effects of climate change, including climate variability and extremes. Vulnerability is a function of the character, magnitude, and rate of climate variation to which a system is exposed, its sensitivity, and its adaptive capacity.

* Definitions adapted from (IPCC 2007; NRC 2007, 2010).

INFORMS THE NATION THE NATIONAL CLIMATE ASSESSMENT SUMMARIZES THE IMPACTS OF CLIMATE CHANGE ON THE UNITED STATES, NOW AND IN THE FUTURE

A team of more than 300 experts guided by a 60-member Federal Advisory Committee produced the report, which was extensively reviewed by the public and experts, including federal agencies and a panel of the National Academy of Sciences.

US Global Change 2014

Improving Emergency Preparation, Response, and Recovery

- "Fort Collins, Colorado, experienced a devastating flood in 1997 that caused loss of life and property. Following that incident, the city implemented a variety of management strategies to mitigate the impacts of floods on life, health, and property in floodplain areas, including floodplain regulations, open space preservation, acquisition of at-risk structures, stormwater capital projects, public education, and flood early warning systems. The Fort Collins Floodplain Management Program is now ranked as one of the top programs nationwide under the FEMA Community Rating System. In September, 2013 another catastrophic flooding event occurred in northern Colorado, causing millions of dollars in property and infrastructure damage. As a result of investments in resilience and mitigation planning, Fort Collins experienced minimal impact, and instead was able to assist neighboring communities in their recovery efforts" (White House 2014).

Implementation of Resource Conservation Strategies

- "Houston has assembled a Water Conservation Task Force to tackle its long-term water needs, which were tested by a severe drought in 2011. The task force is charged with diversifying Houston's water supplies and cutting consumers' water bills through measures like rainwater harvesting, desalination and using greywater and recycled water for irrigation" (Johnson 2013).
- Drought is seen as the most important climate-related risk facing Salt Lake City, Utah, which gets 90 percent of its drinking water from surface sources. "With less rainfall expected in future decades, Salt Lake City will likely face drier conditions and a heightened risk from wildfires. In response, by 2015 the city is working to set aside 10 percent of its surrounding watershed lands and groundwater, and also create roads and trails within the watershed restoration project areas" (Johnson 2013).
- Miami-Dade has several water conservation efforts underway including a 5-year water efficiency plan that will help reduce the demand on its main fresh water sources, like the Biscayne Aquifer (Johnson 2013).
- New York has created a Strategic Gasoline Reserve pilot program to prevent fuel shortages during severe weather and other emergencies. "The $10 million pilot program is part of the state's Fuel NY initiative, launched in response to gasoline supply shortages that occurred during Hurricane Sandy that caused disruptions to emergency response and business operations. The reserve will store three million gallons of gasoline using private industry capacity on Long Island, and it will be used to supplement fuel supplies during an emergency" (Georgetown Climate Center 2014b).

CONCLUSION

The climate has changed and cities are already dealing with the consequences. Faced with more frequent and extreme floods, wildfires, droughts, and heat waves, mayors across America and around the world are not waiting for the scientific or political debates about climate change to be settled. They cannot afford to. "There's simply too much to do and too much at stake," explained former New York City Mayor Bloomberg (Bloomberg 2013).

Progress is being made. More cities are making plans to increase their resilience. They are figuring out how to calculate in the effects of climate change over the coming decades into the planning, investment, and disaster risk reduction decisions they make today. But that planning is still not happening in enough cities—and even where plans have been made, implementation lags.

Bottom line: To do more, cities need more funding, technical, and political support.

REFERENCES

Aylett, A. 2014. Progress and challenges in the urban governance of climate change: Results of a global survey. MIT. Available at https://www.dropbox.com/s/k42x5kejctktdc8/Urban%20Climate%20Governance%20Report.pdf.

Bates, T. Berkeley, California mayor. November 12, 2014. Interviewed by Kim Haddow.

Beamish, R. December 23, 2014. Cities brace for global warming fallout. *Stateline.org, Pew Charitable Trusts*. Available at http://www.pewtrusts.org/en/research-and-analysis/blogs/stateline/2014/12/23/states-cities-brace-for-global-warming-fallout.

Bierbaum, R. et al. June 19, 2012. A comprehensive review of climate adaptation in the United States: More than before, but less than needed. Available at http://link.springer.com/article/10.1007%2Fs11027-012-9423-1#page-1.

Bloomberg, M. New York City mayor. August 26, 2013. Why Sandy forced cities to take the lead on climate change. *CNN*. Available at http://www.cnn.com/2013/08/21/world/europe/bloomberg-why-sandy-force-cities/.

Boltz, F. and A. Granlund. December 10, 2014. Building resilience to the shocks and stresses of climate change. *Just Means*. Available at http://www.justmeans.com/blogs/building-resilience-to-the-shocks-and-stresses-of-climate-change.

Borenstein, S. December 2, 2014. Hotter, weirder: How climate change has changed the earth. Associated Press. Available at http://www.salon.com/2014/12/02/hotter_weirder_how_climate_has_changed_earth/.

Brody, J. June 23, 2014. Too hot to handle. *New York Times*. Available at http://www.well.blogs.nytimes.com/2014/06/23/too-hot-to-handle/?_r=0.

Carmin, J., N. Nadkarni and C. Rhie. 2012. Progress and challenges in urban climate adaptation planning: Results of a global survey. MIT. Available at http://web.mit.edu/jcarmin/www/urbanadapt/Urban%20Adaptation%20Report%20FINAL.pdf.

Chastain, C. et al. 2012. Climate change adaptation: A framework for the City of Philadelphia. Columbia University. Available at http://www.earth.columbia.edu/sitefiles/file/education/capstone/fall2011/Climate%20Change%20Adaptation%20A%20Framework%20for%20the%20City%20of%20Philadeplia_FINAL.pdf.

Cownie, F. Des Moines, Iowa mayor. November 20, 2014. Interviewed by Kim Haddow.

Dattaro, L. December 2, 2013. European cities slow to adapt to climate change, new study says. *The Weather Channel*. Available at http://www.weather.com/science/news/european-cities-slow-adapt-climate-change-new-study-says-20131202.

de Castella, T. February 11, 2014. How does the Thames Barrier stop London flooding? *BBC News Magazine*. Available at http://www.bbc.com/news/magazine-26133660.

Dizikes, P. May 29, 2014. Global survey: Climate change now a mainstream part of city planning. *MIT News Office*. Available at http://newsoffice.mit.edu/2014/global-survey-climate-change-now-mainstream-part-city-planning.

Emergency Management. December 4, 2014. 16 US communities recognized as climate action champions. Available at http://www.emergencymgmt.com/disaster/16-US-Communities-Recognized-Climate-Action-Champions.html.

FEMA. Community rating system fact sheet. Available at http://www.fema.gov/media-library/assets/documents/9998#.

Florida Department of Economic Opportunity. Florida adaptation planning. Available at http://www.floridajobs.org/community-planning-and-development/programs/technical-assistance/community-resiliency/adaptation-planning.

Fritz, A. September 23, 2014. "We have to adapt"—President Obama announces new climate change actions. *Washington Post*. Available at http://www.washingtonpost.com/blogs/capital-weather-gang/wp/2014/09/23/we-have-to-adapt-president-obama-announces-new-climate-change-actions/.

Gallucci, M. June 20, 2013. 6 of the world's most extensive climate adaptation plans. *Inside Climate News*. Available at http://insideclimatenews.org/news/20130620/6-worlds-most-extensive-climate-adaptation-plans.

Georgetown Climate Center. June 20, 2014a. 20 good ideas for promoting climate resilience. Available at http://www.georgetownclimate.org/20-good-ideas-for-promoting-climate-resilience.

Georgetown Climate Center. October 9, 2014b. Find out which US states are making progress preparing for climate change. Available at http://www.georgetownclimate.org/find-out-which-us-states-are-making-progress-in-preparing-for-climate-change.

Georgetown Climate Center. 2015. State and local adaptation plans tracker. Available at http://www.georgetownclimate.org/adaptation/state-and-local-plans.

Grannis, J. et al. July 2014. Lessons from the front line: A synthesis report to the Kresge Foundation. Georgetown Climate Center. http://www.kresge.org/sites/default/files/Preparing%20for%20Climate%20Impacts%20-%20Georgetown%20Climate%20Center.pdf..

Harvard Kennedy School, Shorenstein Center on Media, Politics and Public Policy. October 16, 2014. Global warming, rising seas and coastal cities: Trends, impacts and adaptation strategies. *Journalists Resource*. Available at http://www.journalistsresource.org/studies/environment/climate-change/impact-global-warming-rising-seas-coastal-cities#.

Headwaters Economics. February 2012. Implementing climate change adaptation: Lessons learned from ten examples. Available at http://headwaterseconomics.org/land/reports/climate-adaptation-lessons-learned.

Heartwell, G. Grand Rapids, Michigan mayor. November 14, 2014. Interviewed by Kim Haddow.

ICLEI-Europe. 2014. European cities adapt to climate change. Resilient Cities-ICLEI. Available at http://www.resilient-cities.iclei.org/bonn2014/open-european-day/.

Intergovernmental Panel on Climate Change (IPCC). March 2014. Climate change, 2014: Impacts, adaptation and vulnerability, summary for policymakers. Available at http://www.ipcc-wg2.gov/AR5/report/final-drafts/.

Johnson, T. October 25, 2013. 5 Cities doing big things to reduce their risks from extreme weather. *The Weather Channel.* Available at http://gold.weather.com/news/science/environment/50-cities-reducing-their-risks-extreme-weather-report-20131024.

Kelley, C. and A. Bhattacharyya. October 22, 2013. Storm-ready cities: How climate resilience boosts metro areas and the economy. *The Center for American Progress.* Available at https://www.americanprogress.org/issues/green/report/2013/10/22/77660/storm-ready-cities/.

Knapp, D. May 2014. Survey: US cities report increase in climate change impacts, lag global cities in planning. ICLEI-Local Governments for Sustainability. Available at http://www.icleiusa.org/blog/survey_us_cities_report_increase_in_climate_impacts_lag_in_adaptation_planningworldwide-progress-on-urban-climate-adaptation-planning.

Martin, C. Program Director Sustainable Cities Institute, National League of Cities. November 11, 2014. Interviewed by Kim Haddow.

Morris, A. November 3, 2014. Six dramatic things said about the latest UN climate report. *PBS NewsHour.* Available at http://www.pbs.org/newshour/rundown/six-dramatic-things-said-todays-un-climate-report/.

Plummer, Brad. May 22, 2014a. Should we try to fight rising sea levels—Or abandon the coasts? *Vox.* Available at http://www.vox.com/2014/5/22/5735144/rising-sea-levels-abandoning-the-coasts/.

Plummer, B. October 22, 2014b. Everything you need to know about global warming. *Vox.* Available at http://www.vox.com/cards/global-warming/how-has-global-warming-affected-the-world-so-far#E5499185.

Quinn, P. April 25, 2013. After devastating tornado, town is reborn "green." *USA Today Green Living Magazine.* Available at http://www.usatoday.com/story/news/greenhouse/2013/04/13/greensburg-kansas/2078901/.

Rodin, J. 2014a. *The Resilience Dividend: Being Strong in World Where Things Go Wrong.* Public Affairs, New York.

Rodin, J. June 29, 2014b. Realizing the resilience dividend. Aspen Ideas Institute. Available at http://www.rockefellerfoundation.org/blog/remarks-by-dr-judith-rodin-2014-aspen.

Sheppard, K. December 12, 2014. Most of the coastal US will see over 30 days of flooding due to sea level rise, NOAA projects. *Huffington Post Green.* Available at http://www.huffingtonpost.com/2014/12/18/noaa-sea-level-rise-flooding_n_6350886.html.

Shogren, E. November 2, 2014. Five takeaways from the latest climate change report. *National Geographic.* Available at http://news.nationalgeographic.com/news/2014/11/141102-ipcc-synthesis-report-climate-change-science-environment/.

Stults, M. et al. 2014. Assessing the climate adaptation resource and service landscape. The Kresge Foundation.

Union of Concerned Scientists. October 2014. Encroaching tides: How sea level rise and tidal flooding threaten US East and Gulf Coast communities over the next 30 years. Available at http://www.ucsusa.org/news/press_release/tidal-flooding-0437#.VKMHof8AJA.

United Nations. November 2014. UN and climate change: The science. Available at http://www.un.org/climatechange/the-science/.

US Conference of Mayors. April 2014. Climate mitigation and adaptation action in America's cities. Available at http://www.usmayors.org/pressreleases/uploads/2014/0422-report-climatesurvey.pdf.

US Global Change Research Program (GlobalChange.gov). May 2014. The National Climate Assessment, 2014. Available at http://nca2014.globalchange.gov/.

Vande Bunte, M. January 10, 2014. How Grand Rapids street tax ties into the city's outlook on climate change. *MLive/Grand Rapids Press.* Available at http://www.mlive.com/news/grand-rapids/index.ssf/2014/01/how_streets_tax_in_grand_rapid.html.

West Michigan Environmental Action Council (WMEAC). December 2013. Grand Rapids climate resiliency report. Available at http://wmeac.org/climate-resiliency-report/.

The White House. November 2014. Presidential state, local, tribal leader's taskforce on climate preparedness and resilience: Recommendations for the President. Available at http://www.whitehouse.gov/administration/eop/ceq/initiatives/resilience/taskforce.

The World Bank. 2011. Guide to climate change adaptation in cities. International Bank for Reconstruction and Development. Available at http://www.siteresources.worldbank.org/INTURBANDEVELOPMENT/Resources/336387-1318995974398/GuideClimChangeAdaptCities.pdf.

2 Community Risk Implications

Damon P. Coppola

INTRODUCTION

Hazards, whether of natural, technological, or intentional origin, are the focus of risk management and reduction efforts. *Climate change* is not in and of itself a specific hazard assessed by these efforts, but is rather an ongoing shift in global climatic conditions that collectively influence other existing hazards. Drastic swings in climatic conditions are known to have occurred throughout the course of our planet's existence, although research strongly suggests that observed rates of change in previous eras were significantly less dramatic (i.e., more gradual) than what is presenting today. Furthermore, the evidence collected indicates with overwhelmingly confidence that these accelerated rates of change are still gaining momentum.

Such trends become particularly disconcerting when considered in light of the fact that patterns of human settlement and governance are traditionally predicated on assumptions of static, predictable geologic, meteorological, and hydrological data. A significant shift in any of these factors can result in misalignment of risk management objectives, thus leading to unexpected, disastrous, and even unmanageable conditions that threaten human life, property, and the social fabric itself. It is in this manner that climate change has altered global risk and the methods by which communities must manage it.

In late January 2015, the US Senate voted by a margin of 98 to 1 to affirm its members' agreement that climate change is real (Mascaro 2015). However, the controversy surrounding the human versus natural origin of these recent climate trends continues to simmer despite the late-2014 release of a United Nations Intergovernmental Panel on Climate Change (IPCC) report that found the overwhelming body of research indicates human activity is primarily to blame (and in fact, following the landslide Senate vote accepting the existence of climate change, a subsequent vote seeking to establish agreement with the theory that humans are exacerbating climate change failed to reach a majority).

But to the emergency management community, the question of who is causing these changes is of little consequence. What truly concerns emergency and disaster risk management professionals is that scientists believe climate shifts will continue in the direction they are going, thereby increasing the catastrophic potential of the many meteorological, hydrological, and climatological hazards communities face. A simple online news search reveals that some aspect of almost every recent disaster can be traced back to a change in rainfall, wind currents, air and ocean temperatures, or other climate change effect. Theories concerning the risk influence of climate change that decades ago seemed fantastical and far-fetched are today the mantra of world leaders, major multilateral and international organizations, and private and nonprofit environmental groups alike. Risk management stakeholders in government, nongovernmental, and private sector entities have recognized the influence climate change is having on their risk portfolios and are questioning what needs to be done (and what can be done) to limit exposure and vulnerability within their sphere of responsibility.

Dozens of climate prediction models are presently in use, and together they are providing insight into the changes expected to occur whether in the absence of any effective countermeasures or with moderate to significant intervention (see box "Snapshot of Projected Climate Impacts"). A number of hazard-specific trends have been extracted from these models' outcomes, such as increases in annual average atmospheric temperatures over land surfaces (with great regional variation) as caused by trapping of solar energy by accumulated greenhouse gases. Predictions of increased drought duration and storm intensity have also emerged, as have an increase in the average intensity of tropical cyclones, wetter Asian monsoons, and more frequent and intense midlatitude storms—and this is just a sample.

Many influences on hazard characteristics are caused by complex interactions of two or more climate change effects. For instance, the frequency of cyclonic storm formation has decreased due to decreasing differences between polar and equatorial atmospheric temperatures, while increased water temperatures have increased the magnitude and destructive potential of those storms that do form. On the other hand, for midlatitude storms that affect many of the world's most highly and densely populated urban centers, these decreasing pole to equator temperature differentials have increased both intensity and frequency. With warmer air and ocean temperatures comes increased atmospheric water vapor, and thus, a hotter, more humid environment. These effects are most pronounced at the poles, leading to increased melting of the polar ice caps and a compounded increase in the rate of change. In turn, storms will only become much more intense and more difficult to predict and track.

Other effects of climate change include much higher and lower average temperatures and longer periods of cold and heat waves. Hydrological changes are also affecting the environment, including rising sea levels caused by melting of the world's snow and ice cover at the poles and in glaciers, and a decrease in seawater density. Increased coastal flooding, especially in low-lying countries, has been one of the first and most pronounced negative effects. Predictions by the Intergovernmental Panel on Climate Change have stated that global warming could cause sea levels to rise 0.11–0.77 meters (0.36–2.5 feet) by 2100. In June and July 2010, the governments of Thailand and Viet Nam, respectively, reported projections that both Bangkok and Ho Chi Minh City were facing the prospect of inundation caused by rapidly rising sea levels. In Bangkok, it is estimated that the city will be under water within a century due to a combination of up to 4 inches of land subsidence caused by aquifer drainage and a 3- to 4-centimeter annual rise in the Gulf of Thailand (Associated Press 2007; Bernama 2010). In Viet Nam, it is feared that 61 percent of Ho Chi Minh City, representing land occupied by over 7 million people, will face constant risk of catastrophic flooding due to land falling below the level of the rising seas (Viet Nam News 2010). Countries including the Maldives, Palau, Kiribati, Bangladesh, Tuvalu, Barbados, The Philippines, Solomon Islands, Federated States of Micronesia, Papua New Guinea, and The Cook Islands, among others, face the prospect of losing large areas of land or even their ability to survive as nations if sea-level rise continues at projected rates (Gerken 2012). While the most pronounced impacts seem to be affecting low-lying countries located closer to the tropics, this should offer coastal nations located anywhere else little consolation as no coastal lands are immune from even the most conservative predictions for the twenty-first century.

EFFECT OF CLIMATE CHANGE ON COMMUNITY RISK

Community planning instruments, including emergency operations plans, continuity of operations plans, hazard mitigation plans, disaster recovery plans, and many others, are most effective when predicated on a working knowledge of hazard risk attained through hazard, vulnerability, and risk assessment and analysis. Determination of risk is accomplished by identifying the hazards that exists in the community and then calculating the likelihood of occurrence and the expected impacts and outcomes of each. Patterns of exposure to hazard forces may also be mapped for the purpose of determining which geographic areas within the community and which specific structures and landmarks are at risk of impact. Each of these risk values may be further calibrated by conducting an analysis of the physical, social, economic, and environmental vulnerabilities that exist in the community. This process as described has and continues to go by a range of different terms, including hazard identification/risk assessment (HIRA), threat and hazard identification and risk assessment

(THIRA), and hazard risk assessment (HRA), and their purposes are relatively uniform despite any differences in nomenclature.

Hazard risk is a dynamic factor. Even after assessments have been completed and plans drafted, communities must regularly reassess risk values to address changes that might have occurred in the community in the intervening years. Community change relative to risk can occur for a number of reasons, including significant increases or decreases in the resident and transient populations, the initiation or completion of major development projects, or large-scale changes in land-use patterns. Risk values can also change on account of new information, as is commonly acquired in the aftermath of a major disaster. Data gleaned from the event will impact calculations of the hazard's historical rate of occurrence and realized impacts, thus enabling more accurate calculations of future event likelihood and consequence simply as a factor of increased statistical sampling size and more informed expert analysis. Each of these community- and event-specific factors allows for adjustment in the risk values previously obtained, and likewise improvements to existing risk reduction and disaster response planning instruments.

Consider, for instance, how a community's hazard mitigation plan might require adjustment to address the presence of a sudden and drastic increase in stormwater runoff that is occurring in the aftermath of new building and parking-lot construction that has inadvertently reduced the infiltration rates of the natural ground cover. Or perhaps an adjustment is needed following a marked increase in population specific to a particular geographic area within a community that has occurred as a result of the community's effort to reinvent itself (e.g., as a tourist destination or retirement community).

SNAPSHOT OF PROJECTED CLIMATE IMPACTS (EXCERPT FROM THE PRESIDENT'S STATE, LOCAL AND TRIBAL LEADERS TASK FORCE ON CLIMATE PREPAREDNESS AND RESILIENCE 2014)

- By midcentury, the infrastructure investments needed to combat rising temperatures in the Midwest will require more than $6 billion, and approximately 95% of the electrical infrastructure in the Midwest is susceptible to decreased efficiency due to higher temperatures.
- Across the North Atlantic states, cumulative costs of sea-level rise and associated flood damage may exceed $88 billion by 2100.
- As much as 40% of reef-associated fish may be lost due to massive coral disease outbreaks, associated with higher water temperatures, in the Hawaiian archipelago, impacting $385 million in associated goods and services for Hawaii alone.
- For California and other states across the Southwest, climate change will increase the cost of maintaining and improving drinking water infrastructure by increasing the need for wastewater treatment and water desalination to supplement water supplies; even without the costs of these preparedness measures, California's drinking water system alone will require more than $4 billion in investment per year for the next 10 years.
- In Alaska, thawing and sinking of once frozen ground may cost between $3.6 and $6.1 billion (10% to 20%) above current public infrastructure maintenance over the next 20 years. In more rural parts of Alaska, such permafrost thaw is likely to disrupt community water supplies and sewage systems, posing risks to residents' health.

Source: The White House. 2014. President's State, Local, and Tribal Leaders Task Force on Climate Preparedness and Resilience. Available at http://1.usa.gov/1wWCOo7.

By reexamining risk relative to these changes, plan accuracy is increased as is its relevancy to the community in its current configuration. However, a common flaw exists in this process that, until relatively recently, had little impact on the process. Risk is almost always examined in terms of its present-day condition, even when projections for other community factors (like development patterns) look well into the future (see box "Climate Change Altering Australia's Wildfire Risk"). In other words, even when communities generate long-term development plans that consider what the community might be like years or even decades later, it is likely that these estimates consider each hazard threat in terms of a constant likelihood and a corresponding set of consequences that shift only in terms of their reaction to changes in the makeup of the community itself. With very few exceptions, risk assessments view the natural hazards themselves as being relatively static over time, with risk outcomes therefore adjusted only in relation to the mitigation actions taken to minimize them (or in relation to the actions of the community that serve to exacerbate them).

What would happen to these assessments if the climate were to change? In the face of irrefutable climate change, the traditional risk assessment approach becomes invalidated. An increase in the number and severity of hurricanes, surpluses and shortages of precipitation, sustained record-breaking high and low temperatures, warming ocean waters, and many other effects each reveal that use of the traditional planning assumption fixing natural hazard risk at static nontemporal levels is no longer valid nor advisable (see box "Climate Change Influence on Risk"). Natural hazard likelihood is often considered a fixed constant in the equation Risk = Likelihood × Consequence. But flood map inundation designations are expanding and contracting like never before. Storm surges are pushing further and further inland. Hurricane winds are inflicting stronger and more regular pressures on buildings, the environment, infrastructure, and other community features. Snow is accumulating to ever-increasing depths, impacting roofs and roads in ways never seen before encountered, and while some areas are experiencing more rain than has ever been encountered or expected, others are enduring shortages that have spanned much longer than ever imagined possible.

CLIMATE CHANGE ALTERING AUSTRALIA'S WILDFIRE RISK

Australia is considered the most developed country to have already sustained significant impacts directly linked to climate change. Factors related to its geographic location, the physical characteristics of its land and coastal areas, its climate, and its hazard profile (among other factors) are to blame. One of the most obvious of those effects relates to the country's rising average annual temperature. The year 2013 was the hottest on record, while 2014 was the third hottest. These figures are in line with longer-term trends that show the continent is warming faster than the global average, and these figures have corresponded to a steady drying of much of the content. Together, the heat and corresponding drought have led to severe wildfire risk.

Wildfires have long been a part of Australia's risk profile, but as climate change alters temperature and rainfall patterns throughout the country, it is not just fire magnitude that has changed—wildfire locations have changed as well. Researchers at the Australia National University Fenner School of Environment & Society found that the areas of greatest exposure and vulnerability are shifting in a southeasterly direction, which is where the largest population centers exist. The fire seasons are also starting earlier in the year, and the behaviors and characteristics of the fires are making them much more dangerous. This phenomenon has been so great as to require an adjustment to the national fire warning system wherein a new qualifier—"Catastrophic"— has been added above the former limit of "Extreme."

Source: Einhorn, B. 2015. Australia is Burning, and Climate Change is Making it Worse. *Bloomberg Business Week*. Global Economics. Available at http://buswk.co/1ujFU0A.

CLIMATE CHANGE INFLUENCE ON RISK

The following is an excerpt from the IPCC climate change report released in 2014 explaining how climate change interacts with and amplifies preexisting risks:

> There are two components of risk: the probability of adverse events occurring and the impact or consequences of those events. Climate change increases the probability of several types of harmful events that societies and ecosystems already face, as well as the associated risks. For example, people in many regions have long faced threats associated with weather-related events such as extreme temperatures and heavy precipitation (which can trigger flooding). Climate change will increase the likelihood of these two types of extremes as well as others. Climate change means that impacts already affecting coastal areas, such as erosion and loss of property in damaging storms, will become more extensive due to sea level rise. In many areas, climate change increases the already high risks to people living in poverty or to people suffering from food insecurity or inadequate water supplies. Finally, climate and weather already pose risks for a wide range of economic sectors, including agriculture, fisheries, and forestry: climate change increases these risks for much of the world. Climate change can amplify risks in many ways, including through indirect interactions with other risks. These are often not considered in projections of climate change impacts. For example, hotter weather contributes to increased amounts of ground level ozone (smog) in polluted areas, exacerbating an existing threat to human health, particularly for the elderly and the very young and those already in poor health. Also, efforts to mitigate or adapt to climate change can have negative as well as positive effects. For example, government policies encouraging expansion of biofuel production from maize have recently contributed to higher food prices for many, increasing food insecurity for populations already at risk, and threatening the livelihoods of those like the urban poor who are struggling with the inherent risks of poverty. Increased tapping of water resources for crop irrigation in one region in response to water shortages related to climate change can increase risks to adjacent areas that share those water resources. Climate change impacts can also reverberate by damaging critical infrastructure such as power generation, transportation, or health care systems.

Source: IPCC. 2014. Climate Change 2014: Impacts, Adaptation, and Vulnerability. Available at http://bit.ly/1DHtHeF.

The problem with relying on short-sighted assumptions that fail to account for Earth's changing climate is more than a simple increase in the inaccuracy of plans. In fact, short-sighted risk assessments have in several instances been blamed for inadvertently increasing population risk above what would have otherwise existed in the absence of such assessments. Consider, for example, a community with a sizeable amount of property located within the 100-year floodplain. Construction in flood-prone identified areas should remain relatively limited on account of the recognition of seasonal or periodic flooding that occurs, which also results in a general awareness of flood exposure boundaries. Imagine that a risk assessment determines that at current risk levels, building levees of height X would provide enough protection in the floodplain to reduce risk to 500-year flood levels (or a 0.2 percent chance of flooding in any given year). If those plans move forward, the floodplain designation is thus removed, and perceptions of risk likewise decrease, thereby leading to a go-ahead for increased development. With a new assurance of safety, it may take little time for developers to build in the former floodplain, protected by nothing other than faith that the levees will perform as designed. Now imagine that changes in climatic patterns associated with regional precipitation result in higher-than-normal peak rates of discharge on the river, such that the protection offered by the levees is reduced first to 100-year flood levels (1 percent chance of flooding in any given year), and eventually to 50-year levels (2 percent chance of flooding in any given year).

For those people and structures located in the shadow of the levees, risk levels will have increased 10-fold to untenable levels which fail to reflect the original assumptions that motivated the increase in development. Parallels can be found in pre-Katrina New Orleans, where residents were encouraged to settle in low-lying areas within the city limits that would have otherwise been considered undesirable (or perhaps even unbuildable) were the levees never constructed. Many coastal communities in New Jersey that lie behind seawalls were similarly devastated by Hurricane Sandy's unexpected impacts (Mills 2012; Young 2013).

In altering natural hazards' meteorological and hydrological drivers, climate change has shuffled the deck for risk managers. Scientists agree that communities in the United States, as is true elsewhere in the world, will begin experiencing much greater disaster impacts from more widespread disaster events, and these changes will continue to increase in their scope and scale *even if recommended actions to address climate change are taken.* In other words, irrespective of the causes of climate change and regardless of the political actions that are taken to combat ongoing anthropogenic contributors, communities should expect to see changes in their hazard profiles in the years and decades to come—and in some cases those changes will be severe and lasting. Communities will see whole districts previously identified as being "safe" reclassified to fall within the 100-year floodplain. Towns that rarely see a cumulative foot of snow throughout the course of a year will face multiple events that each drop 24 to 36 inches of snow or more. Coastal communities are already facing much larger and more powerful hurricanes, higher storm surges, and more severe winter storms, while those neighborhoods falling close to the wildland-fire interface have seen the threat from wildfires increase (see box "Measuring the Effect of Climate Change on Risk").

MEASURING THE EFFECT OF CLIMATE CHANGE ON RISK

The following excerpt from the report *A Sign of Things to Come* describes the influence of climate change effects on nations' hazard risk portfolios.

> The influence of greenhouse gases on the climate system is unequivocal. We know that global temperatures rose during the 20th century due to human emissions, and are very likely to rise in future. Understanding the influence of greenhouse gases and global warming on extreme weather events is more difficult. This is partly because extreme events are, by definition, rare, and so data are limited, and partly because of natural variability in the climate system. Extreme weather has always occurred, and natural variability will continue to influence weather in future. However, scientists expect that emissions from fossil fuels will alter the frequency and intensity of extreme weather events, and there is an increasing amount of evidence to support this.

There are two related areas of enquiry that can help us to understand the link between weather and climate change:

1. Trends in extreme events, about which we can draw some conclusions from basic physics, historical observations, and model experiments exploring future climate scenarios;
2. While it will never be possible to confidently state that an event would not have occurred without human-induced climate change, "probabilistic event attribution studies" (PEA), that consider whether and to what extent climate change altered the magnitude of and the risk of such an event occurring, can be conducted for certain types of extreme event.

Source: Coghlan, C. et al. 2014. *A Sign of Things to Come*. Oxfam Research Reports, September.

For several communities, climate-induced shifts in risk have already brought about disastrous consequences. A December 2014 Woods Hole Oceanographic Institution report found that the California communities affected by the 2012–2014 drought sustained the worst stretch of dry conditions ever experienced in the preceding 1200 years that the study examined, and that these impacts were the result of warmer temperatures "creating 'hot' droughts that compound the severity of the conditions created by an absence of rainfall." The report further states that, "[t]here is no doubt that we are entering a new era where human-wrought changes to the climate system will become important for determining the severity of droughts and their consequences for coupled human and natural systems" (Cameron 2014). The year 2014 was also determined to have been the hottest year on record worldwide, including in many parts of the United States. Moreover, empirical data extending back more than 160 years indicates that all of the 14 years that have passed since the turn of the century are counted among the 15 hottest years on record (see Figure 2.1), and these effects are already leading to secondary impacts. For instance, vector-borne and communicable diseases that haven't been seen in the United States for decades or even longer, including dengue fever, have started to appear as conditions have become more favorable to their spread (Maron 2013). Algal blooms and changes in ocean pH have also corresponded to these changes, and it is likely that there are many other novel effects for which the link to climate change has just not yet been recognized.

Without knowing how hazard risk is changing and how these changes impact community risk, communities' planning products will quickly become irrelevant, inaccurate, and possibly detrimental to long-term community resilience. Mitigation plans will seek the wrong goals. Response plans will underestimate event requirements. Recovery plans will fail to provide long-term sustainability. And development plans will encourage population expansion and construction where neither should ever occur.

A 2014 study conducted by the government of Australia highlighted that nation's attempt to understand and likewise estimate the increased cost of disasters that are likely to be sustained as a result of changes to climate-related hazard risk patterns. In the report produced as an outcome of this project, the authors determined that financial losses tied to natural-disaster events will exceed $1 trillion (in that country) by the year 2100 and could even reach $1.4 trillion if climate effects exceed predictions (Australian Business Review 2014; Sturmer 2014). Skeptics need only to look to the World Meteorological Organization's 2014 report, which showed that the worldwide cost of climate-related disasters totaled almost $500 billion in just the 5 years preceding the report, which amounts to more than three times the total sustained across the entire 1970s even when accounting for inflation (Byanyima 2014).

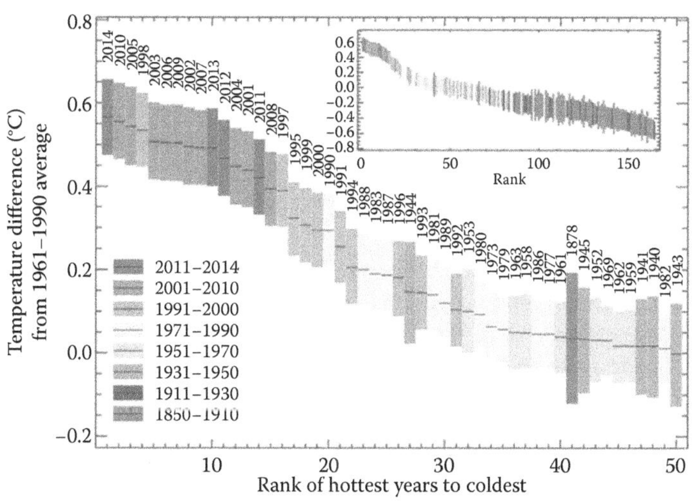

FIGURE 2.1 Global average temperatures by year. (From UK Met Office. 2014. Human Influence Important Factor in Possible Global and UK Temperature Records. UK Met Office website. Available at http://bit.ly/1UFIY3y (accessed September 2015).)

HOW CLIMATE CHANGE WILL IMPACT DIFFERENT PARTS OF THE WORLD

The following is an excerpt from UNISDR Concept Note 01:

Climate change will affect all countries, but people in the poorest countries and poor people in richer countries are more likely to suffer the most. They tend to live in high risk areas such as unstable slopes and flood plains, and often cannot afford well-built houses. Many of them depend on climate-sensitive sectors, such as agriculture, and have little or no means to cope with climate change, for example owing to low savings, no property insurance and poor access to public services. Climate change is expected to reduce already low incomes and increase illness and death rates in many developing countries. Africa, small island states, and the Asian and African mega-deltas are likely to be particularly affected by climate change. According to the IPCC:

- Africa is particularly vulnerable to the effects of climate change because of multiple stresses and low adaptive capacities, arising from endemic poverty, weak institutions, and complex disasters and associated conflicts. Drought will continue to be a primary concern for many African populations. The frequency of weather- and climate-related disasters has increased since the 1970s, and the Sahel and Southern Africa have become drier during the twentieth century. Water supplies and agricultural production will become even more severely diminished. By 2020, in some African countries agricultural yields could be reduced by as much as 50%. By the 2080s, the area of arid and semiarid land in Africa will likely increase by 5–8%.
- Asia's sustainable development will be challenged as climate change compounds the pressures that rapid urbanization, industrialization, and economic development have placed on natural resources. One of the main issues will be the availability of adequate fresh water, which by the 2050s will be a concern for possibly more than one billion people. The continued melting of glaciers in the Himalayan region is projected to increase flooding and rock avalanches and to adversely affect water resources in the next two to three decades. Asia's coastal areas, and especially its heavily populated delta regions, will become even more prone to increased flooding because of both rising sea levels and river flooding.
- Australia and New Zealand may face more frequent extreme events such as heat waves, droughts, fires, floods, landslides and storm surges. These conditions will increase stresses on water supplies and agriculture and will change natural ecosystems. Less seasonal snow cover and shrinking glaciers will create additional problems. Coastal areas are projected to be threatened by the consequences of rising sea levels and increasingly severe and more frequent storms and coastal flooding by 2050.
- Europe will need to cope with retreating glaciers and extend of permafrost, reduced precipitation in Southern Europe and the possibility of more droughts in some areas, as well as increased risk of flash floods. Higher temperatures and heat waves will increase health risks and increase the frequency and severity of wildfires. Reduced forest area and agricultural productivity and greater vulnerability of low-lying coastal areas to rising sea levels are likely. Many economic sectors will be affected. For Southern Europe, less water will reduce hydropower potential, tourism and, in general, crop production.
- Latin America's changes in precipitation patterns and the disappearance of glaciers will significantly reduce the amount of water available for human consumption,

agriculture and energy generation. In drier areas, climate change is expected to lead to increased salinization and desertification of agricultural land. The productivity of some crops and livestock will decrease, with adverse consequences for food security. Rising sea levels will cause increased risk of flooding in low-lying coastal areas.

- North America will experience further decrease of mountain snow due to rising temperatures leading to increased winter flooding and reduced summer flows, and to altered seasonal availability of water. Many areas are expected to experience more, longer and hotter heat waves, with a greater potential for adverse health impacts. Sustained higher temperatures also will increase the risk of forest fires. Coastal communities will be increasingly threatened if the intensity of tropical storms increases.
- Polar Regions are likely to experience reductions in the thickness of glaciers and the extent of ice sheets, and changes in natural ecosystems. The impacts on human communities in the Arctic will be reflected in changes in infrastructure and traditional or indigenous ways of living.
- Small island states, coastal systems and other low-lying areas are especially vulnerable to the effects of climate change, rising sea levels and extreme weather events. Millions of people are likely to be affected by floods, storm surges, erosion and other coastal hazards every year due to rising sea levels by the 2080s, particularly in the large deltas of Asia and Africa and the small island states. By mid-century, reduced water resources are expected in many small islands, e.g., in the Caribbean and Pacific.

Source: UN Office for Disaster Risk Reduction (UNISDR). 2008. Climate Change and Disaster Risk Reduction. Briefing Note 01, United Nations. Available at http://bit.ly/1ACI1lY.

American communities must also recognize the influence that climate-related emergency and disaster event impacts in other countries—both neighboring and far afield—can have in terms of shifting their own risk profiles (see box "How Climate Change Will Impact Different Parts of the World"). A study conducted by Coghlan et al. (2014) highlighted the global impacts of hazards and events affected by climate change in their study *A Sign of Things to Come*. The report's authors explain the mechanisms by which a heat wave in Russia caused local, national, and international impacts beginning with a dramatic reduction in the Russian wheat harvest. The low output reverberated outward as national food supplies became depleted, in turn resulting in negative financial impacts for the farming communities. To address shortages, the government of Russia banned the export of wheat, thereby triggering global shortages that resulted in worldwide wheat price increases. Analysts subsequently linked these price increases to widespread upticks in global unrest, inclusive of the events that transpired in the nations affected by the Arab Spring—almost all of which remain dependent on Russian wheat. These effects were notably pronounced in Pakistan, which that same year concurrently faced higher-than-normal monsoon-related rainfall that severely damaged crops, killed livestock, and caused shortages in local markets, in addition to causing health-related problems and damaging housing stock and infrastructure (Coghlan et al. 2014).

American communities will need to develop an understanding of new and increased risk factors that extend beyond what has traditionally been prepared for and mitigated against, even if the most conservative estimates of climate-related risk adjustments prove accurate. Even when accounting for inflation, the number of billion-dollar disasters has already begun its steady rise, as has total average annual cost of disasters (see Figure 2.2). Martin Parry, a climate scientist with IPCC, characterized the problem back in 2007 at the release of the first comprehensive report on

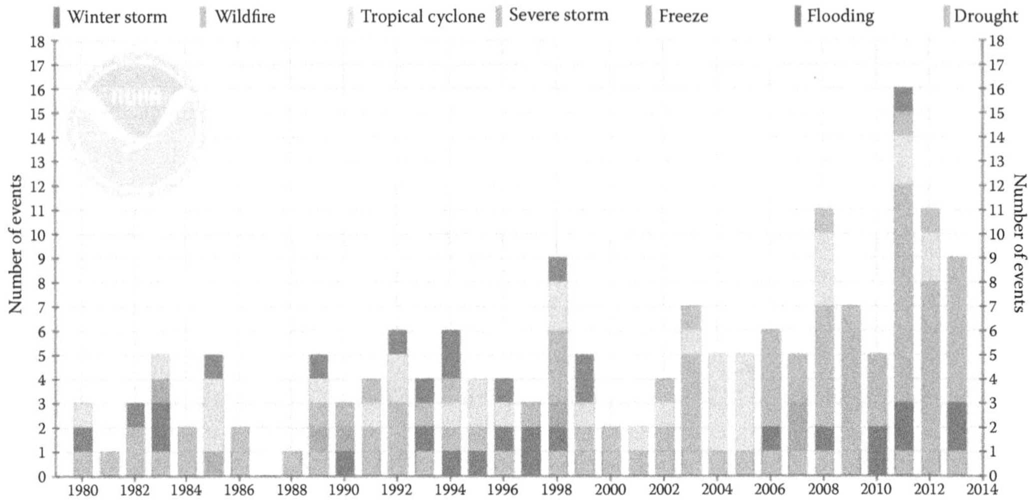

FIGURE 2.2 Billion dollar weather and climate related disasters by year and hazard type. (From NOAA. 2014. Billion Dollar Weather/Climate Disasters: Time Series. NOAA National Climatic Data Center. Available at http://1.usa.gov/1BJgX3h.)

the anticipated global impacts of climate change in stating that, "We are all used to talking about these impacts coming in the lifetimes of our children and grandchildren. Now we know that it is us" (Adam 2007).

The November 2014 *President's State, Local, and Tribal Leaders Task Force on Climate Preparedness and Resilience* report supports this concern, recommending that the US government take immediate steps to support risk reduction for American communities by performing the following:

- Provid[ing] actionable data and information on climate change impacts and related tools and assistance to support decision-making
- [Making] climate-smart planning and investment decisions at a regional, state, tribal, territorial, and local level [by providing] decision-makers [with] access to the best available information about climate impacts in a user-friendly and accessible format
- Building on successful efforts like the National Oceanic and Atmospheric Administration's (NOAA) Regional and Integrated Sciences and Assessments program [so that] more can be done to provide authoritative, consistent, and relevant information and tools to help inform planning and decision making at all levels (White House 2014).

Climate change impacts disasters of hydrological, meteorological, and climatological origin. Each of these is described next.

- *Hydrological hazards* are those associated with either a severe excess or a lack of water. This could include coastal flooding, riverine flooding, flash flooding, heavy snowfall, avalanche, mudslide, and drought.
- *Meteorological hazards* are those that are related to atmospheric weather conditions. These hazards are the result of interactions between precipitation, temperature, wind speed, humidity, and other more complex drivers. Meteorological hazards include tropical cyclones, tornadoes, strong winds (including derechos, straight-line winds, and other variants), ice storms, severe winter storms, hailstorms, frost, sandstorms, wildfires, thunderstorms, heat waves, extreme cold, and fog, to name a few.

- *Climatological hazards* are related to meteorological hazards, but include the factor of time and relate to patterns. Climatological hazards include shifting annual average temperatures, long-term changes in precipitation, El Niño and La Niña, soil salination, desertification, soil and coastal erosion, and others (drought is often considered a climatological hazard).

These three hazard groupings are just a subset of the larger category of natural hazards, which also includes tectonic hazards, mass-movement hazards, biological/health-related hazards, and other more obscure threats like meteors, animal attacks, and space weather. Nonetheless, during the 20-year period spanning from 1988 to 2007, researchers determined that 76 percent of all disaster events were hydrological, meteorological, or climatological in nature, and these events accounted for 45 percent of the deaths and 79 percent of the economic losses caused by natural hazards in that span of time (UNISDR 2008).

The climate change threat is not limited to these three hazard groups. Many of the hazards they include result in situations or conditions conducive to or directly supportive of secondary hazards that are equally dangerous. For instance, increases in average annual temperatures, as have been occurring in the southwestern United States and elsewhere, creates a hospitable environment for rodent and insect vector-borne diseases, some of which have not been seen in decades or ever (including dengue fever, encephalitis, yellow fever, Lyme disease, and malaria (IPCC 2014; Morin, Comrie, and Ernst 2013). Increased human exposure to environmental allergens, such as pollen, has also been associated with changes in temperature and precipitation patterns, and has caused an increased public health threat to populations that are not accustomed to these conditions (Radford 2014).

Communities will also contend with new or increasing technological and industrial hazards caused by changes in climate and atmospheric patterns. For instance, smog and haze, often exacerbated by air-trapping temperature inversions caused by temperature inversions, are but one example (Bailey et al. 2011). Climate change is also associated with increased long-term exposure to the carbon dioxide, carbon monoxide, and ozone that is increasing in concentration in the lower atmosphere (IPCC 2014). Even increased airline turbulence, which has damaged several planes in recent years and poses a general threat to commercial and cargo aviation, has been tied to the instability in atmospheric air currents caused by climate change. These effects are expected to increase dramatically over the next 50–100 years, resulting in travel delays, damage to aircraft, and possibly the loss of lives in the event of an accident (Subbaraman 2014).

The collective impact of climate change on the global hazard portfolio is so great, in fact, that it poses a sizable threat to the nation's security. In March 2014, 16 retired US generals released a report that explained how changing global climates pose multiple risks to national security. Their findings are extensive and include warnings that

- Increased frequency of extreme weather events will cause undue strain on US military resources, stretching them beyond their intended capacity
- Climate change will cause civil unrest in unstable regions of the world
- Rising sea levels will put people and food supplies in increased danger, and many may seek refuge in the United States
- Melting polar ice will set in motion global posturing for control of newly uncovered resources, including trade routes
- Several different climate-related effects will increase the threats facing the nation's critical infrastructure, including that of the power grid (Briggs 2014)

The Pentagon's Quadrennial Defense Review (QDR), which considers the most important issues facing the military, echoed these concerns about climate change, and went as far as to state that climate change remains a major driver of global terrorism (Davenport 2014) (see box "Defense Assessment of the Security Threats Posed by Climate Change"). In the 2015 State of the Union Speech, President Obama stated that, "No challenge—no challenge—poses a greater threat to future generations than climate change" (Neuhauser 2015).

DEFENSE ASSESSMENT OF THE SECURITY THREATS POSED BY CLIMATE CHANGE

"Climate change poses another significant challenge for the United States and the world at large. As greenhouse gas emissions increase, sea levels are rising, average global temperatures are increasing, and severe weather patterns are accelerating. These changes, coupled with other global dynamics, including growing, urbanizing, more affluent populations, and substantial economic growth in India, China, Brazil, and other nations, will devastate homes, land, and infrastructure. Climate change may exacerbate water scarcity and lead to sharp increases in food costs. The pressures caused by climate change will influence resource competition while placing additional burdens on economies, societies, and governance institutions around the world. These effects are threat multipliers that will aggravate stressors abroad such as poverty, environmental degradation, political instability, and social tensions—conditions that can enable terrorist activity and other forms of violence."

Source: Department of Defense. 2014. Quadrennial Defense Review 2014. Available at http://1.usa.gov/1u8C5eB.

And finally, climate change is itself resulting in additional compounding drivers of negative climatological trends. The increased temperatures at the planet's poles are causing a melting of permafrost, which in turn results in the release of millions of tons of the greenhouse gas methane. Methane in the atmosphere traps heat, causing a chain reaction many scientists feel to be the greatest climate-related threat of all (DiConsiglio 2015). The loss of polar ice, which reflects solar energy back out to space, is revealing darker energy-absorbing ocean waters as it recedes, further intensifying the warming of ocean surface temperatures. These are only a small sample of the known compounding effects, and it follows that additional exacerbating effects are happening or are yet to begin that have simply not yet been recognized or discovered. What we do know is that there is not much working against the current trends.

INFLUENCE OF CLIMATE CHANGE ON COMMUNITY HAZARDS

Community risk managers must incorporate temporal hazard trends and predictions as influenced by climate change effects if community planning instruments are to remain accurate over time. Climate change affects the existing hazard portfolio in a number of ways, and it also results in the addition of new hazards that have not been seen in certain areas or even anywhere in the nation's 50 states and territories (see box "Observed and Projected Climate Change Impacts in Different Regions of the United States"). The following sections present these changes in relation to the first- and second-order effects of climate change; in other words, it explains how the climate is changing and how this change affects one or more hazards. The warming effect of air and water are first-order climate change effects, while other effects (such as increases in precipitation or storm intensity) result from these first-order effects. It is important to note that for some hazards, multiple climate change effects will influence the hazard, and as such the hazard may appear in more than one of these categories. For instance, the flood hazard is influenced in different ways by increased air temperatures, increased ocean temperatures, and increased precipitation. Also, one must keep in mind that effects will differ depending on where a community is located and the time of year. Where appropriate, differentiation per these factors is provided, but there may be differences that exist beyond those listed in this text. For a full, detailed listing of these effects, the authors recommend that the reader consults the following two resources:

- US National Climate Assessment. 2014. Climate Change Impacts in the United States. US Global Change Research Program (http://1.usa.gov/1L4S6wO)
- Intergovernmental Panel on Climate Change (IPCC). 2014. Fifth Assessment Report (AR5) (http://bit.ly/1xHUfFv)

OBSERVED AND PROJECTED CLIMATE CHANGE IMPACTS IN THE DIFFERENT REGIONS OF THE UNITED STATES

- *Northeast:* Communities are affected by heat waves, more extreme precipitation events, and coastal flooding due to sea level rise and storm surge.
- *Southeast and Caribbean:* Decreased water availability, exacerbated by population growth and land-use change, causes increased competition for water. There are increased risks associated with extreme events such as hurricanes.
- *Midwest:* Longer growing seasons and rising carbon dioxide levels increase yields of some crops, although these benefits have already been offset in some instances by occurrence of extreme events such as heat waves, droughts, and floods.
- *Great Plains:* Rising temperatures lead to increased demand for water and energy and impacts on agricultural practices.
- *Southwest:* Drought and increased warming foster wildfires and increased competition for scarce water resources for people and ecosystems.
- *Northwest:* Changes in the timing of streamflow related to earlier snowmelt reduce the supply of water in summer, causing far-reaching ecological and socioeconomic consequences.
- *Alaska:* Rapidly receding summer sea ice, shrinking glaciers, and thawing permafrost cause damage to infrastructure and major changes to ecosystems. Impacts to Alaska Native communities increase.
- *Hawaii and the Pacific Islands:* Increasingly constrained freshwater supplies, coupled with increased temperatures, stress both people and ecosystems and decrease food and water security.
- *Coasts:* Coastal lifelines, such as water supply infrastructure and evacuation routes, are increasingly vulnerable to higher sea levels and storm surges, inland flooding, and other climate-related changes.
- *Oceans:* The oceans are currently absorbing about a quarter of human-caused carbon dioxide emissions to the atmosphere and over 90 percent of the heat associated with global warming, leading to ocean acidification and the alteration of marine ecosystems.

Source: US National Climate Change Assessment. 2014. Climate Change Impacts in the United States. US Global Climate Change Research Program. Available at http://NCA2014.globalchange.gov.

SEA-LEVEL RISE

Sea levels rise and fall over extended periods of time as a result of the melting and freezing of mountain glaciers and ice sheets (called eustatic sea-level rise). There are other factors that influence this phenomenon to varying degrees, including changes in ocean temperatures (which affect density and are called thermosteric effects), changes in salinity (halsosteric effects), the volume of water retained in lakes, rivers, aquifers, and the atmosphere, and changes in the shape of the sea floor (isosteric changes), to name a few. Land masses can also rise and fall relative to the ocean, due to subsidence, expansion, or tectonic movement (termed "relative sea-level change") (Yoskowitz, Gilbeaut, and McKenzie 2009). However, of all of these effects, none are as impactful as the melting of ice, which is a direct second-order effect of rising global temperatures.

Historic models of mean sea level (MSL) differ significantly from each other prior to the mid-nineteenth century on account of the fact that proxy data is required in the estimations. As such there is general disagreement about how much MSLs have risen and fallen over the course of Earth's history. There is a general consensus, however, that these levels have existed at both higher and

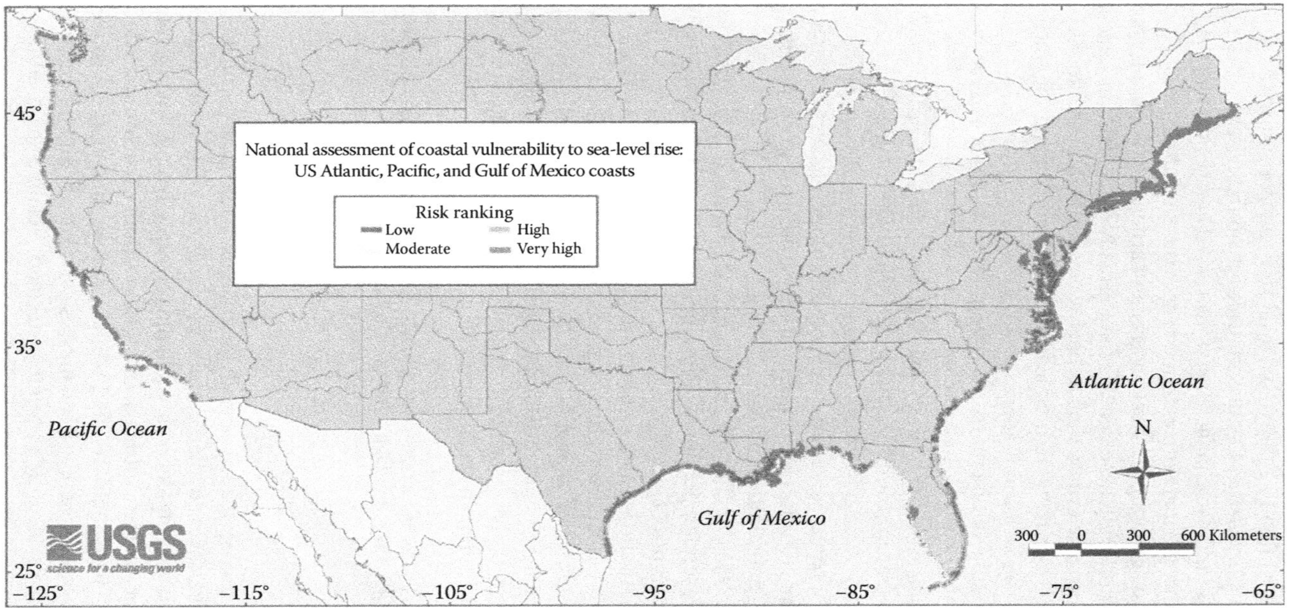

FIGURE 2.3 Map showing coastal vulnerability to sea-level rise in the 48 contiguous United States. (From Thieler, E.R., J. Williams, and E. Hammar-Klose. 2014. National Assessment of Coastal Vulnerability to Sea Level Rise. Woods Hole Field Center. USGS. Available at http://woodshole.er.usgs.gov/project-pages/cvi/.)

lower levels than are observed today, and at times this increase or decrease has reached 100 meters or more (and the majority of the time they have been much higher than today). However, the fact that sea levels have existed at levels much higher than they are today, even before humans walked the planet, is of little consolation given that our societies have since elected to settle in low-lying coastal land situated just meters above the mean high tide.

Empirical MSL data indicates that the sea is again on an upward trajectory, and studies completed in early 2015 found that the actual rate of rise between 1901 and 1990 was about 1.2 millimeters per year. However, measurements spanning the period from 1990–2014 show that the rate, which had previously been gradually increasing, has suddenly increased its speed, and the rate during this more recent period averaged 3 millimeters per year. That is more than double what was observed over the full course of the twentieth century (Maritime Executive 2015). Even if this rate was to remain constant at 3 millimeters per year, we would see almost 1 foot of increase by the year 2100, and if the rates of increase were to continue compounding in the manner they have during the past 25 years, we could experience MSL rises as great as 2 meters or more. This would most certainly lead to catastrophic changes of such great magnitude as to alter the world map (Figure 2.3).

While the rising and falling of the sea is in fact in line with natural cycles and may be of no real long-term consequence to the planet, it is certainly a concern of those who have settled in low-lying coastal zones likely to be inundated. In addition to bringing these structures and infrastructure closer to the inundation risk, the threats associated with storm surge effects, variable and extreme tides, wave action, and coastal erosion each increase as well (see box "Rising Seas Likely to Flood US History") (Figure 2.4).

For coastal communities, flood maps provide insight into the nature of risk that will accompany sea-level rise. Flood zones will grow and the structures inside those zones will see increased damages as a result of more frequent flood events. However, there are other effects that must be considered, such as the destruction of environmental protections like wetlands, mangroves, coral reefs, and other natural barriers. Any change in sea level can result in changes in sedimentation processes, which in turn can result in the movement of beaches and barrier islands. Coastal currents can also shift,

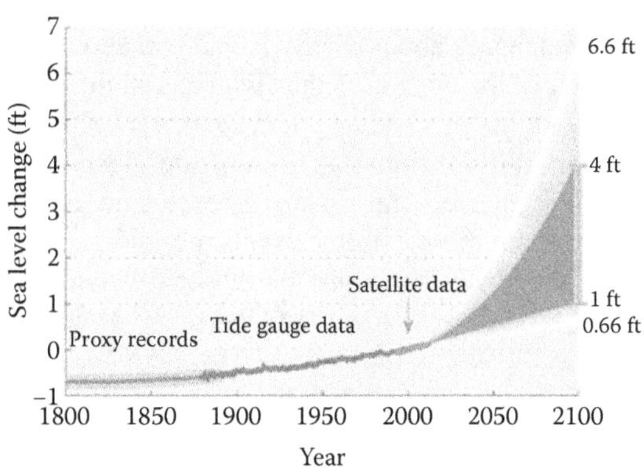

FIGURE 2.4 Past and projected sea-level rise estimated, observed, and possible future amounts of global sea level rise from 1800 to 2100 relative to the year 2000. Estimates from proxy data (e.g., sediment records) are shown between the dates of 1800–1890, with the lighter band showing uncertainty, from tide gauge data between 1880–2009, and from satellite observation from 1993 to the present day. The future scenarios range from 0.66 to 6.6 feet in 2100. These scenarios are not based on climate model simulations, but rather reflect the range of possible scenarios based on other scientific studies. Where the graph extends past 2015 shows the currently projected range of sea level rise of 1 to 4 feet by 2100, which falls within the larger risk-based scenario range. The large projected range reflects uncertainty about how glaciers and ice sheets will react to the warming ocean, the warming atmosphere, and changing winds and currents. As seen in the observations, there are year-to-year variations in the trend. (From US National Climate Change Assessment. 2014. Climate Change Impacts in the United States. US Global Climate Change Research Program. Available at http://NCA2014.globalchange.gov.)

RISING SEAS LIKELY TO FLOOD US HISTORY

The following is an excerpt from an article by Seth Borenstein that appeared in the *Washington Post* on September 23, 2007:

Ultimately, rising seas will likely swamp the first American settlement in Jamestown, Virginia, as well as the Florida launch pad that sent the first American into orbit, many climate scientists are predicting. In about a century, some of the places that make America what it is may be slowly erased. Global warming—through a combination of melting glaciers, disappearing ice sheets and warmer waters expanding—is expected to cause oceans to rise by one meter, or about 39 inches. It will happen regardless of any future actions to curb greenhouse gases, several leading scientists say. And it will reshape the nation.

Rising waters will lap at the foundations of old money Wall Street and the new money towers of Silicon Valley. They will swamp the locations of big city airports and major interstate highways.

Storm surges worsened by sea level rise will flood the waterfront getaways of rich politicians—the Bushes' Kennebunkport and John Edwards' place on the Outer Banks. And gone will be many of the beaches in Texas and Florida favored by budget-conscious students on Spring Break.

That's the troubling outlook projected by coastal maps reviewed by The Associated Press. The maps, created by scientists at the University of Arizona, are based on data from the US Geological Survey. Few of the more than two dozen climate experts interviewed disagree with the one-meter projection. Some believe it could happen in 50 years, others say 100, and still others say 150.

Sea level rise is "the thing that I'm most concerned about as a scientist," says Benjamin Santer, a climate physicist at the Lawrence Livermore National Laboratory in California. "We're going to get a meter and there's nothing we can do about it," said University of Victoria climatologist Andrew Weaver, a lead author of the February report from the Intergovernmental Panel on Climate Change in Paris. "It's going to happen no matter what—the question is when."

Sea level rise "has consequences about where people live and what they care about," said Donald Boesch, a University of Maryland scientist who has studied the issue. "We're going to be into this big national debate about what we protect and at what cost." And it's not just a rising ocean that is the problem. With it comes an even greater danger of storm surge, from hurricanes, winter storms and regular coastal storms, Boesch said. Sea level rise means higher and more frequent flooding from these extreme events, he said.

All told, one meter of sea level rise in just the lower 48 states would put about 25,000 square miles under water, according to Jonathan Overpeck, director of the Institute for the Study of Planet Earth at the University of Arizona. That's an area the size of West Virginia. The amount of lost land is even greater when Hawaii and Alaska are included, Overpeck said. The Environmental Protection Agency's calculation projects a land loss of about 22,000 square miles. The EPA, which studied only the Eastern and Gulf coasts, found that Louisiana, Florida, North Carolina, Texas and South Carolina would lose the most land. But even inland areas like Pennsylvania and the District of Columbia also have slivers of at-risk land, according to the EPA.

This past summer's flooding of subways in New York could become far more regular, even an everyday occurrence, with the projected sea rise, other scientists said. And New Orleans' Katrina experience and the daily loss of Louisiana wetlands—which serve as a barrier that weakens hurricanes—are previews of what's to come there. Florida faces a serious public health risk from rising salt water tainting drinking water wells, said Joel Scheraga, the EPA's director of global change research. And the farm-rich San Joaquin Delta in California faces serious salt water flooding problems, other experts said. "Sea level rise is going to have more general impact to the population and the infrastructure than almost anything else that I can think of," said S. Jeffress Williams, a US Geological Survey coastal geologist in Woods Hole, Massachusetts.

Even John Christy at the University of Alabama in Huntsville, a scientist often quoted by global warming skeptics, said he figures the seas will rise at least 16 inches by the end of the century. But he tells people to prepare for a rise of about three feet just in case. Williams says it's "not unreasonable at all" to expect that much in 100 years. "We've had a third of a meter in the last century." The change will be a gradual process, one that is so slow it will be easy to ignore for a while. "It's like sticking your finger in a pot of water on a burner and you turn the heat on," Williams said. "You kind of get used to it."

Source: Borenstein, S. 2007. Rising Seas Likely to Flood US History. *Washington Post*, September 23. Available at http://wapo.st/1DJpUNS.

thereby exacerbating erosion effects and changing marine habitats (which can be devastating to fishing and tourism-based economies). Rising sea levels also contribute to the salinization of freshwater resources, such as aquifers and wells, and can cause salinization of coastal farmland. Yoskowitz, Gilbeaut, and McKenzie (2009) discuss several instances where sea-level rise has led to mass migrations in different regions of the world, including as many as 17 million people in Bangladesh since the 1950s and 4.3 million people in the Philippines between 1970 and 1990. There is no reason to believe this is outside the realm of possibility for coastal American communities.

CHANGES IN SEA SURFACE TEMPERATURE AND pH (ACIDITY)

Changing atmospheric temperatures have remained the central focus of the media on the topic of climate change. However, sea temperatures are also rising, and the impacts in terms of global risk are as great if not greater. Ocean temperatures actually rise in response to rising air temperatures, absorbing much of that heat and dampening the extent to which the air temperature actually changes. Over the past century, observations of sea temperature have indicated a much lower rate of increase, totaling about .1 degree Celsius (or about .18 degrees Fahrenheit.) However, the marine environment can hold a much greater amount of energy given its density, and it is much more sensitive to change. Moreover, the impacts of a rise in sea temperature are often farther reaching than most people realize (Figure 2.5).

To begin with, warmer water equates to more water due to the isothermal effects discussed in the preceding section on sea-level rise. As sea temperatures increase, water molecules move slightly farther away from each other and the water becomes less dense. Warmer waters therefore translate to inundated land and increased flooding.

Warmer water also fuels several types of storms, including cyclonic storms (hurricanes, cyclones, and typhoons). Hurricanes require warm seawater to thrive, as well as the large quantities of vaporized seawater that warmer oceans produce. Increased water vapor in the air also leads to increases in the number and severity of periods of heavy rainfall and snow, and can change storm tracks leading to extreme weather in areas that do not traditionally experience such weather (EPA 2015a).

Warmer sea temperatures, which have been shown to extend to depths up to 700 meters (NASA 2014), cause coral to die as a result of the chemical changes that occur in warmer waters (including a decrease in oxygen concentrations). Bleaching of coral reefs throughout the world has been attributed to ocean warming (LeGault 2014). Warmer waters also disturb the balance in microscopic sea life, resulting in an increase in zooplankton, which feed on phytoplankton. The result of this shift is an overall decrease in phytoplankton, which serves as a reliable sink for the greenhouse gas carbon dioxide (Biello 2009).

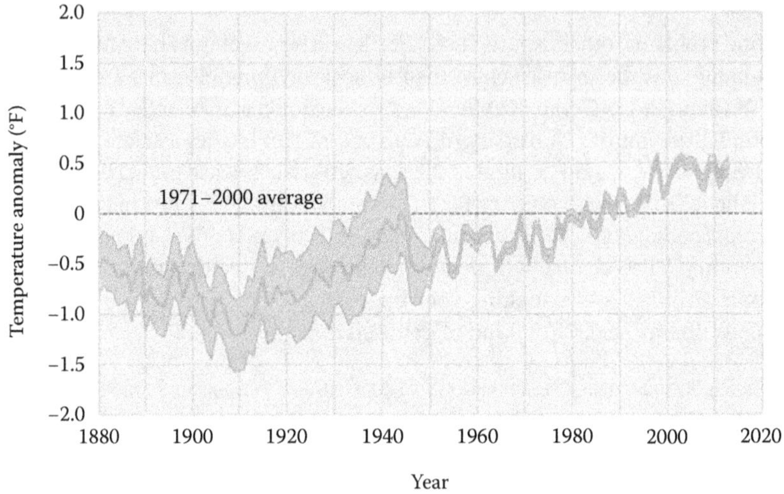

FIGURE 2.5 Average global sea surface temperature 1880–2013. (From NOAA. 2014. Average Global Sea Surface Temperature, 1880–2013. NOAA website. Available at http://1.usa.gov/15pv7fQ.)

Research has also shown that increases in ocean temperature can result in changes or even proportional decreases in ocean life due to alternations of the marine food chain. Fish and shellfish breeding cycles are also highly sensitive to even a single degree of change, and marine diseases and predators can bloom beyond control in warmer waters (Gray 2015). For communities dependent on the fisheries industry, including shellfish, lower yields will translate to negative economic impacts.

Ocean acidification, the result of increases in dissolved carbon dioxide, also results in impacts to human populations in a number of ways that affect risk profiles. Oceans have since the industrial revolution been absorbing about a third of carbon dioxide emissions, which then move to the deep ocean where they are stored as carbonic acid. Because carbon dioxide levels have increased significantly (31% since 1750) with the advent of combustion engines and the burning of fossil fuels for power generation, the amount of dissolved carbon dioxide in the oceans has increased to an extent that pH levels have steadily fallen (NOAA 2010). Lower pH corresponds to greater acidity, which in turn affects the abilities of several marine species to thrive, including coral, plankton, shellfish, algae, and calcium-dependent invertebrates. Long-term negative impacts to plankton populations threaten to destabilize the entire marine food chain, which would have cascading effects on fish, seabirds, and marine mammals. For fishing communities and communities dependent on marine tourism, the economic impacts would be disastrous (NOAA 2010).

INCREASED, DECREASED, AND UNPREDICTABLE PRECIPITATION LEVELS

All creatures, and likewise all communities, rely on reliable access to water for their survival. Water stores exist in many forms, including rivers, lakes, reservoirs, aquifers, groundwater, and man-made storage tanks. The original source of all of this water is precipitation in the form of rain or snow. In addition to ensuring steady access to drinking water, community water resources are used to feed crops, enable sanitation and cooking, support industrial production and manufacturing, transport goods, and enable plant and animal life. The economic dependence on water is extreme given that the two greatest uses of water by sheer volume, accounting for more than three-quarters of all consumption, are thermoelectric power generation (41%) and irrigation (37%) (EPA 2015b).

Communities and their associated economies have each developed on the basis of an assumed and relatively static quantity and schedule of precipitation. Too little rain, or drought, can result in a loss of crops and livestock, increase wildfire risk, and restrict certain activities (e.g., river freight).

Too much water leads to floods and flash floods and causes injuries, deaths, and destruction of property and the environment. Water storage mechanisms, such as reservoirs and dams, have enabled populations to endure periods of comparatively high or low rainfall, but these mechanisms have their limits of protection and climate change is pushing those limits.

In fact, climate change is affecting each of the factors associated with water delivery and retention. Precipitation is increasing in some areas and decreasing in others as a result of increases in both average global air temperatures and average global sea surface temperatures. Both of these factors contribute to an increase in the amount of water vapor that is created and stored in the atmosphere. As atmospheric water moves over land, changes in temperature and pressure can cause that water to fall to the ground as rain or snow.

Throughout the United States and in several other nations around the world, average annual precipitation rates have steadily increased over the past century. Since 1901, global precipitation has increased at an average rate of 0.2 percent per decade, while precipitation in the contiguous 48 states has increased at a rate of 0.5 percent per decade. Of course, these rates generalize all areas, but locally there are many examples where much higher rates of increase and where decreases in rates have occurred (e.g., Hawaii and in states located in the Southwest). Because changing wind patterns have shifted the location of atmospheric water, rain is moving in less predictable patterns, leaving some areas with more water and others with less, further exacerbating regional disparities. Perhaps the cruelest part of the equation is that some areas will see less rain overall, and the rain which does come falls in heavy bursts that fails to absorb into the soil, damaging infrastructure and property that lies in its path, and evaporates more quickly because of increased temperatures. Figures 2.6 to 2.9 illustrate where rates of precipitation have changed and by what extent.

Increased precipitation that exceeds the soil absorption rate will cause runoff. Communities manage stormwater runoff by constructing sewer systems. These systems have maximum flow capacities, and when rates of runoff exceed them flash flooding is the result. Despite its short-lived presence, flash flooding is extremely destructive, as witnessed in many communities impacted by Hurricane Irene in Vermont in 2011 (see box "Impact of Increased Precipitation in Vermont"). It can be logistically difficult and extremely costly to increase maximum flow capacities of systems designed for outdated hydrological data, yet many communities with decades-old systems are confronting the limits of their once-adequate systems.

Floods remain the most damaging and costly hazard in the United States, and the overall increase in total precipitation in terms of both rising annual averages and increased frequency of extreme events only adds to the weight of these statistics. Community flood mitigation measures, including culverts, canals, gullies, channels, and other mechanisms are not designed to meet many of the future precipitation estimates. Unstable slopes, like those which turned to mud in Snohomish County, Washington, during the 2014 Oso mudslide, will be more likely to reach the point of saturation and subsequent failure. Roads, bridges, railways, and other transportation networks in the path of natural routes of runoff will experience higher levels of erosion and more severe flood damage. Where runoff collects and compounds in rivers and streams, risk will quickly travel and extend far beyond those areas directly affected by the increased rains. Higher-than-normal flood levels on major rivers reduce or eliminate the natural flood buffer provided by these conduits, thereby increasing the frequency and severity of flooding in downstream areas.

Changes in precipitation coupled with the climatological effects associated with the seasons have resulted in an increase in winter storms. Winter storms have increased in both their frequency and intensity over the past 60 years. Record-breaking snowfalls in areas like the Washington, DC, metropolitan area have seen accumulations in single or closely-timed events that exceed typical annual snowfalls, as occurred in the 2009–2010 season. Events like these can collapse roofs that were constructed to withstand much lower accumulations, close city streets for days, and shutter businesses and government (as occurred with the federal government during a 3.5-day shutdown that followed a 2010 blizzard). Many cities that rarely experienced heavy snowfall events lack the equipment (e.g., plows), supplies (e.g., salt), and human resources to quickly and efficiently clear snow-covered

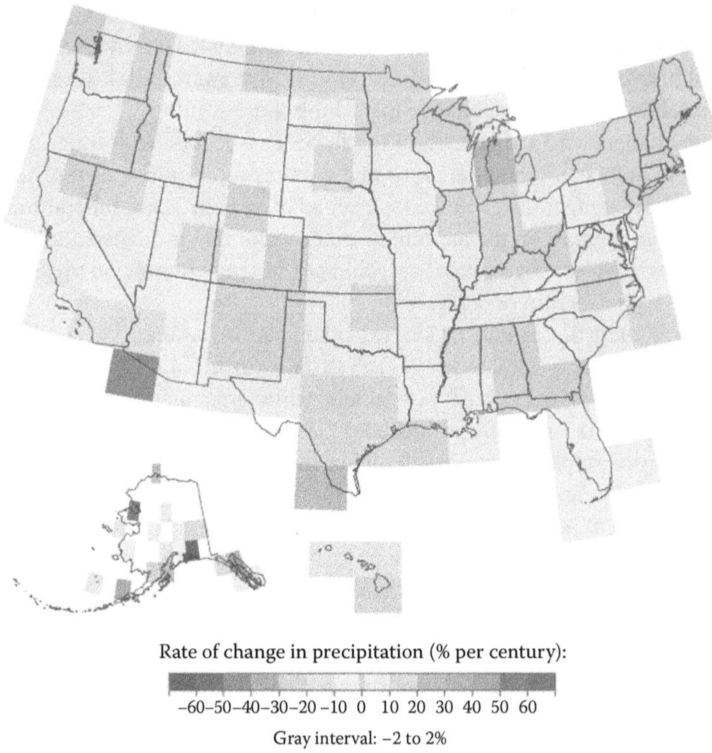

FIGURE 2.6 Rate of change in precipitation (reported as a percent change per 100 years). (From EPA. 2013. Rate of Precipitation Change in the United States, 1901–2012. Climate Change Indicators in the United States. US and Global Precipitation. Available at http://1.usa.gov/1um16sm.)

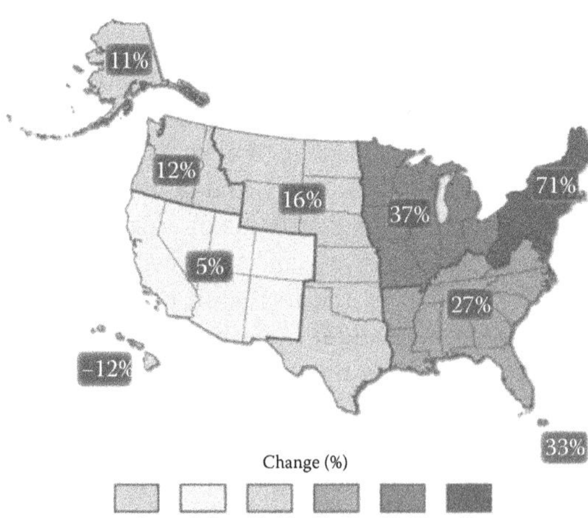

FIGURE 2.7 Observed changes in heavy precipitation from 1958–2012. (From US National Climate Change Assessment. 2014. Climate Change Impacts in the United States. US Global Climate Change Research Program. Available at http://NCA2014.globalchange.gov.)

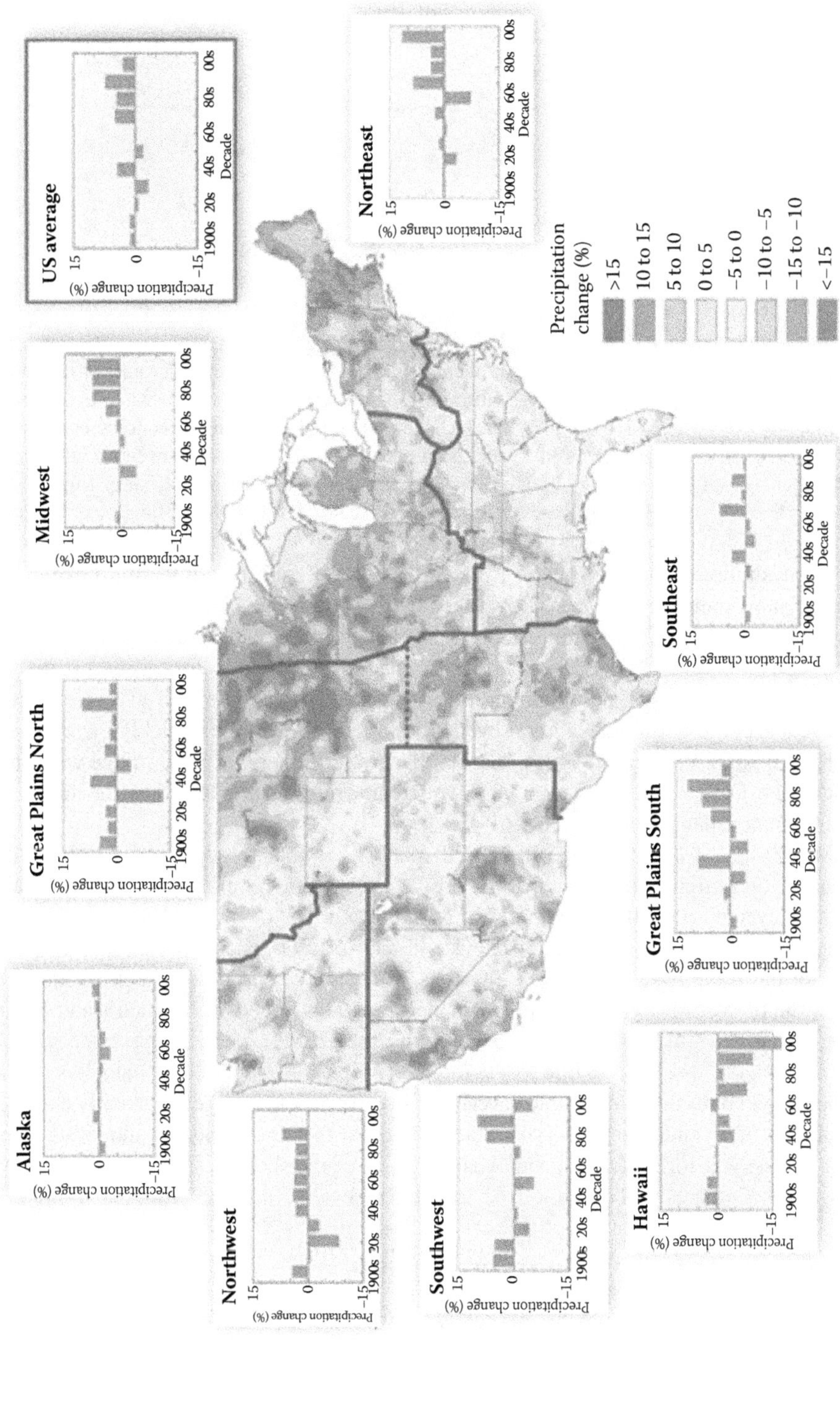

FIGURE 2.8 Observed changes in average annual precipitation for 1991–2012 compared to the 1901–1960 average. (From US National Climate Change Assessment. 2014. Climate Change Impacts in the United States. US Global Climate Change Research Program. Available at http://NCA2014.globalchange.gov.)

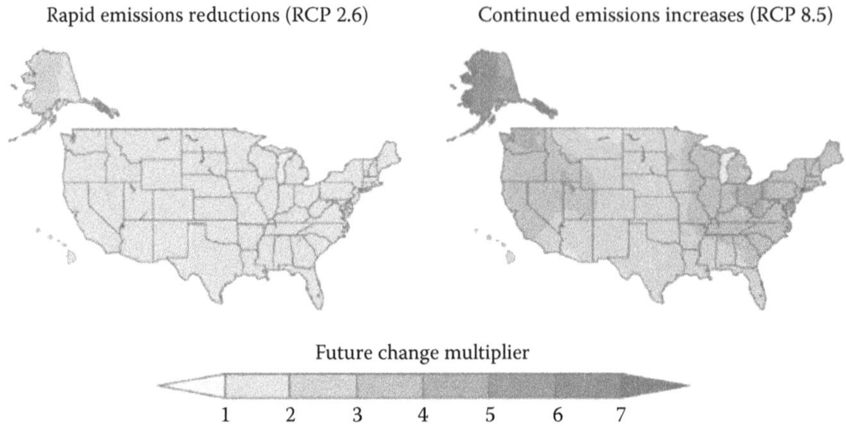

FIGURE 2.9 Projected change in heavy precipitation events. Maps show the increase in frequency of extreme daily precipitation events (a daily amount that now occurs once in 20 years) by the later part of this century (2081–2100) compared to the later part of last century (1981–2000). Such extreme events are projected to occur more frequently everywhere in the United States. Under more rapid emissions reduction scenarios, these events would occur nearly twice as often. With increasing emissions, however, these events could occur up to five times as often. (From US National Climate Change Assessment. 2014. Climate Change Impacts in the United States. US Global Climate Change Research Program. Available at http://NCA2014.globalchange.gov.)

streets and railways following heavy storms. Even Buffalo, New York, one of the most experienced cities in terms of snow removal operations, was taken by surprise when a late 2014 event dropped an astonishing 5–7 feet of snow. This extreme depth led to collapsed roofs despite some of the nation's most stringent construction codes and resulted in negative health impacts for those unprepared to clear the snow from their own property, among other impacts. Heavier snowfall averages and individual snow event accumulations can lead to severe flooding if temperatures quickly increase, especially when accompanied by rainfall. This particular hazard will likely increase in line with annual snowfalls and will affect communities even far downstream from those affected by the heavy snow (US National Climate Change Assessment 2014).

An interesting impact of the change in rainfall patterns is that in some areas, total rainfall remains the same or similar across the span of the year, yet this volume of water is now falling in fewer, heavier events rather than gradually in a higher number of minor downpours. Because heavier storms result in greater runoff and likewise less absorption, changes such as these have a dramatic impact on groundwater resources. In areas with varied topography, such as those with steeper slopes, the problem is exacerbated ever further due to the speed with which water escapes. Pakistan, which has a highly mountainous topography, is already experiencing extended water shortages on account of this effect. While that country's risk profile is notable in that it has included a dramatic increase in both the number and severity of flooding events, it is concurrently experiencing rapid depletion of groundwater reserves. Scientists point to the decrease in moderate rainfall events that more easily absorb into the ground as the cause of the shortages, and they believe with climate change these impacts will only increase for the foreseeable future (Shaikh and Tunio 2014).

INCREASING AVERAGE ANNUAL TEMPERATURES

It was increasing global temperatures that first elevated the risks associated with climate change to the forefront of the public agenda. Early understanding of the problem was somewhat limited with regard to the more complex factors that feed climate change trends, and as a result this phenomenon was first termed "global warming." Since that time, broader understanding of the complex interactions between temperature, precipitation, wind, and atmosphere, among other factors, has resulted

IMPACT OF INCREASED PRECIPITATION IN VERMONT

Rainfall monitoring indicates that the number of extreme rain events and damaging floods experienced by the State of Vermont are both increasing at a steady rate. The state expects these increases to pose a major challenge to communities moving in coming years. Floods have always occurred on Vermont's rivers, but the frequency, intensity, and timing of these events, on which the state's mitigation programs and development patterns have been based, are changing in uncertain ways (see Figure 2.10).

Tropical Storm Irene drenched Vermont in unusually heavy rain and wind over the course of two days (August 28–29) in 2011. While throughout the state rainfall averages of 3–5 inches were commonly encountered, several areas—many of which included communities with steep slopes and complex topography—endured more than 7 inches of rain in this one event. The runoff that resulted was extremely heavy and in some places resulted in catastrophic damages as had never before been seen in the state. Water from throughout the watershed continued to collect in rivers and streams, from the Mad River valley south to the Deerfield River, and impacting many communities along the way. In total, 10 of the state's 17 major rivers experienced record or near record flood levels, and 225 municipalities were affected.

A sample of the impacts that resulted from this brief event includes:

Transportation Sector

- More than 500 miles of highways and about 200 highway bridges were damaged
- More than 2000 municipal road segments, more than 280 bridges, and approximately 960 culverts were damaged
- More than 200 miles of railway track and six rail bridges in the state-owned rail system were damaged

Buildings and Infrastructure

- 73,000 customers lost power, 45 percent for more than 24 hours
- 4954 properties were damaged, 1500 seriously
- At least 12 municipalities experienced over $1 million in infrastructure losses

FIGURE 2.10 Rochester, Vermont, September 13, 2011: A bridge connected to Route 100 was destroyed but quickly replaced with a pedestrian walkway. Tropical Storm Irene took away many bridges and culverts as it moved throughout Vermont and beyond. (Photo by Angela Drexel/FEMA. http://1.usa.gov/1ALrYPg.)

Fisheries

- Significant damage was sustained by the White River National fish hatchery in Bethel, with 10 percent of hatchery fish killed and thousands of other fish contaminated.
- $500,000 in damages at the State Roxbury Fish Hatchery.

Wastewater Treatment

- Seventeen municipal wastewater treatment facilities (WWTFs) reported compromised operations, with issues ranging from pump station overflows to incomplete processing of sewage.

Forests

- High flows and saturated ground conditions undermined tree roots, and floating debris injured tree stems. A brief duration of standing water at most locations prevented further near-term tree damage; however, great amounts of accumulated sediment and debris in some streamside forests or establishment of invasive plants may inhibit tree growth over time.
- Aerial surveys found 9213 acres with trees exhibiting flood damage symptoms from both spring and Irene-related flooding.

Agriculture

- Farm fields and barns were washed out or covered with flood sediments and debris; more than 450 farms filed Farm Loss claims with the US Department of Agriculture (USDA) and roughly 20,000 acres of farmland were affected.
- Food advisories forced farmers to throw away food crops that may have been contaminated by flood waters.
- More than $10 million in crop losses statewide (see Figure 2.11).

Source: Vermont Agency of Natural Resources. 2012. Tropical Storm Irene: By the Numbers. Climate Change Team. http://bit.ly/1yHkiAS.

FIGURE 2.11 Brandon, Vermont, September 11, 2011: Main Street of Brandon showing damages that Tropical Storm Irene left behind. President Obama declared Disaster for the State of Vermont to supplement state and local recovery efforts in the area affected by the storm. (Photo by Angela Drexel. http://1.usa.gov/1BDqGLF.)

in a rebranding of the issue. Yet it is the increasing global temperature that is perhaps the driver behind all other effects.

The year 2014 was established in early 2015 to have been the hottest since scientific measurement of average annual temperatures began in the late nineteenth century. The National Aeronautics and Space Administration (NASA) and NOAA together released a report on these findings, which also states that 9 of the 10 hottest years on record occurred since 2000, and the 10th took place in 1998 (which means that the 10 hottest years on record occurred in the last 16 years) (Foran 2015).

Despite these record-breaking occurrences, the changes that have been observed in the United States are still lower than has occurred in many other parts of the world. But that fact should bring little consolation, as even a small variance in annual temperatures can have drastic secondary effects (and larger changes occurring in distant countries and continents will still result in both tangible and intangible indirect impacts to the United States). Temperatures in the United States will continue to rise, however, with the lowest projections calling for a 2-degrees-Celsius rise by the end of the twenty-first century and more severe projections claiming as much as 4 degrees. The greatest changes are expected to occur in the country's northern parts, including Alaska, where mean annual temperatures could rise as much as 6 degrees Celsius over present-day rates (Figure 2.12).

This increase in temperature expected to occur over the course of the century may seem small—a mere degree or two—but it is important to remember that these are annualized averages that do not convey the scope of day-to-day readings (which also differ considerably from what was experienced a century ago). In fact there is an increasing frequency in the number of periods where heat reaches and remains at much higher-than-typical levels (called extreme heat or a heat wave) (see box "Description of the Change in Heat Wave Frequency and Severity"). Because of the way that the climate has altered, heat waves will involve higher temperatures, last longer in duration, and will be more frequent in their occurrence (IPCC 2013). Projections of the number of days considered "extremely hot" show a continued increase over the majority of the United States, worsening as the year 2100 approaches. Rising summer temperatures coupled with reductions in soil moisture will lead to more sustained heat-wave events, especially in the western and central regions. Temperature extremes that occurred on average only once every 20 years (5% annual likelihood) in the 1950–1979 period are expected to occur at least seven out of every 10 years (70% annual likelihood) by 2035 (US National Climate Change Assessment 2014).

Extreme heat most significantly affects humans and other living things, although stresses on electrical infrastructure related to intense demand caused by widespread use of air conditioners can cause secondary, exacerbating disasters. Power outages leave people without any option for cooling down, and negative health effects skyrocket. Extreme heat can also increase the risk from wildfires

FIGURE 2.12 Lovelock, Nevada, February 7, 2014: State park rangers burned weeds on the exposed lake bed of the Rye Patch Reservoir in Nevada, which was at 3.5 percent capacity amid a drought that has caused the worst water shortage the region has faced in more than a century. (Photo by Max Whittaker, FEMA. http://1.usa.gov/1L1QS5w.)

DESCRIPTION OF THE CHANGE IN HEAT WAVE FREQUENCY AND SEVERITY

The following is an excerpt from the 2014 US National Climate Assessment Report *Climate Change Impacts in the United States*:

> Heat waves are periods of abnormally hot weather lasting days to weeks. Heat waves have generally become more frequent across the US in recent decades, with western regions (including Alaska) setting records for numbers of these events in the 2000s. Tree ring data suggests that the drought over the last decade in the western US represents the driest conditions in at least 1200 years (Kiger 2014) (see Figure 2.13). Most other regions in the country had their highest number of short-duration heat waves in the 1930s, when the multi-year severe drought of the Dust Bowl period, combined with deleterious land-use practices, contributed to the intense summer heat through depletion of soil moisture and reduction of the moderating effects of evaporation. However, the recent prolonged (multi-month) extreme heat has been unprecedented since the start of reliable instrumental records in 1895. The recent heat waves and droughts in Texas (2011) and the Midwest (2012) set records for highest monthly average temperatures, exceeding in some cases records set in the 1930s, including the highest monthly contiguous US temperature on record (July 2012, breaking the July 1936 record) and the hottest summers on record in several

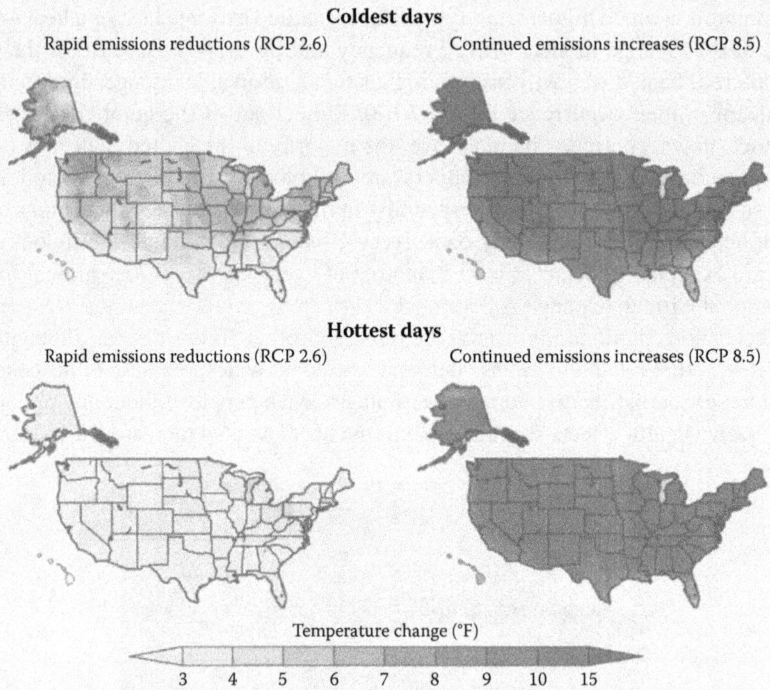

FIGURE 2.13 Projected temperature change of hottest and coldest days. Change in surface air temperature at the end of this century (2081–2100) relative to the turn of the last century (1986–2005) on the coldest and hottest days under a scenario that assumes a rapid reduction in heat trapping gases (RCP 2.6) and a scenario that assumes continued increases in these gases (RCP 8.5). This figure shows estimated changes in the average temperature of the hottest and coldest days in each 20-year period. In other words, the hottest days will get even hotter, and the coldest days will be less cold. (From US National Climate Change Assessment. 2014. Climate Change Impacts in the United States. US Global Climate Change Research Program. Available at http://NCA2014.globalchange.gov.)

states (New Mexico, Texas, Oklahoma, and Louisiana in 2011 and Colorado and Wyoming in 2012). For the spring and summer months, 2012 had the second largest area of record-setting monthly average temperatures, including a 26-state area from Wyoming to the East Coast. The summer (June–August) temperatures of 2012 ranked in the hottest 10% of the 118-year period of record in 28 states covering the Rocky Mountain states, the Great Plains, the Upper Midwest, and the Northeast. The new records included both hot daytime maximum temperatures and warm nighttime minimum temperatures. Corresponding with this increase in extreme heat, the number of extreme cold waves has reached the lowest levels on record (since 1895).

when conditions are very dry. Between 1979 and 2001 excessive heat exposure was attributed to almost 10,000 deaths in the United States. That accounts for more deaths than from hurricanes, lightning, tornadoes, floods, and earthquakes in the same period combined. Climate change has resulted in more severe and longer periods of extreme heat, raising the average annual death toll from heat waves to over 600 in the United States, and some predict this number will triple by the end of the century, thus resulting in approximately 150,000 heat-related deaths over the course of the century (CDC 2013; Coppola 2015; Foster 2013; Sheppard 2012). The elderly, children, and people with certain medical conditions such as heart disease are most vulnerable to this hazard (ISDR 2008). However, even young and healthy individuals can succumb to heat if they participate in strenuous physical activities during hot weather.

Interestingly enough, climate change, which is often singularly linked to a warming of the planet, will result in much lower temperatures—at times extreme—in parts of the country and the world. It might seem counterintuitive that a warming planet might cause colder weather, but scientists believe the phenomenon is caused by the melting of the polar ice caps, which in turn exposes darker oceans that more readily absorb solar radiation (and likewise increase warming effects, which disrupt arctic jet streams by pushing them southward). The earth isn't getting colder, it's just that the cold arctic air is being pushed southward as the poles continue to warm. The term "polar vortex," which describes this southwardly trend of cold airstreams, has come into more regular use as this hazard has increased in its likelihood in recent years (Walsh 2014).

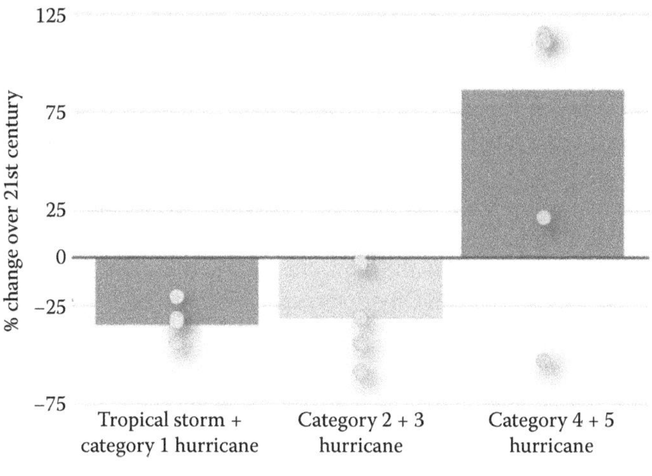

FIGURE 2.14 Climate change impact on hurricane frequency and intensity. Climate models project there will be fewer weak- to moderate-strength Atlantic hurricanes as surface temperatures rise this century. However, the models predict that a greater number of the hurricanes that do form will tend to strengthen to category 4 and 5 hurricanes. The bars in this graph show the average results from 18 different models. The dots on each bar show a range of results from 4 of the 18 different models. (From Scott, M. 2010. Will Hurricanes Change as the World Warms? Climate.Gov. NOAA. Available at http://1.usa.gov/1EuqHQf.)

Extreme cold days, which do little to offset the upward trajectory of annual global temperatures, can be life-threatening in a number of ways. They result in a higher incidence of injury and death from hypothermia and frostbite, especially when associated with power outages or for vulnerable populations that may not have the cognitive or financial abilities to protect themselves from the elements. They can also damage property, whether from frozen pipes, transportation accidents, or other means. They also cause dangerous ice dams in rivers, which can result in catastrophic flooding when they break or as they rapidly melt in springtime (Figure 2.14).

STRONGER (YET LESS FREQUENT) STORMS

The first real hint to the emergency management community that climate change may be more than simply an environmental issue came when scientists began linking a possible increase in severe weather with observed increases in global average temperatures. Initially, a higher frequency in stronger, damaging storms were blamed almost exclusively on the El Niño Southern Oscillation (ENSO or El Niño) wherein equatorial winds slow down, thereby causing changes in ocean temperatures associated with tropical storm formation. Cyclonic storms are fueled by warm, moist air, especially over warm bodies of water, which is why even the moderate increases in ocean temperatures are resulting in more destructive storms. Interestingly enough, as the climate has warmed there has been an associated decrease in the total number of tropical storms that are generated each year, which is postulated to be the result of lower temperature differences between the poles and the tropics. However, those that form do so in a manner that enables the storage of greater energy, placing them squarely within the highly destructive Category 3, 4, and 5 ranges on the Saffir-Simpson Hurricane Wind Scale (Cimons 2013; Elsner, Kossin, and Jagger 2008) (Figure 2.15).

During the winter months, outside of the typical hurricane season, the observed changes are a little different. In these colder periods, especially in the northern parts of the country, storms are gradually increasing in both strength and intensity, as tracked for the past 60 years. The typical storm track has been moving north, which has led to an associated decrease in storm frequency and intensity in the southern parts of the country (see Figure 2.16).

The impact of warmer land surface temperatures on convective storms, including hurricanes, is much less clearly understood. While it is assumed that because climate change exacerbates many of the conditions that are conducive to hurricanes, thunderstorms, and other convective events, there is insufficient data to draw any significant conclusions. Predictions, however, are in favor of increases in each of these events over time (Riobek 2005).

Any increases in storm intensity are going to result in increased statistical risk. Hurricanes Sandy and Katrina are commonly held up as harbingers of what's to come. But really, these storms are small compared to what is happening in Southeast Asia where supertyphoons like Haiyan, which struck the Philippines in 2013, are reaching intensities previously thought to be impossible. The Joint Typhoon Warning Center (JTWC) estimated Haiyan's 1-minute sustained winds to be 195 miles per hour at landfall, which is the highest ever recorded. The event caused a storm surge in the Islands of Leyte and in Tacloban City that reached as high as 17 feet in some areas. Rainfall reached almost a foot in less than 12 hours, and all low-lying areas were flooded. By the time the storm had passed, approximately 90 percent of Tacloban was obliterated, over 6300 people were dead, 28,000 injured, and another 1000 missing. The storm caused over $2 billion in damages in a country where damage tallies of that magnitude are seldom reached. Were a storm of this intensity to strike the United States, the results could be similar if not greater. One of the most terrifying scenarios that the Federal Emergency Management Agency (FEMA) Catastrophic Planning Program prepares for is a Category 5 hurricane striking the Lake Okeechobee area and breaching the Herbert Hoover Dike in Florida. A scenario like this could result in the flooding of much of South Florida, which as modeled would result in fewer fatalities than resulted from Hurricane Katrina but much greater economic damages (over $450 billion) (Reynolds and Shelly 2013). But the problem is not unique to South Florida. In fact, all coastal communities have to consider how

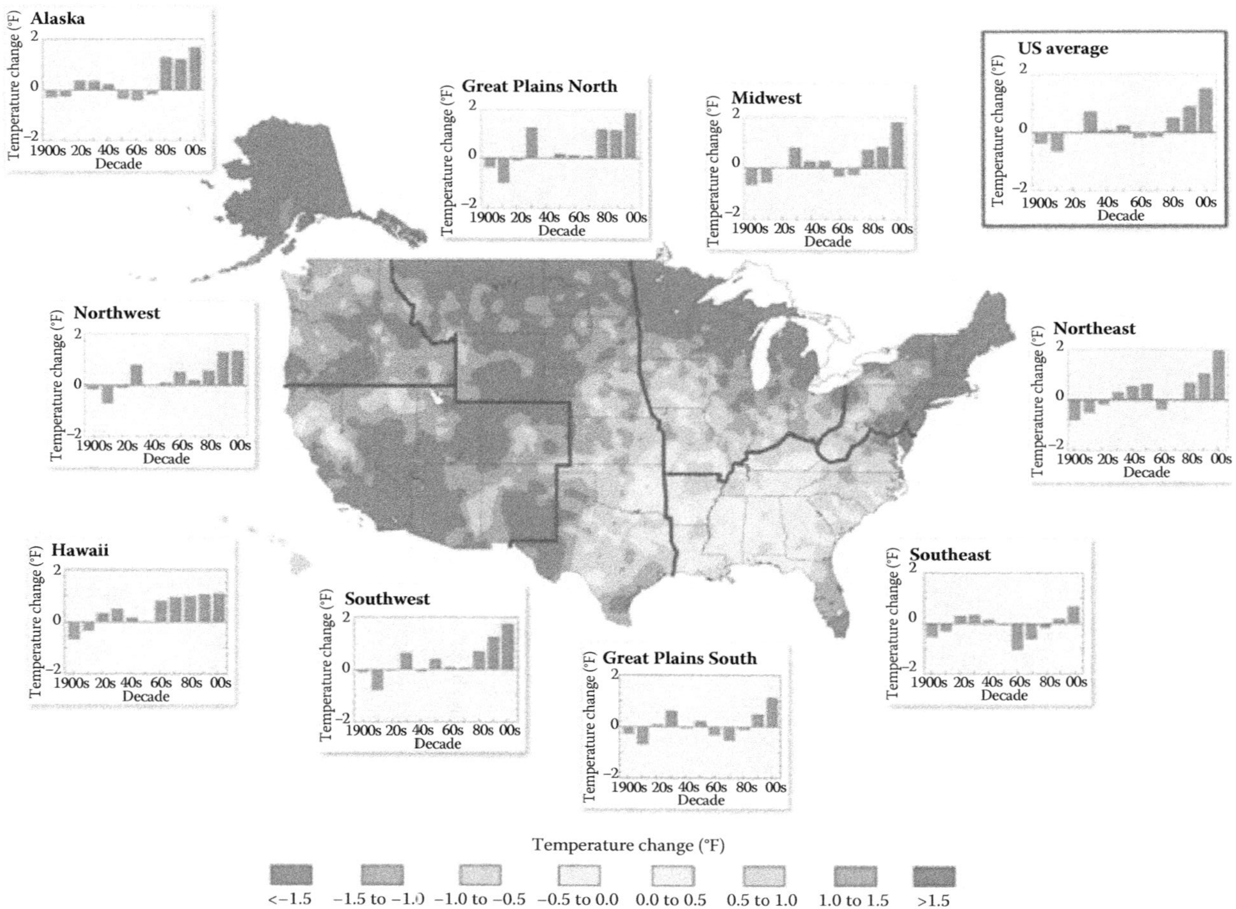

FIGURE 2.15 Changes in average temperatures between 1991–2012 as compared to the 1901–1960 temperatures in the contiguous United States, and to the 1951–1980 average for Hawaii and Alaska. (From US National Climate Change Assessment. 2014. Climate Change Impacts in the United States. US Global Climate Change Research Program. Available at http://NCA2014.globalchange.gov.)

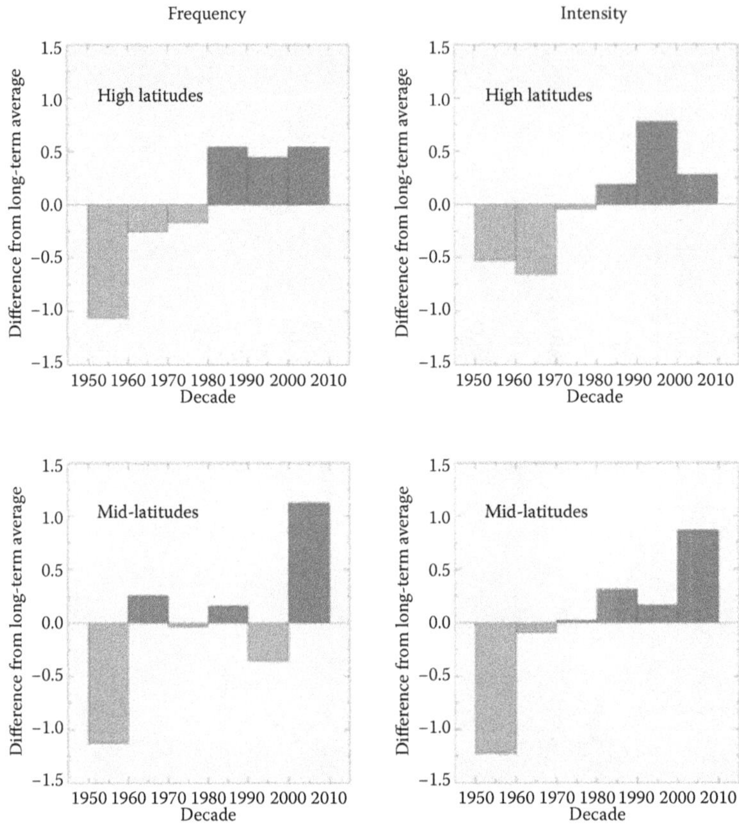

FIGURE 2.16 Variation of storm frequency and intensity in the cold season (November–March). Variation of winter storm frequency and intensity during the cold season (November–March) for high latitudes (60°N–90°N) and mid-latitudes (30°N–60°N) of the Northern Hemisphere over the period 1949–2010. The bar for each decade represents the difference from the long-term average. Storm frequencies have increased in middle and high latitudes, and storm intensities have increased in middle latitudes. (From US National Climate Change Assessment. 2014. Climate Change Impacts in the United States. US Global Climate Change Research Program. Available at http://NCA2014.globalchange.gov.)

increases in typical tropical storm and hurricane intensities might change their coastal flood zone designations, and more importantly, how damaging a Category 3, 4, or 5 hurricane would be to their infrastructure, their economy, and their residents.

HEALTH-RELATED ISSUES

Human health will face a multifront onslaught as a result of the varied conditions associated with climate change. Health and life safety are affected both directly and indirectly, and vulnerabilities will differ according to the specifics of each threat, many of which are described herein.

Perhaps the most obvious risk to health is that of exposure to heat. As was previously discussed, climate change is rooted in the increased tendency for solar energy to be absorbed by the land and sea, held within the atmosphere, and reflected back to earth. Heat waves and record high temperatures stress humans by causing dehydration, heat stress, heat stroke, and death. The elderly and the very young are most at risk from the physical effects of heat-related hazards, and these same groups are also most likely to be exposed to extreme heat due to social vulnerabilities that typify them.

Climate change will also result in an increase in waterborne diseases due to increasing water temperatures, increased incidence of floodwaters, increased routes of pollutants into waterways, and more hospitable conditions for certain waterborne diseases (e.g., cholera and schistosomiasis). Increased rainfall has led to overflows and even failures of water treatment facilities, resulting in the release of untreated waste into waterways. Boil-water orders caused by raw sewage releases may become more common as heavier rainfall events and heavy spring snowmelts rise in frequency. Rising water temperatures, especially in coastal areas, can lead to an increase in bacterial growth, which can cause a number of gastrointestinal illnesses. Water shortages can likewise cause increased incidence of waterborne illness due to the concentrating effects of chemicals and biological materials in shrinking water reserves (National Institutes of Health 2015).

Many disease vectors, such as mosquitoes, ticks, mites, and rodents, benefit from the changes in climate that are occurring. As such, certain vector-borne diseases are likely to increase as climate effects are magnified. Tropical diseases like malaria, yellow fever, and dengue fever are transmitted by mosquitoes that reproduce much more successfully in warmer climates. These diseases, which were eradicated in the United States years ago when such vectors were much easier to control, are again threatening the southern states as they move northward with rising temperatures. Were these diseases to result in outbreaks in the United States, the public health system might not at first be adequately prepared to manage an epidemic. Once conditions became hospitable to the disease vectors, it may be difficult to eradicate them from all areas, especially outside of city centers (Physicians for Social Responsibility 2008). Diseases that are already endemic to the United States, including West Nile virus and Lyme disease, will likely spread far beyond their current range as temperatures increase the hospitable zone for the associated vectors (see Figure 2.17).

It has also been postulated that mental-health-related cases will increase as a result of climate change due to the stress caused by disasters and physical discomfort. Disasters have long been known to cause psychosocial impacts on those exposed to them, similar to posttraumatic stress disorder. These impacts can last for months or years following the event and require long-term treatment. Increased population exposure to floods, hurricanes, wildfires, and other climate-related disasters will likewise increase the number of people experiencing these systems. Chronic effects of climate change, such as longer periods of extreme heat or cold, water shortages, power outages, and other effects can result in increased stress-related illnesses and disorders.

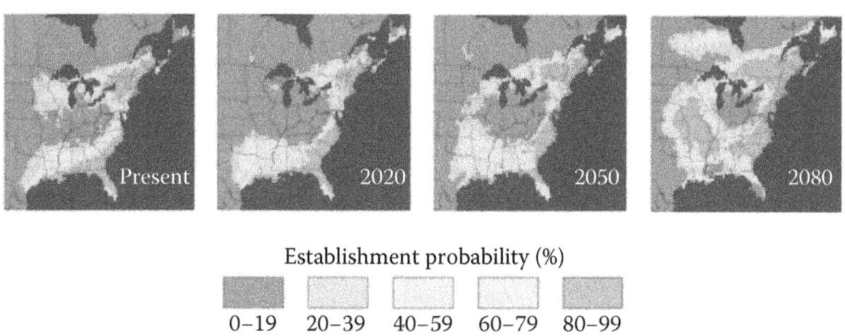

FIGURE 2.17 Projected changes in tick habitat, 2015–2080. The maps show the current and projected probability of establishment of populations of the deer tick (the species that transmits Lyme disease). Projections are shown for 2020, 2050, and 2080. The projected expansion of tick habitat includes much of the eastern half of the country by 2080. Note that in some areas around the Gulf Coast, the probability of tick population establishment is projected to decrease by 2080. (From US National Climate Change Assessment. 2014. Climate Change Impacts in the United States. US Global Climate Change Research Program. Available at http://NCA2014.globalchange.gov.)

Climate change has often been associated with respiratory illnesses, due to the fact that many experts link increases in airborne pollutants (namely the greenhouse gases linked to fossil fuel burning) with both the cause of climate change and the triggering of asthma. Of course, this is not a cause-and-effect equation. Rather, the two issues can only be said to be occurring in conjunction with each other (i.e., it's not climate change that is causing these health problems, it is pollution). That being said, there is now growing evidence that many common allergens, which also cause respiratory problems like asthma and hay fever, are in fact exacerbated by the *effects* of climate change; namely, increased temperature and rainfall. Mold and pollen affect millions of Americans (some estimates place grass allergies alone at over 20% of the US population) (Mooney 2014), and a recent study has found that warmer temperatures could lead to as much as a 200 percent increase in the concentration of pollen in the air (Albertine et al. 2014). While conclusive links to climate change have not been established, it is interesting to note that the prevalence of asthma (the percentage of people who have ever been diagnosed with asthma and still have asthma) throughout the United States increased from 7.3 percent in 2001 to 8.4 percent in 2010 (US National Climate Change Assessment 2014). Dust from prolonged drought and increased convection, as occurred during the Dust Bowl years (1934, 1936, and 1939–1940), and smoke from increased wildfires, are both additional factors related to potential increases in respiratory illnesses associated with climate change effects.

POWER FAILURE

The primary and secondary hazards associated with climate change will continue to impact both human and built environments. When infrastructure is affected, additional secondary hazards can arise, resulting in severe physical and economic risk to the populations served. Perhaps the most likely scenario, which has already manifested in many American communities, is that power failures will become a more common occurrence.

There are a number of mechanisms by which climate change affects both the generation and the transmission of electrical power. At the most basic level, increased temperatures drive up demand for a number of power-intensive machines and services. Cooling is by far the greatest power consumer. The experience of the United Arab Emirates (UAE), where summer temperatures climb to upward of 120 degrees Fahrenheit, is indicative of the demand that cooling can impose on a nation's electrical infrastructure and energy sector. It has been estimated that upward of 70 percent of all energy consumed in the UAE is dedicated to cooling in the commercial and residential sectors, and as a result power failures have become common during extended hot periods when there is excess demand (Pamuk 2011). Globally, cooling already accounts for half of the world's electrical consumption, amounting to 10 trillion kilowatts per year (Cox 2012).

Power failures also occur when the energy infrastructure is impacted by severe storms, flooding, or other hazards, many of which are exacerbated by climate change. Many recent hurricanes have resulted in downed power transmission lines, flooded generation facilities, and disruption in the extraction of fossil fuels or energy supply chain. Ice storms, blizzards, and tornadoes can inflict many similar impacts (see box "Climate Change Threats to Electrical Power").

Power loss can be devastating to the individuals, businesses, and communities served. They are devastating to health when they occur in the middle of extreme temperatures (hot or cold) because vulnerable elderly, poor, and very young people can die of exposure. They can also prevent access to critical lifesaving and life-sustaining equipment. Power failure also has devastating economic impacts both to individual businesses that lose revenue and face closure and to local economies that depend on the taxes and fees they collect that are generated by successful businesses. The 2003 Northeast Blackout, which affected over 50 million people in the United States and Canada, caused over $6 billion in economic costs (and affected water supplies and transportation systems) (US Government Accountability Office 2003). It is devastating to the other components of infrastructure that depend on electricity to function, such as hospitals, communications systems, water treatment facilities, and more. In fact, much of the flooding that occurred in New Orleans following

CLIMATE CHANGE THREATS TO ELECTRICAL POWER

The Union of Concerned Scientists issued a report entitled *Power Failure: How Climate Change Puts Our Electricity at Risk, and What We Can Do* in April 2014 that describes the increasing threat to the nation's electricity infrastructure. This report summarized the threat as follows:

- *Accelerating sea level rise:* Today some 100 electric facilities in the contiguous US, including power plants and substations, are sited within four feet of local high tide. Global average sea level has increased eight inches since 1880, and is projected to rise up to an additional 6.6 feet over the course of this century, greatly increasing coastal flooding risks.
- *Increasing wildfires:* Wildfires can directly damage transmission poles and other electricity infrastructure, but the greatest risk comes from smoke and particulate matter, which can ionize the air, create an electrical pathway away from transmission lines, and shut down the lines. Climate change affects multiple factors that increase wildfire risk; the average number of large wildfires in the western US rose from 140 in the 1980s to 250 between 2000 and 2012.
- *More frequent and intense heat waves:* Climate change is already bringing more intense, more frequent, and longer lasting heat waves in North America. These periods of extreme heat decrease the efficiency of power plants during periods when electricity demand is highest, placing additional stresses on the electricity system.
- *Droughts and reduced water supplies:* The electricity sector is highly dependent on water for cooling. Power production accounts for the single largest share of all freshwater withdrawals in the US. As temperatures continue to rise, droughts and reduced water supplies are likely to become the norm in some regions, increasing the risk to the power sector.
- *Elevated water temperatures:* Higher air temperatures warm the water in rivers and reservoirs used by power plants for their cooling needs. If the temperature of incoming water is too hot, or if the temperature of the discharge water is too high, power plants must dial back production or shut down temporarily, as has occurred at numerous coal and nuclear power plants over the past decade.

Source: Union of Concerned Scientists. 2014. Power Failure: How Climate Change Puts Our Electricity at Risk, and What We Can Do, April. Available at http://bit.ly/15zfAsT.

Hurricane Katrina was linked to or at least exacerbated by the failure of several drainage pumps, which occurred after power to the city was lost.

CONCLUSION

Communities have no choice but to consider the temporal nature of their hazard profile given the long-term threat of climate change. It is no longer feasible to calculate static likelihood and consequence factors when assessing risk, even for those hazards that have no obvious association with climate change, such as landslides. The effects of climate change are varied and complex, and without due consideration emergency plans and risk reduction policies will both become insufficient as these trends persist. As models of climate shifts become more accurate, this task will become easier. Recent reports, including the *Fifth Climate Assessment of the IPCC* and the US National Climate Assessment report *Climate Change Impacts in the United States*, offer a fairly comprehensive view into the future of risk in America. It is contingent on not only the emergency management community, but all community stakeholders, to take notice and respond accordingly.

REFERENCES

Adam, D. 2007. How Climate Change Will Affect the World. *The Guardian*, September 19. Available at http://bit.ly/1AS40nX.

Albertine, J.M., W.J. Manning, M. DaCosta, K.A. Stinson, M.L. Muilenberg, and C.A. Rogers. 2014. Projected Carbon Dioxide to Increase Grass Pollen and Allergen Exposure Despite Higher Ozone Levels. *PLoS One* 9(11): e111712. Available at http://bit.ly/15vcyq2.

Associated Press. 2007. Thai Official: Bangkok Will Be Underwater in 20 Years. *Fox News* website, October 23. Available at http://fxn.ws/1B4z9Hi.

Australian Business Review. 2014. CSIRO Warns of "$1 Trillion" Natural Disasters Exposure, December 14. Available at http://bit.ly/1vR015o.

Bailey, A., T.N. Chase, J.J. Cassano, and D. Noone. 2011. Changing Temperature Inversion Characteristics in the US Southwest and Relationships to Large-Scale Atmospheric Circulation. *Journal of Applied Meteorology and Climatology* 50(6): 1307–1323. Available at http://bit.ly/1u8xFUW.

Bernama. 2010. Researchers Warn Bangkok under Threat to Sink, Bernama.com July 23. Available at http://bit.ly/1Irn0NR.

Biello, D. 2009. How Will Warmer Oceans Affect Sea Life? *Scientific American*, August 25. Available at http://bit.ly/1xPb6oP.

Borenstein, S. 2007. Rising Seas Likely to Flood US History. *Washington Post*, September 23. Available at http://wapo.st/1DJpUNS.

Briggs, B. 2014. Four-Star Warning: Generals Dub Climate Change a Security Risk. *NBC News*, May 14. Available at http://nbcnews.to/1C9Jn6f.

Byanyima, W. 2014. Climate-Related Disasters Have Cost Half a Trillion Dollars Since World Leaders Met to Discuss Climate Change. *Oxfami*, September 19. Available at http://bit.ly/1qQH72Z.

Cameron, C. 2014. Research Shows California Drought is the Worst in 1200 Years. *Inhabitat*, December 8. Available at http://bit.ly/1zdQQ52.

CDC. 2013. Heat related deaths after an extreme heat event: Four states, 2012, and United States, 1999–2009. *Morbidity and Mortality Weekly Report* 62(22): 433–436.

Cimons, M. 2013. Perfect Storm: Climate Change and Hurricanes. *Live Science*, April 5. Available at http://bit.ly/1yum4F7.

Coghlan, C., M. Muzammil, J. Ingram, J. Vervoort, F. Otto, and R. James. 2014. A Sign of Things to Come? Oxfam Research Reports, September. Available at https://www.oxfam.org/en/research/sign-things-come.

Coppola, D. 2015. *Introduction to International Disaster Management.* Elsevier, Burlington, MA.

Cox, S. 2012. Climate Risks Heat Up as World Switches on Air Conditioning. *The Guardian*, July 10. Available at http://bit.ly/1D85f2A.

Davenport, C. 2014. Climate Change Deemed Growing Security Threat by Military Researchers. *The New York Times*, May 13. Available at http://nyti.ms/1IMGPNC.

Department of Defense. 2014. Quadrennial Defense Review 2014. Available at http://1.usa.gov/1u8C5eB.

DiConsiglio, J. 2015. As Alaskan Permafrost Melts, Chemistry Professor J. Houston Miller and His Students are Designing New Technology and Preparing for an Excursion to the Ground Zero of Global Warming. George Washington University, Columbian College of Arts & Sciences, January 14. Available at http://bit.ly/1FZlYe7.

Einhorn, B. 2015. Australia is Burning, and Climate Change is Making it Worse. *Bloomberg Business Week.* Global Economics. Available at http://buswk.co/1ujFU0A.

Elsner, J.B., J.P. Kossin, and T.H. Jagger. 2008. The Increasing Intensity of the Strongest Tropical Cyclones. *Nature* 455: 92–95. Available at http://bit.ly/1CEf0Vy.

EPA. 2013. Rate of Precipitation Change in the United States, 1901–2012. Climate Change Indicators in the United States. US and Global Precipitation. Available at http://1.usa.gov/1um16sm.

EPA. 2015a. Sea Surface Temperatures. Climate Change. *EPA* website. Available at http://1.usa.gov/15pv7fQ (accessed January 20, 2015).

EPA. 2015b. Water in Daily Life. Water Sense. *EPA* website. Available at http://1.usa.gov/1J6hM8v (accessed January 20, 2015).

Foran, C. 2015. Agencies: 2014 was the Hottest Year on Record. *National Journal*, January 16. Available at http://bit.ly/1xwNE0s.

Foster, J. 2013. Deaths from Heat Waves May Increase Ten Times by Mid-Century. *Climate Progress*, November 8. Available at http://bit.ly/1xwXX4K.

Gerken, D. 2012. Sea Level Rise Threatens Countries, Regions, and Cities Around the World. *Huffington Post*, August 26. Available at http://huff.to/1n7HBw9.

Gray, R. 2015. Study Shows Parasites Could Become More Common as Temperatures and Sea Levels Rise. *Daily Mail*, January 13. Available at http://dailym.ai/1wn83DE.

IPCC. 2013. Summary for Policymakers. In T.F. Stocker, D. Qin, G.-K. Plattner, M. Tignor, S.K. Allen, J. Boschung, A. Nauels, Y. Xia, V. Bex, and P.M. Midgley (eds.). *Climate Change 2013: The Physical Science Basis. Contribution of Working Group I to Firth Assessment Report of the Intergovernmental Panel on Climate Change*. Cambridge University Press, Cambridge, UK.

IPCC. 2014. Climate Change 2014: Impacts, Adaptation, and Vulnerability. Available at http://bit.ly/1DHtHeF.

ISDR. 2008. Briefing Note 1: Climate Change and DRR. Geneva, September. Available at http://bit.ly/1zqSooP.

Kiger, P. 2014. California Drought is Worst in 1200 Years. *Discovery News*, December 5. Available at http://bit.ly/1EtVxsn.

LeGault, D. 2014. Researchers Say Rise in Sea Temperatures May Cause Coral Killings Worldwide. *Capital Berg*, December 23. Available at http://bit.ly/1zt3A9T.

Maritime Executive. 2015. Sea Level Rising Faster Than Thought. MarEx Staff, January 18. Available at http://bit.ly/1GmOpTa.

Maron, D.F. 2013. Dengue Fever Makes Inroads Into the US. *Scientific American*, November 15. Available at http://bit.ly/1uLNiSd.

Mascaro, L. 2015. Climate Change is Real and Not a Hoax, Senate Overwhelmingly Decides. *Los Angeles Times*, January 21. Available at http://lat.ms/1L4ssID.

Mills, A. 2012. Jersey Shore Development Failures Exposed by Hurricane Sandy. *Huffington Post*, December 12. Available at http://huff.to/1FUi8Tn.

Mooney, C. 2014. Global Warming Could Make Your Pollen Allergies Much Worse. *The Washington Post*, November 5. Available at http://wapo.st/1JbIbU9.

Morin, C.W., A.C. Comrie, and K. Ernst. 2013. Climate and Dengue Transmission: Evidence and Implications. *Environmental Health Perspectives* 121(11–12): 1264–1272. Available at http://1.usa.gov/1C2LlY1.

NASA. 2014. NASA Finds Earth's Abyss Has Not Warmed. NASA Press Release, October 6. Available at http://1.usa.gov/1DWZxEc.

National Institutes of Health. 2015. Waterborne Diseases and Climate Change. Global and Environmental Health. *NIH* website. Available at http://1.usa.gov/1yPULaf (accessed January 22, 2015).

Neuhauser, A. 2015. Obama: No "Greater Threat" than Climate Change. *US News*, January 20. Available at http://bit.ly/1yLOIly.

NOAA. 2010. Changing Climate, Changing Sanctuaries. National Marine Sanctuaries. News & Events. *NOAA* website. Available at http://1.usa.gov/1AFqJkJ.

Pamuk, H. 2011. UAE's Mission Impossible: Cooling the Desert. *Reuters*, July 13. Available at http://reut.rs/1ze0GX2.

Physicians for Social Responsibility. 2008. Health Implications of Global Warming: Vector Borne and Water Borne Diseases. Fact Sheet. Available at http://bit.ly/1un50M0.

Radford, T. 2014. How Climate Change Could Increase Pollen Levels by 200%. *Eco Watch*, November 9. Available at http://bit.ly/1IzQVDw.

Reynolds, D., and J. Shelly. 2013. The Day the Dike Breaks. *Risk & Insurance*, August 1. Available at http://bit.ly/1Gz9CcE.

Riobek, H. 2005. The Rising Cost of Natural Hazards. *NASA Earth Observatory*, March 28. Available at http://1.usa.gov/15C8CEs.

Scott, M. 2010. Will Hurricanes Change as the World Warms? *Climate.Gov. NOAA*. Available at http://1.usa.gov/1EuqHQf.

Shaikh, S., and S. Tunio. 2014. Flooding Could Worsen Pakistan's Water Shortage, Experts Warn. *Reuters*, November 3. Available at http://reut.rs/1tvAdSd.

Sheppard, K. 2012. Climate Change Will More than Triple Annual US Heat Death Toll. *Mother Jones*, May 24.

Sturmer, J. 2014. CSIRO Says Climate Change and Poor Planning Could Cost Australia Over a Trillion Dollars, December 10. Available at http://ab.co/1zqNGHB.

Subbaraman, N. 2014. Fasten Seat Belts: Climate Change Could Mean More Turbulence. *NBC News*, February 20. Available at http://nbcnews.to/1zfMRqo.

Thieler, E.R., J. Williams, and E. Hammar-Klose. 2014. National Assessment of Coastal Vulnerability to Sea Level Rise. Woods Hole Field Center. USGS. Available at http://woodshole.er.usgs.gov/project-pages/cvi/.

US Government Accountability Office. 2003. Electricity Restructuring: 2003 Blackout Identifies Crisis and Opportunity for the Electricity Sector. GAO-04-204, November 18. Available at http://1.usa.gov/1zMpy7P.

US National Climate Change Assessment. 2014. Climate Change Impacts in the United States. US Global Climate Change Research Program. Available at http://NCA2014.globalchange.gov.

UN Office for Disaster Risk Reduction (UNISDR). 2008. Climate Change and Disaster Risk Reduction. Briefing Note 01, United Nations. Available at http://bit.ly/1ACI1lY.

Union of Concerned Scientists. 2014. Power Failure: How Climate Change Puts Our Electricity at Risk, and What We Can Do, April. Available at http://bit.ly/15zfAsT.

Viet Nam News. 2010. HCM City Mulls Climate Change Threat. *Viet Nam News* website, July 13. Available at http://bit.ly/1C3cjPo.

Walsh, B. 2014. Climate Change Might Just Be Driving the Historic Cold Snap. *Time*, January 6. Available at http://ti.me/1y01EyR.

The White House. 2014. President's State, Local, and Tribal Leaders Task Force on Climate Preparedness and Resilience. Available at http://1.usa.gov/1wWCOo7.

Yoskowitz, D.W., J. Gilbeaut, and A. McKenzie. 2009. The Socio-Economic Impact of Sea Level Rise in the Galveston Bay Region. Environmental Defense Fund Report. Harte Research Institute. Available at http://bit.ly/1EkdI3u.

Young, R. 2013. A Year After Sandy, The Wrong Policy on Rebuilding the Coast. *Yale Environment 360*, October 31. Available at http://bit.ly/1Ae3sG5.

3 The Stages of Adaptive Planning

INTRODUCTION

Planners and emergency managers are currently at the forefront of hazard risk reduction efforts at the community level in the United States. Many of these same people are leading the way as communities seek to deal with climate change and become more resilient. This chapter presents two essays, one by a community planner with years of experience working in hazard risk reduction and one authored by a local emergency manager who has toiled for years in the world of community-based disaster mitigation. The two essays examine how the fields of planning and emergency management can help guide the integration of climate change adaptation planning and actions into building resilient communities.

The first essay examines the evolution of the integration of hazard mitigation planning into comprehensive community plans and how this trend can serve as a guide to communities as they integrate climate change adaptation actions into their comprehensive plans. As James Schwab notes in his essay, "Although young and evolving, the system and methodology for integrating climate adaptation into both local hazard mitigation plans, under the federal Disaster Mitigation Act, and local comprehensive plans through hazard-related and other elements, is clearly taking hold in those jurisdictions with the largest stake in surviving the impacts that may have some connection to global climate change. With or without federal or state legislative encouragement, they are tapping a variety of scientific, governmental, and institutional resources to experiment and innovate with meaningful approaches to this challenge amid a number of uncertainties."

The second essay addresses climate change adaptation from the view of the emergency manager and examines what local emergency managers can do to help make their communities more resilient. Eric Holdeman notes in his essay, "While it may appear that no one is listening, the emergency manager must continue to provide information on hazards and risks to their individual communities. Building coalitions of people and organizations with mutual interests in climate adaptation remains one of your best tools for success."

We hope that these two essays will provide some guidance to community leaders as to what role community planners and local emergency managers can play in helping communities understand their risks from climate change and to identify and implement those adaptation actions that will protect their citizens, homes, businesses, and natural resources in the future.

PLANNING AND CLIMATE CHANGE: CREATING RESILIENCE IN US COMMUNITIES

James Schwab, AICP
Hazards Planning Center for the American Planning Association (APA)

INTRODUCTION

Planning at the community and regional level is essentially the practice of developing a vision for that community's future welfare and development, embodying that vision in an adopted plan, and implementing that vision through regulations, incentives, public policies, and administrative actions. Part of that process involves issues of land use, but other parts relate to economic development, public investment, infrastructure, and environmental policy, as well as other tools and techniques that may be available for the community to accomplish its stated goals. The exact form and focus of a community's plan will often depend on the statutory authorities it has to pursue those goals, as well as the authorities delegated to neighboring, superior, or special jurisdictions that may cooperate with or hinder the vision behind the plans. For instance, a community's goal of compact growth may rely to a significant extent on the cooperation of an independent school district in its school siting policies, or on a regional water board or transit authority in locating infrastructure or developing new bus or rail lines. The challenge is not necessarily that the various jurisdictions want to obstruct change, but that each has its own institutional dynamics and objectives.

The planning movement in the United States dates to the early twentieth century, while the issue of climate change and its impacts on urban growth is a relative newcomer. While it is obvious that local political systems and even planning functions have evolved over the past century largely without reference to climate change, and in many cases without much reference to natural hazards, the origins of modern planning nevertheless provide some basis for optimism about core values. The early twentieth-century planners paid considerable attention to public health because pollution, industrial hazards, and disease were prevalent urban problems in the late nineteenth century. Modern sewer and water infrastructure, among other public investments, were meant to alleviate many of those problems. Thus, in anticipating health and damage impacts from climate change and natural disasters, planners can in effect see themselves extending their original mission to address new threats to the public welfare.

Nonetheless, progress in the public interest has seldom come easily, especially when it conflicts with powerful vested interests, as the politicization of the discussion of climate change has demonstrated. Even without reference to climate change, hazard mitigation as a public policy priority has often faced challenges, both subtle and direct, from real estate and development interests that may find coastal, riverfront, or mountainside properties both attractive to buyers and lucrative for sellers. On the other hand, a growing body of federal (and sometimes state) public policy is providing a counterweight of sorts to such pressures. Community planning often finds itself in the middle of such pressures, trying to establish a public consensus that will lead to implementable solutions. Those solutions vary, and work at various levels, so that some fall directly within the realm of one local government or another, while others may require a good deal of political persuasion and regional cooperation in order to make a difference. This essay discusses the most important ways in which planning may be able to make those changes happen.

Mitigation versus Adaptation

Before pursuing the question of how planning is addressing the impact of climate change in the form of increased extreme weather events, thus accentuating the need for effective and forward-looking hazard mitigation, it is important to distinguish this emphasis from "climate mitigation" to avoid conceptual confusion. It is perhaps unfortunate for the understanding of the general public that we use the same words in such disparate ways. In the world of climate action planning, mitigation refers to the use of public policy mechanisms and strategies to reduce emissions of greenhouse gases that produce global warming and related climate changes. The idea is for the local jurisdiction, region, or state to do its part to slow the process of overall global climate change.

In contrast, adaptation refers to those strategies aimed at coping with or responding to the impacts of climate change that are already happening. In the world of hazard mitigation planning, that adaptation is a new wrinkle in what is already known as mitigation. To avoid that confusion in a book that discusses both types of planning, Boswell et al. (2012) chose to refer to climate mitigation as emissions reductions or reduction strategies.

Both types of climate action planning have their imperatives. It is the argument here that they also should not be done in isolation from each other, but inform each other. One challenge is the organizational and policy development structure that local governments choose to use in developing such plans. Some communities have chosen to place climate action plans within some sort of framework emphasizing sustainability or sustainable development, which has often lacked any focus on hazard mitigation. In more optimistic times, there may have been a belief that a shift to green development would forestall the impacts of climate change, but that is no longer a tenable position as the studies accumulate that demonstrate that the world is already on a trajectory with too much accumulation of greenhouse gases in the atmosphere and far too little effort to constrain that momentum. A world that does not prepare for accelerated impacts of global climate change may be a world headed for more catastrophes traceable at least indirectly to an increasingly volatile climate. The problem for local planners is that the level of certainty on this point that may be possible on the global or even continental scale is much harder to pin down in a local context.

Climate adaptation, on the other hand, necessarily finds a home, if it is taken into consideration at all, in the context of local and state hazard mitigation planning. This framework has been driven since 2000 by the federal Disaster Mitigation Act, which makes eligibility for federal mitigation grants contingent on adoption and federal approval of a mitigation plan that complies with the law and accompanying FEMA regulations. Often, because FEMA is first and foremost an emergency management entity, states and local jurisdictions assign the development of such plans to their own emergency managers, although planners are increasingly involved. Thus, the two sides of local climate action planning are set up to operate in somewhat separate bureaucratic environments with very different goals. Merging or coordinating these two sets of strategies becomes dependent on local political leadership or city managers, for whom neither issue may be a high priority.

In a planning environment informed by integrative thinking, there are plenty of areas in which cobenefits apply and the two strategies are mutually advantageous. For instance, urban forestry and green infrastructure programs can serve to mitigate urban heat island impacts, thus also mitigating the impacts of urban heat wave emergencies. Urban forestry in riparian buffers can also anchor soil and reduce urban stormwater runoff, reducing some flooding impacts. Both also have beneficial effects with regard to greenhouse gas emissions. Likewise, permeable paving reduces urban heat island impacts compared to asphalt pavements in typical parking lots while also improving the absorptive capacity of the ground and improving stormwater management. There are numerous other examples of such mutually beneficial strategies for urban development, but these suffice to make the point that it is far better to think about climate mitigation and adaptation as a package deal than to isolate one from the other in siloed plans, and certainly better than to debate which strategy is more important.

As we will see, however, experience remains a good teacher. Communities directly affected by recent weather-related major disasters have been prone to make significant policy changes.

The Rise of Resilience

I have personally been prone to say that nothing is less green, less sustainable, and more wasteful than the level of destruction witnessed along the Gulf Coast after Hurricane Katrina. At a New Partners for Smart Growth conference in Albuquerque, New Mexico, in 2009, I used my closing slide in a presentation to raise the question, "If it ain't safe, how smart is it?" I do believe that smart growth activists have been catching on to the significance of hazard mitigation as a factor in smart growth, though I have more doubts about the willingness of New Urbanism to accept this emphasis. APA's report, *Hazard Mitigation: Integrating Best Practices into Planning* (Schwab 2010), included a Safe Growth Audit. The audit was also described in an article in *Zoning Practice* (Godschalk 2009).

In recent years, the concept of resilience has gained increasing impetus as a public policy imperative as the nation has faced growing losses from major natural disasters like Hurricanes Katrina and Sandy, and the Midwest floods of 2008. The discussion of resilience began somewhat earlier, appearing in planning literature by the late 1990s, but events and the growing evidence of human-induced climate change have propelled the concept to the center of the discussion of disaster response and recovery.

The National Academies of Science (2012) has defined resilience as "the ability to prepare and plan for, absorb, recover from, and more successfully adapt to adverse events." The applicability of this concept to planning for hazard mitigation or disaster recovery seems clear. What is perhaps less clear is how this concept relates to the concept of sustainability, generally defined since the report of the United Nations' Bruntland Commission as meeting "the needs of the present without compromising the ability of future generations to meet their own needs" (World Commission on Environment and Development 1987). That definition stresses a long-term view of economic development that is different from, but not necessarily in conflict with, the more immediate concern with the ability of a community to withstand and even prosper from adverse events. It is important that planners and others involved in both types of climate action planning, and in hazards planning more generally, understand how these two concepts relate to each other. For that reason, APA offered, in its newest report on postdisaster recovery, the following explanation.

Simply put, it is suggested here that planners and their communities should strive for *both* resilience and sustainability. They should also understand that, in truly thinking about the welfare of the proverbial seventh generation whose abilities to meet their own needs should not be impaired, they should go about the business of developing the culture of sustainability that contains within it the seeds of a culture of preparedness to ensure that potential current disasters do not foreclose those future opportunities. Infrastructure built to withstand foreseeable disasters within the current generation or even the next will consume fewer resources, and borrow less environmental capacity from future generations, by not needing to be rebuilt (or at least by being rebuilt less extensively) every time a disaster occurs. Moreover, long-term sustainability that builds serious resilience in the face of increasing natural hazard threats as a result of climate change may actually increase sustainability for future generations. Civic and institutional capacity to understand and confront these issues may perhaps be the most significant heritage we can bequeath to those who come after us (Schwab 2014).

Moving toward Integrated Hazards Planning

Given the difference in impetus between climate mitigation, which involves strategies for reductions of greenhouse gas emissions, and planning for climate adaptation, which involves hazard mitigation planning with an eye to the ways in which a more volatile climate may exacerbate existing weather-related hazards, it is not surprising that these two strands of activity have taken somewhat different paths. Moreover, it is also clear that planning related to climate adaptation has been a somewhat later development in the planning field than planning to reduce greenhouse gases. The latter is considerably easier to quantify and identify, and the strategies, while politically challenging, are in many cases relatively straightforward: hybrid buses and cars, more efficient appliances, increased reliance on transit, and so on. Quantifying and isolating the specific impacts on a local jurisdiction of global climate change, in contrast, has been fraught with uncertainty, typified by predictions within fairly wide bands of impacts—for example, varying amounts of predicted sea-level rise or increased propensity for suffering long-term drought. Teasing out how much of a current hazardous event is a result of climate change versus a product of random variation in weather patterns can be exceedingly difficult. As a result, making the public case for adaptation is often less than a rewarding endeavor.

Nonetheless, there are trends within the practice of hazard mitigation planning that hold promise. The most promising trend may be the idea that hazard mitigation should be more thoroughly integrated into the overall community planning process. APA described how this should include a range of planning activities, beginning with community visioning and goal setting, then incorporating hazards elements or considerations into not only local comprehensive plans, but area and functional plans as well, and then carried through with implementation tools such as subdivision and zoning ordinances, capital improvement plans, and development activities such as site plan review (Schwab 2010). FEMA, which funded that project, has followed up with a series of guidance manuals at both the national and regional levels that have continued to stress the need for such integration (e.g., see FEMA 2013; FEMA Region X 2013). The subject has also gained attention in various academic and practical settings.*

It is important to lay out the statutory framework that pervades this subject at both the federal and state levels because that framework introduces both policy incentives to achieve such integration and some practical problems for effective implementation. The linchpin of this framework is the federal Disaster Mitigation Act of 2000 (DMA), which established the principle that FEMA would no longer provide grants for hazard mitigation projects to states and communities that had not prepared hazard mitigation plans, gained federal approval for such plans as compliant with the act and FEMA regulations, and formally adopted the plan. At the local level, that meant adoption by the governing body, usually a city council or county board. In recent years, FEMA has estimated that more than 20,000 jurisdictions, which can include special districts and state universities, have adopted local hazard mitigation plans (LHMPs). The number of actual plans may be considerably lower, however, because jurisdictions can participate in and adopt multijurisdictional plans. In many states, these are coordinated by counties or regional entities such as councils of governments or regional planning agencies. The crucial question in such cases is the linkage between the plan and local implementation through ordinances and policies.

The impetus for the APA study and many similar critiques of the process was that many LHMPs became the projects of local emergency management agencies in stand-alone plans that had little bearing on local development policies expressed in comprehensive plans. To the extent that one plan failed to inform the other, glaring weaknesses in the use of land-use policy to address hazard mitigation persisted, with development continuing to use problematic sites with potential for flooding, landslides, or coastal erosion, to name a few possibilities. As a result, the right questions were not being asked as part of the local development approval process. In some cases, particularly in more rural or politically conservative states, zoning and building codes might be entirely lacking, thus begging the question of how future development could be made less hazardous. Certainly, this situation prevailed in much of Mississippi and rural Louisiana at the time of Hurricanes Katrina and Rita, and in Texas when Ike struck the Galveston area in 2008.†

Thus, the idea has grown that while the LHMP is a useful document, it would be far more useful if the information and recommendations it contains were incorporated into a hazards element in the local comprehensive (or master or general) plan, especially if that element were cross-referenced with specific opportunities for hazard mitigation identified in other commonly used elements such as land use, transportation, housing, and economic development. Unlike LHMPs, comprehensive plans typically under state law have some level of legal standing as municipal policy documents, with many states requiring some degree of consistency between zoning and development decisions and the adopted comprehensive plan. Therefore, this integration goes much farther than the LHMP alone to ensure timely and effective implementation of the policies it embraces. Put simply, things might actually get done.

Curiously, in November 2008, the voters of the city of New Orleans adopted a charter amendment that required zoning and development decisions to be consistent with a master plan the city had not yet completed. (It adopted the new plan two years later.) The new master plan, unlike those that had preceded Hurricane Katrina, contained significant details about hazard mitigation in future development. Catastrophe provided a vital lesson in the value of such integrated planning.

---

* The theme of integrated hazard mitigation planning, for instance, is increasingly common in presentations at local, state, and national planning and hazard-related conferences. I have presented this topic several times myself, sometimes in concert with others, either federal officials or consultants.

† Zoning and building codes were essentially nonexistent in the Bolivar Peninsula in rural Galveston County, which was badly devastated in Hurricane Ike, losing about three-fourths of its housing stock.

Even more impressively, Iowa in May 2010 adopted the Iowa Smart Planning Act, which did not mandate planning but did contain 10 smart growth principles, basically mirroring those long in circulation in the smart growth movement nationally, but also recommending 13 specific elements for inclusion in local comprehensive plans. This was the first time Iowa had specifically defined what a local comprehensive plan should contain. One of those items was a natural hazards element, which the new law said should be prepared with reference to the local hazard mitigation plan prepared for FEMA. For the first time, state law enshrined the principle of tying the LHMP to the comprehensive plan. This legislation was prepared and advanced by the Rebuild Iowa Office which was established after the 2008 floods that engulfed cities like Cedar Rapids. The law was the product of collaboration by the state with the US Environmental Protection Agency (EPA) and FEMA under a memorandum of understanding between FEMA and EPA's Office of Smart Growth (now the Office of Sustainable Communities). The law has proven to be a valuable incentive for convincing cities all across the state to include a natural hazards element in their local plans.

The Iowa law also highlights an issue that has existed in some states since the adoption of DMA. Even before that federal law, about 10 states had various requirements for including some type of hazard-related element in local comprehensive plans, almost all in states that mandated local comprehensive plans. There are some noteworthy patterns in the nature of these state mandates. California, Arizona, and Nevada all have safety elements; the California requirement has shifted from an original seismic emphasis to a nearly all-hazards focus as wildfires and floods have become part of the expected analysis. California has also made climate change a mandatory consideration in required general plans. Florida and North Carolina have had a historic emphasis on coastal issues, while Oregon's Goal 7 is a rather comprehensive inventory of potential hazards. In other words, states requiring treatment of hazards in local comprehensive plans logically do so based on the state's historic hazards profile.

The relationship of the required safety element to the LHMP was a subject of some concern to the California Governor's Office of Emergency Services (OES) in the years immediately after Hurricane Katrina. The agency engaged a team at the California Polytechnic State University (Cal Poly) to examine practices in this regard while also conducting a content analysis of LHMPs in more than 500 California jurisdictions. A survey of local planners made clear that simply having a required safety element did not guarantee cross-pollination with LHMPs or their preparers because roughly one-third of those surveyed were unable to say what relationship existed between the two plans. Fortunately, a substantial minority either indicated that the two plans were identical (3%), suggesting that the same work was done only once, that the LHMP was adopted as part of the safety element (12%), or at least that the safety element referenced the LHMP (23%), meaning that the information was shared and reused. Nonetheless, a disturbing number of jurisdictions still seemed to prepare entirely unrelated plans that both dealt with hazard mitigation in one way or another (Boswell et al. 2008). If California, generally deemed as being in the forefront on such issues, faced such challenges, one can only imagine what problems exist in more permissive states with regard to integration. OES has subsequently undertaken to provide incentives to spur higher levels of integration within the state, presumably with improved results.

The arena of plan integration that is far less well established, and also lacks the statutory basis of hazard mitigation, is that of planning for postdisaster recovery. To date, only one state has ever enacted a mandate for local governments to engage in pre-event planning for recovery after disasters. Florida did so in 1993 for coastal counties and their municipalities but subsequently failed to develop guidance that would make clear to them precisely what was expected in what the statute labeled Post-Disaster Redevelopment Plans (PDRPs). About 15 years later, however, in part spurred by the experience of four hurricanes criss-crossing the state within 6 weeks in the fall of 2004 (1 year before Katrina), the Florida Department of Community Affairs (DCA), with support from FEMA and the National Oceanic and Atmospheric Administration (NOAA), supported development of prototype plans in five counties and one municipality (Schwab 2009b). Based on the experiences in those communities, DCA subsequently issued guidance on the preparation of PDRPs (Florida DCA/Florida Division of Emergency Management 2010). By 2011, however, a more conservative administration under Gov. Rick Scott de-mandated the PDRPs, which nonetheless are encouraged as a good idea. DCA was also converted into what is now the Florida Department of Economic Opportunity, with a reorientation of its earlier mission to oversee growth management on a statewide basis. Notably, however, major jurisdictions such as Hillsborough County, which contains the Tampa Bay area, have stood by their adopted PDRPs. Even more importantly, Hillsborough County (2010) used its PDRP to elaborate a policy establishing what it calls Priority Redevelopment Areas, which are essentially less hazard-prone parts of the county to which redevelopment would be steered after a disaster, away from more vulnerable disaster-stricken areas. This remains a courageous public policy move that is as yet untested in battle. The county acknowledges that, through sheer dumb luck, it has escaped impact from a major hurricane for more than half a century. It also states plainly in its plan that such impact is simply a matter of time.

The existence of such efforts has had a noticeable impact, in a way. The National Disaster Recovery Framework (NDRF) (FEMA 2011) and associated FEMA publications certainly encourage communities to give serious consideration to predisaster planning for recovery. The NDRF, however, is purely an administrative framework and lacks the force of law of DMA. Outside Florida, then, in the absence of federal or state requirements or monetary incentives, local adoption of such plans or their integration into a larger planning framework is almost entirely the result of local initiative and the decisions of local officials. One benefit of such initiatives, however, is that they usually indicate significant buy-in by local citizens and stakeholders and a higher level of commitment than plans that result from mandates. At the same time, they remain rare.

Climate Adaptation in the Mitigation Plan

Climate adaptation, as noted earlier, involves adopting and implementing hazard mitigation measures with potentially adverse impacts from climate change in mind. Those measures can be undertaken either before a disaster as part of what some would call peacetime or blue skies planning, or as part of a recovery plan following a disaster. In the United States, the latter option provides access to certain pots of federal money, such as the Hazard Mitigation Grant Program, which are available only as part of a presidential disaster declaration. It is important either way to keep in mind that the aftermath of one disaster is the prelude to the next because any natural event that has occurred once in a particular region can obviously occur again. The troubling wrinkle with climate change is its disruptive attributes. Historic projections of the frequency and intensity of weather-related disasters is no longer adequate because the changing climate warps the probabilities in the direction of more frequent extreme events. The issue for planners is how to take the information that is available in this regard and incorporate it into effective planning and policy making to protect the community. How does a community anticipate a future that begins to diverge from historic trends? How do planners translate what often remains rather imprecise technical data into defensible prescriptions for how and where people are allowed to build or where the community should locate critical facilities?

Boswell et al. (2012, 154) note two key differences in climate adaptation strategies from those addressed in natural hazard mitigation. One is that climate adaptation must address ecosystem changes that are not normally part of hazard mitigation, which tends to focus instead on the intermittent events that are typically defined as hazards. The other is the inability to use history alone as a predictive factor in future occurrences of episodic events because of their changing frequency and severity. Adaptive strategies become an opportunity to enhance community resilience.

A few examples may suffice to highlight the challenges in this regard. We can anticipate, for instance, that most residential development can be expected to last at least 50 years, given that many of our cities have substantial amounts of housing stock that it in fact is much older, sometimes more than 100 years old. Within the ranges of projected sea-level rise along any particular stretch of coast, how restrictive should a community be in establishing setbacks? It is easy enough to argue with the North Carolina legislature's inane decision to limit the state's Department of Natural Resources to using historical data rather than projected sea-level rise in establishing such setbacks in state coastal regulations, but exactly what is the best projection to use? Likewise, if we can expect higher temperatures to lead to a greater propensity for wildfires, how does that affect building code and zoning regulations for the wildland-urban interface? In a chapter in the earlier edition of this book, I advocated building a greater margin of safety into our standards on several such issues, but the time may have arrived for being much more precise in how we integrate climate data into local hazard mitigation planning (Schwab 2009a).

On the national scale, these measurements and projections will likely reach higher levels of resolution over time, but how do the resources of federal agencies and research institutions reach local and regional planners in a form that they can effectively utilize in crafting climate-conscious mitigation policy? The problem of making the needed data readily available and comprehensible to these planners is not small. Outside the largest jurisdictions, which tend to have greater resources to invest in such planning, many communities are already straining to absorb such data and incorporate it effectively. Scientific agencies like NOAA and the US Geological Survey should have a role in providing such technical assistance and tools, and so should state agencies and institutions of higher education. For example, the State of California has taken significant steps including an interactive website using state-generated projections to help local governments develop their strategies (Boswell et al. 2012). It should also be noted that Florida produced an addendum to postdisaster redevelopment planning guidance dealing specifically with climate adaptation (Florida DEO/DEM 2013). Unfortunately, most states have not yet developed such capabilities, and several have state legislatures or administrations that refuse to concede that a problem even exists. In those states, local governments pursuing such strategies must rely on federal, institutional, and academic resources.

There may also be some role for regional planning agencies. Although they do not have the scientific resources of some state and federal agencies, they can play a role in translating for local governments within their domain the means of planning appropriately for climate adaptation. For instance, the Chicago Metropolitan Agency for Planning, seeking to carve out a meaningful role in assisting local government, produced a climate adaptation guidebook (CMAP 2013). It has subsequently also been working to assist with resilience planning as a result of the National Disaster Resilience Competition announced by the US Department of Housing and Urban Development in September 2014.*

NOAA has undertaken some significant steps in that direction with efforts like Digital Coast, whose website† offers a range of tools, resources, and data to support coastal planning, some of which, like the Sea Level Rise Viewer, are clearly climate-related. This NOAA effort began in 2009, however, an indication of just how recently such resources have even started to become available. One notable feature is that the program includes a national partnership with eight (originally five) national organizations that serve, in part, as a conduit to the user communities for such services.‡ One interesting pilot case involves cooperation between FEMA and NOAA in supporting an effort in Monterey County, California, to incorporate the use of Digital Coast data in developing the county's local hazard mitigation plan, with the help of AECOM, a consulting firm heavily involved in mitigation planning.

One way or another, there are ways to bridge the gap between end users of such information for local and regional planning and the producers of such data, although it may take time. The need to do so is certainly clear.

---

* The author has met with CMAP personnel in informal consultations concerning this endeavor.

† http://coast.noaa.gov/digitalcoast/.

‡ For example, see the APA web pages detailing its connection to the partnership (https://www.planning.org/research/digitalcoast/project/). These include an extensive needs assessment survey of APA members regarding their own needs for data and tools to facilitate coastal planning.

Regional Approaches

Climate change, like most natural hazards, clearly has impacts that transcend municipal or even county boundaries. Thus, it makes sense for counties and municipalities to cooperate on regional solutions wherever possible, and a few regions within the United States have taken noteworthy steps in that direction, even as individual local governments take those actions that depend on municipal authorities, such as using zoning and subdivision controls to steer development out of harm's way.

One of the most noteworthy regional efforts toward climate adaptation has taken place in one of the most deeply impacted areas—southeast Florida. Including Miami, the Everglades, and the Florida Keys, this region is ground zero in many respects with regard to the perils of long-term sea-level rise and the potential increased severity of coastal storms. Florida experiences more hurricanes than any other state in the nation because of its high exposure on a peninsula fronting both the Atlantic Ocean and the Gulf of Mexico. Floridians are almost instinctively aware that the next hurricane is not a question of if, but when. Yet there are billions of dollars of real estate and millions of human lives at stake with each event.

Those risks had everything to do with motivating the political leadership of four Florida counties (Monroe, Dade, Broward, and Palm Beach) to come together in creating the Southeast Florida Regional Climate Compact.* Since its inception, the regional compact has produced a number of documents addressing regional problems including a regional climate action plan (Southeast Florida Regional Climate Change Compact Counties 2012). It serves as an excellent example of a consolidated climate action plan, addressing both climate mitigation and adaptation at a regional level and demonstrating a remarkable level of intergovernmental cooperation. The plan contains recommendations with regard to infrastructure, water supply, transportation, and natural systems, but also with regard to risk reduction and emergency management. On the latter point, it addresses the need to incorporate climate change and sea-level rise into local hazard mitigation plans. It has been hailed as one of the most progressive efforts of its type in the nation. Excerpts from the Compact's website describing their activities are presented in the box "The Southeast Florida Regional Climate Change Compact."

Local Approaches

A couple of major examples of the integration of climate adaptation strategies into local comprehensive and hazard mitigation planning should suffice to illustrate that this concept is gaining substantial momentum, at least in communities that have serious reason for concern.

New York City was already engaged in such planning even before Hurricane Sandy through PlaNYC and other initiatives. Sandy and the recovery planning initiatives that followed, coupled with the need to comply with changes in federal flood insurance standards, certainly cemented that commitment but did not produce it (Schwab 2013). In the year after Sandy, as the city struggled to rebuild, its planning department completed two major studies examining the challenges of making its waterfront districts more resilient while retaining the urban aesthetics of densely built neighborhoods and commercial districts.

One study emphasized the need for diverse solutions, including both structural protection for those areas with major sunk investment in infrastructure, such as transportation facilities, and green infrastructure, such as restoration of tidal wetlands, for those areas where such approaches were feasible. New York City, in short, faced the hard fact that no one solution can work everywhere in the most densely populated urban region in the nation, and that flexibility will be the key to success (New York City Department of City Planning 2013a).

The other study concerned itself with maintaining attractive, livable, walkable neighborhoods in areas near the city's more than 500 miles of waterfronts. It provides planners and builders with

* See http://www.southeastfloridaclimatecompact.org/.

THE SOUTHEAST FLORIDA REGIONAL CLIMATE CHANGE COMPACT

The Southeast Florida Regional Climate Change Compact was executed by Broward, Miami-Dade, Monroe, and Palm Beach Counties in January 2010 to coordinate mitigation and adaptation activities across county lines. The Compact represents a new form of regional climate governance designed to allow local governments to set the agenda for adaptation while providing an efficient means for state and federal agencies to engage with technical assistance and support.

At the conclusion of the first Southeast Florida Regional Climate Leadership Summit on October 23, 2009, four County representatives—Broward County Commissioner Kristin Jacobs, Palm Beach County Commissioner Shelley Vana, Miami-Dade County Commissioner Katy Sorenson, and Monroe County Mayor George Neugent—signed the Southeast Florida Regional Climate Change Compact committing to take it back to their County Commissions for review and adoption.

The Compact calls for the Counties to work cooperatively to

- Develop annual Legislative Programs and jointly advocate for state and federal policies and funding
- Dedicate staff time and resources to create a Southeast Florida Regional Climate Action Plan to include mitigation and adaptation strategies
- Meet annually in Regional Climate Summits to mark progress and identify emerging issues

Since 2009, the four Compact Counties have accomplished much together and have also worked with a growing number of federal, state, regional, municipal, nonprofit, academic, and private sector partners. Widely recognized as one of the nation's leading examples of regional-scale climate action, the Compact continues to serve as the mechanism to collaborate for climate adaptation.

The Compact members and participants look forward to another productive year of regional collaboration and will continue to seek opportunities to bring additional funding and resources to the region for climate change mitigation and adaptation activities. Activities for the coming year include

- Implementation of the RCAP with the Yale Cultural Cognition Project, support from the Institute of Sustainable Communities and funding from the Kresge Foundation
- Engagement of the region's 109 municipal partners through the Mayors' Climate Action Pledge
- Advocacy for the 2014 State and Federal Energy and Climate Legislative Program
- Supporting the Adaptation Action Area pilot project in partnership with Fort Lauderdale, Broward County, South Florida Regional Planning Commission, and Florida Department of Economic Opportunity

Each of the Compact partners is moving forward to implement greenhouse gas mitigation and climate adaptation strategies. Working with municipal partners, key strategies are advancing to create a more economically competitive and climate-resilient community.

Source: The Southeast Florida Regional Climate Change Compact website. Accessed February 2, 2015. http://www.southeastfloridaclimatecompact.org/who-we-are/ and https://southeastfloridaclimatecompact.files.wordpress.com/2014/05/compact-1-page-flyer-ia-final-sa.pdf.

design alternatives for protecting the active street life of dense urban neighborhoods while also protecting homes and businesses from flooding, accounting for the possibilities of increased coastal storm activity and common urban flooding threats (New York City Department of City Planning 2013b). In both cases, the city was hopeful that its efforts might also provide a pathway to new solutions for smaller cities with similar problems but fewer resources for innovative planning (Schwab 2013).

The City of Baltimore, on the other hand, has used the 5-year update of its all-hazards mitigation plan for FEMA as the entry point for launching its Disaster Preparedness and Planning Project (DP3), whose web page states up front that, "In an effort to plan for existing hazards and prepare for predicted hazards due to climate change, the Office of Sustainability will develop and implement an integrated project to provide the City with a comprehensive system for addressing existing and future impacts."* The effort culminated in a new plan released in late 2013 that addresses impacts to natural systems, infrastructure, buildings, and public services from a variety of climate change-induced hazards (City of Baltimore 2013).

Conclusion

Although young and evolving, the system and methodology for integrating climate adaptation into both local hazard mitigation plans, under the federal Disaster Mitigation Act, and local comprehensive plans through hazard-related and other elements, is clearly taking hold in those jurisdictions with the largest stake in surviving the impacts that may have some connection to global climate change. With or without federal or state legislative encouragement, they are tapping a variety of scientific, governmental, and institutional resources to experiment and innovate with meaningful approaches to this challenge amid a number of uncertainties. The odds are that as the snowball of local government with such planning accumulates, more communities facing up to the demands of protecting their citizens and businesses from disruption by climate-induced disasters will avail themselves of the lessons of others, sustaining the momentum that already exists. Indeed, in due course, they may have little choice.

CLIMATE ADAPTATION FOR EMERGENCY MANAGERS

Eric Holdeman
Center for Regional Disaster Resilience

Playing Catch-Up

While the rest of the world has been thinking about the warming climate and planet for some time, here in the United States we are just now starting to play catch-up on what we need to be doing to deal with projected issues that will impact emergency managers everywhere. In reality we have spotted the world a 10-year head start in thinking about and reacting to climate change issues. As late as 2008, the National Weather Service (NWS) was not allowed to officially or unofficially write about or discus climate change and its potential impacts. In an interview in January 2009, the Federal Emergency Management Agency (FEMA) headquarters was asked what they had done or were doing about climate change and climate adaptation. At that point, they had done nothing and the topic was not even on the radar screen at their national headquarters.

Since that time and with the onset of the new President Obama Administration, it became politically acceptable for federal officials to write about, discuss, and present publically on the topic of climate change. The United States military have described climate change as one of the greatest destabilizing forces in the world and a significant threat to our national security. The President's

* http://www.baltimoresustainability.org/disaster-preparedness-and-planning-project#sthash.HMC5mcC8.dpuf.

2014 US Climate Action Report highlighted the existing impact of severe weather and attributed these impacts to climate change here in the United States. This impact is being measured in terms of the size of disasters and their scope, severity, and financial costs to respond and recover from them.

Hazards Increasing

The year 2014 was the warmest year on record for modern record keeping. These warming global temperatures are projected to bring humanity a whole host of natural disasters and related impacts to our critical infrastructures. There isn't one geographical area of the nation that will be spared some impacts from climate change. Coastal areas will see and are beginning to see significant impacts from rising seas. While the projections vary considerably, the waters are already inching upward. Even conservative estimates for future ocean levels will cause significant issues around the world. There are coastal areas that are already having high-tide flood events on a regular basis. Extreme tides called king tides accent today what the impact of permanently higher sea levels will be. While "normal" weather conditions might still make things bearable for now, combining sea rise with other weather phenomenon like hurricanes is going to make those types of events much more damaging. Take Super Storm Sandy for instance; it was a flood event and not a wind event. The storm was huge and it came ashore at high tide, making the storm surge much more destructive. Tied to the increasing risk for physical damages to our coastal cities is the fact that population densities continue to increase in our coastal areas. All coastal communities are expected to have significant population growth in the coming decades. This added population growth will amplify the impacts of storm destruction on critical infrastructures that have already exceeded their design capacities and their expected life cycles.

Moisture, Too Much or Too Little

Severe weather is predicted to become more frequent and destructive. Warmer air holds more moisture, which means record-breaking rainfalls and snowstorms. In the past few years we have seen huge rainfall events across the nation. Jurisdictions like the State of Minnesota and the Province of Alberta and the cities of Seattle, Colorado Springs, and Phoenix are communities that have had deluges of rain that generated significant flooding events. Is it possible that Tornado Alley will become more like a tornado boulevard with wider swaths of the nation being impacted? In 2014, Washington State experienced a rare tornado.

While there are areas of the nation that will have record amounts of rain and snow, other sections of the country will be living with continued drought conditions. Snowpack has been the expected standard for many parts in the western states that are supporting growing populations, hydroelectric projects, and irrigation for our nation's food basket. California's current water emergency can be directly credited to a lack of snowpack.

Terminology Defined

It would be good to define some terminology. In some political circles and geographical areas the terms "climate change" and "global warming" are not received well. Both of these terms are used somewhat interchangeably by many people. There are segments of our society who believe that climate change proponents are using voodoo science and computer models to contort what is happening in nature. One way to get around getting sucked up into the debate about the causes of climate change is to talk about the warming climate using the term "climate variability." Many who oppose the notion of the changing climate being attributable to human causes such as carbon emissions do accept the notion that our world has gone through various periods of warming and cooling such as the ice age. The term "climate variability" speaks to this issue. No matter what the cause, we need to deal with the consequences of a warming climate. Even today there are segments of the population who believe that the earth is really cooling and this emphasis on global warming and humans' role in it is something that has been concocted in order to exert more government control over personal liberties.

In climate change terms the word "mitigation" is about controlling carbon emissions. Cap and trade is one of the tools some countries are using to reduce carbon emissions. If you are involved in a process of taking steps to deal with the changing climate, the term used is "climate adaptation." In emergency management terms, "climate adaptation" may look more like our typical disaster mitigation, but sticking with "climate adaptation" will help you stay out of hot water. Another warning is required. For purists who are seeking to reverse global warming, the thought of focusing on climate adaptation is like surrendering to the inevitable. They fear that a move to climate adaptation will negate all the efforts at climate mitigation by reducing carbon emissions.

The Politics of Climate Change

Climate change can have a polarizing effect on people and organizations. Much of the angst that emanates from climate change comes when the causes of the warming climate are attributed to human-caused carbon emissions. People quickly separate into different camps. The energy industry is reluctant to take the full blame for a changing climate, but in fact, fossil fuels are critical to our functioning economy. Much of our nation's electricity comes from coal-fueled power plants and our system of roads and rails that move goods and people are dependent on carbon-based fuels.

Some businesses fear the cost of increased regulation and higher taxes to pay for implementing more expensive alternative fuels or the systems to further scrub pollutants from the air. Elected officials have their different constituencies that expect them to represent their points of view on the issue of climate change. Those opposed to taking any action to reduce carbon emissions want to delay any implementation of carbon reduction strategies until there is conclusive and irrefutable proof that there is a direct correlation between the warming climate and carbon emissions coming from direct human interaction with the planet.

This minefield of opinions is one that many emergency managers are afraid to step into. One strategy to consider is to concentrate on the aspect of climate change that has a direct impact on emergency manager functions and responsibilities, this being climate adaptation, and failing that, consequence management of events as they occur. The one saving grace is that many people are in agreement that the weather has changed and we are seeing more impacts from severe weather. This generally accepted view of a changing climate can be leveraged to allow you to focus on the areas of climate change that fall within the responsibility of emergency managers.

The Emergency Manager's Role

In the last 25 years, additional hazards have been added to the inventory of things that emergency managers need to be prepared to deal with. Before the turn of the century, terrorism in all its forms, including weapons of mass destruction, were an afterthought if considered at all. Solar storms and what they can do to our modern electronic society were not a topic at conferences and seminars. The likelihood of an asteroid strike on earth, which while we acknowledged as possible, was dismissed as one of those things you could not plan for. All of the above are now part of the lexicon of emergency managers and dealing with climate change is going to top that list.

Climate change is not likely to bring us a totally new category of disaster. Instead, it will be the frequency, severity, and wider impacts of natural disasters. Climate change may exacerbate some disasters so that they could morph and become compounding events, one building on the other.

Rain and snow. Warmer air holds more moisture, and when released in the form of rain and snow events will become more dramatic and sudden. Already we have seen rainfall totals from some thunderstorms that have surpassed the amount of rain that might fall during a hurricane or tropical storm event. Flash flooding can be expected in places that have never experienced that type of flooding before. Additionally, continued urbanization replaces soils that absorb rainfall with impermeable surfaces. Even areas without streams and rivers may experience surface-water flooding due to drainage systems being undersized for the levels of rainfall that might be experienced in the future.

Drought. Interestingly, while some areas of the nation may experience record amounts of precipitation, there will be others where long-term and debilitating drought becomes the norm. Arid parts of the West and Southwest have seen huge population growth in the last 40 years. The water supply to support this population is already being strained. Water conservation measures are becoming more rigorous in some areas. A significant debate will emerge as to the best uses for water. Western agriculture depends on man-made irrigation systems. Declining rainfall and snowpack can significantly decrease the amount of water available for this irrigation for use by municipalities to support their general population and commercial industry. Water flow to support endangered fish populations will have to be balanced against the other needs of a region.

While dam construction has significantly decreased in recent decades, there may be a resurgence as people realize that additional water storage capacity is needed for flood events that can be episodic and also be used as a reservoir to support water and hydro power needs in a region. These dams will need emergency plans to support their operation that will need to be integrated with other area disaster plans.

Wildland fires. The federal budget for fighting wildland fires has ballooned in recent years. Drought has fueled this growth in the size and costs of fires. The State of California now has a year-long fire season that becomes much more active months before what used to be their peak season. Contributing to the natural fire risks is the destruction of forests by insects that were never endemic to the region. An example of this is the pine beetle that has ravaged huge swaths of forestland in the Pacific Northwest. The pine beetle was normally contained because of severe winters that would kill off large portions of the population feasting on trees. Because of milder winters the beetle has thrived and greatly expanded its reach in pine forests extending beyond the United States into Canada.

Consequence Management

The traditional role of emergency management in most people's minds remains consequence management, which is the disaster response function of coordinating agency responses to disasters as they are detected and unfold. This role will continue in the future. The change that can be expected is that the frequency and severity of disasters will increase. The increasing impacts of natural hazards will focus more attention on the discipline of emergency management and its leaders. Federal, state, and local emergency managers will be in the spotlight more often and perhaps in circumstances where the resources of a community are totally overwhelmed and the systems that are in place are no match for the size and scale of the event and the damages incurred.

Warning

Part of this response role will include warning people and organizations of impending events. The greatest partnership in this respect is with the National Weather Service from whom we get the majority of our weather warnings. However, warnings require more than hazard detection. Systems for distributing the warning must be put into place and tested on a regular basis to ensure they are functional and that a warning can be issued in a timely manner to populations who are threatened. Today the majority of those warning systems come from the twentieth century. Newer technologies are available that use modern telecommunications systems like smartphones and tablets to provide a warning and also instructions that can be used by people to protect themselves. The last challenge in the warning process is to get people to react to the warning message and immediately follow the instructions being provided. There are plenty of sociological studies which show that people do not immediately act but check with others as to what they are doing. This process is called milling and we need to find ways to shorten that process. One way to do that is with utilizing social media before an event to help build trust with the general public so that your warnings have more credibility and the reaction time by the public is shortened.

Adaptation

Emergency managers will need to be proponents for finding ways to adapt to increased natural hazard risks. The process for doing this will require determining what assets to defend, protect, or abandon. Every community will have to do its own homework as to the value of specific infrastructure or properties. In some cases the decision to protect certain pieces of real estate will be a no-brainer. For large urban areas like New York City, the primary actions will be to defend the critical infrastructure because moving it or abandoning it to the elements is not an option. The particular challenge in New York will be deciding to what degree different aspects of the infrastructure will be defended. For instance, following Hurricane Sandy the decision has already been made to seek ways to seal off the New York City subway systems from future flooding. The debate is already in full swing about the need to build a seawall to keep waters out of the urban area. How high should that wall be and most importantly, who will pay for that level of protection? Should the project move forward there are sure to be compromises because of the costs involved and uncertainty about when the next large event will occur.

On the opposite end of the scale will be the challenge of determining which properties don't have the same high value and for which the cost-benefit analysis doesn't add up to protecting properties with limited community-wide benefit. Unfortunately for some, this equates to individual homeowners who find themselves facing repeated losses, typically due to flooding, and running out of personal resources, and on top of that facing the governments who may or may not be able to support them. Repetitive losses have yet to be meaningfully addressed via our disaster recovery mechanisms.

Emergency managers will become even more dependent on science and technology to provide the information for making adaptation choices before events and also in the aftermath as we pick up the pieces during the recovery phase. The timing for making most choices on climate adaptation will only come after a large-scale disaster. It is then, and usually only then, that there will be the political willpower to make hard choices. You have the best opportunity to make advances in disaster resilience in the immediate aftermath of a disaster. As the months and years go by the motivation and energy to make significant changes will fade rapidly. Emergency managers will need to work with their community partners in advance of a large disaster to envision how the community might be changed for the better by protecting some portions and abandoning others. Abandoning means using techniques, such as creating green space, when and where it is not fiscally possible to defend every inch of land.

Besides the science community, a broader constellation of partners will need to be engaged. Large-scale disasters are regional in nature. They cross jurisdictional borders and responding and recovering from disasters will require partnerships between governments and private and nonprofit sectors. These partnerships must be nurtured and renewed over time as people rotate in and out of positions of influence. It is a continuing challenge, but much time must be invested in gaining and maintaining a wide swath of personal relationships that cut across the physical, economic, and social spectrums.

Moving Forward

Black swan events are those disasters that seem rare, but in hindsight you can find that evidence existed to know that these were possible. It is just that the majority of people function in their daily lives unaware of the dangers that surround them. In part this comes from our perception of risk that is based little on statistics and more on our personal feelings and our collective ability to measure time only in human lifespan terms. While it may appear that no one is listening, the emergency manager must continue to provide information on hazards and risks to their individual communities. Building coalitions of people and organizations with mutual interests in climate adaptation remains one of your best tools for success.

You can expect that senior political leadership will want public assurances that we are ready for any disaster that might impact your community. Given the potential for truly catastrophic disasters occurring in coastal areas from hurricanes, it is unlikely that any government has the resources to adequately respond in a timely manner to individual citizen needs. Therefore, if you live in a high-risk area, rather than promising what you are unable to deliver, it is better to accentuate what preparations are being made by government, but then also to place some of that preparedness burden on the individual. We all want someone to take the responsibility for disaster preparedness from us. Do not let others off the hook for what they need to be doing for their own personal and organizational welfare.

Emergency managers have gotten used to the federal government providing guidance, policies, and funding for a whole host of issues. In the case of climate change and the need for climate adaptation you can expect that it will be states and cities that will lead the way, as they already have been doing on these topics. The politics of national government and Washington, DC, will provide significant obstacles to any forward movement on anything other than conceptual strategies. This is a case for doing it yourself and not waiting for guidance and especially funding.

Expect the Unexpected

Insurance companies have always used past losses to measure and estimate what their future losses might be. They used these figures to set their customers' premiums as they seek to have the necessary resources to meet future loss estimates. Today they acknowledge that the past is not necessarily a good guide to what losses may be in the future. They are struggling to have a better system to prognosticate about the future.

The above is also true for emergency managers. We are seeing extremes across the board. Seven feet of snow in Buffalo from one storm; 10 inches of rain in 1 hour in the South without being part of a tropical storm. We are dealing with what will be a new normal where we don't know what normal might be in the future. The bottom line: Expect the unexpected and be prepared for all eventualities. Doing so will reduce your chance of getting caught flatfooted in responding to a disaster. Remember that today's disasters may be tomorrow's catastrophes.

REFERENCES

Boswell, M.R., A.L. Greve, and T.L. Seale. 2012. *Local Climate Action Planning*. Washington, DC: Island Press.

Boswell, M.R., K.C. Topping, W. Siembieda et al. 2008. *Local Hazard Mitigation Planning in California: A Report on the Implementation of LHMPs under DMA*. Sacramento, CA: California Governor's Office of Emergency Services.

Chicago Metropolitan Agency for Planning (CMAP). 2013. Climate Adaptation Guidebook for Municipalities in the Chicago Region, June. Available at http://www.cmap.illinois.gov/climate-adaptation.

City of Baltimore. 2013. Disaster Planning and Preparedness Project. Available at https://www.dropbox.com/sh/c32rmuy0ov8xtq7/NDibFOfTbD.

Federal Emergency Management Agency (FEMA). 2011. *National Disaster Recovery Framework: Strengthening Disaster Recovery for the Nation*. Washington, DC: FEMA, September.

______. 2013. Integrating Hazard Mitigation into Local Planning: Case Studies and Tools for Community Officials. Available at https://www.fema.gov/media-library/assets/documents/31372?id=7130.

FEMA Region X. 2013. *Integrating the Local Natural Hazard Mitigation Plan into a Community's Comprehensive Plan*. Seattle, WA: FEMA Region X. Available at http://www.fema.gov/media-library-data/1388432170894-6f744a8afa8929171dc62d96da067b9a/FEMA-X-IntegratingLocalMitigation.pdf.

Florida Department of Community Affairs (DCA)/Florida Department of Emergency Management. 2010. *Post-Disaster Redevelopment Planning: A Guide for Florida Communities*. Tallahassee, FL: Florida DCA. Available at http://www.floridadisaster.org/Recovery/documents/Post%20Disaster%20Redevelopment%20Planning%20Guidebook%20Lo.pdf.

Florida Department of Economic Opportunity (DEO)/Florida Division of Emergency Management (DEM). 2013. Post-Disaster Redevelopment Planning: Addressing Adaptation during Long-Term Recovery. Available at http://www.floridajobs.org/fdcp/dcp/PDRP/Files/PDRPSeaLeveRiseAddendum.pdf.

Godschalk, D.R. 2009. Safe Growth Audits. *Zoning Practice*, October.

Hillsborough County, Florida. 2010. *Hillsborough County Countywide Post-Disaster Redevelopment Plan.* Tampa, FL: Hillsborough County.

National Academies of Science. 2012. *Disaster Resilience: A National Imperative.* Washington, DC: National Academies Press.

New York City Department of City Planning. 2013a. Coastal Climate Resilience: Urban Waterfront Adaptive Strategies. Available at http://stormsmart.org/uploads/qs/nyc-urban-waterfront-adaptive-strategies.pdf.

_____. 2013b. Designing for Flood Risk. Available at http://www.nyc.gov/html/dcp/html/sustainable_communities/sustain_com6.shtml.

Schwab, J. 2009a. The Role of Planning in Reducing Impacts of Global Warming. In Bullock, Jane A., George D. Haddow, and Kim S. Haddow (eds.), *Global Warming, Natural Hazards, and Emergency Management.* Boca Raton, FL: CRC Press.

_____. 2009b. Winds of Change. *Planning*, October 24–29.

Schwab, J. (ed.). 2010. *Hazard Mitigation: Integrating Best Practices into Planning.* Planning Advisory Service Report No. 560. Chicago: American Planning Association.

Schwab, J. 2013. High and Dry on the Waterfront. *Zoning Practice*, November. Issue 11. Available at https://www.massport.com/media/266323/2013-November_Zoning-Practice.pdf.

_____. 2014. *Planning for Post-Disaster Recovery: Next Generation.* Planning Advisory Service Report No. 576. Chicago: American Planning Association.

Southeast Florida Regional Climate Change Compact Counties. 2012. A Region Responds to a Changing Climate, October. Available at http://www.southeastfloridaclimatecompact.org//wp-content/uploads/2014/09/regional-climate-action-plan-final-ada-compliant.pdf.

World Commission on Environment and Development (1987). *Our Common Future.* Oxford: Oxford University Press.

4 The Role of Local, State, and Federal Governments in Climate Change Adaptation

INTRODUCTION

This chapter includes three essays focused on government programs and actions in the fields of climate change adaptation and disaster risk reduction.

The first essay examines ongoing federal, state, and local government programs designed to address climate change and provides background information concerning what government agencies across the country are doing relative to climate change adaptation.

The second essay examines federal strategies for risk management that provide a potential Roadmap for how communities can address the challenges of climate change and design adaptation programs and actions that will reduce the impacts of climate change.

The third essay examines the results of a "research project [that] was jointly undertaken between The Kresge Foundation and graduate students at the University of Michigan to: (1) assess the current landscape of climate-adaptation resources and services available to support local and regional adaptation; (2) assess how well these resources align with the needs of local practitioners; and (3) identify strategic opportunities for moving the local adaptation field forward. This piece profiles the results from the first phase of this analysis, highlighting the types of adaptation-related resources and services provided by 85 primarily nonprofit and boundary organizations that support local or regional climate adaptation efforts."

RECENT ACTIONS THAT FEDERAL, STATE, AND LOCAL GOVERNMENTS ARE ENGAGED IN RELATIVE TO THE ISSUES ARISING FROM THE IMPACTS OF CLIMATE CHANGE

Jane A. Bullock
Bullock & Haddow LLC

INTRODUCTION

The intent of this essay is to highlight the recent actions that federal, state, and local governments are engaged in relative to the issues arising from the impacts of climate change.

There is a wide diversity in acknowledgment of the issues of climate change and significant differences in how agencies and communities at all levels of government have approached this problem. The increase is in acknowledgment of the problems brought on by climate change and the commitment to action to address these problems has not necessarily been driven by scientific evidence but by the more practical experiences on natural disasters and politics.

The Obama Administration came into office committed to the fact that climate change was real, and that we were already experiencing its impacts, nationally and internationally, and the United States needed to exercise leadership in this area especially with the developing nations of the world such as India and China. Domestically, the United States faced a series of unprecedented natural disasters, ranging from Hurricane Katrina to increasingly severe wildfires and floods and then Superstorm Sandy that ravaged the East Coast but especially the vulnerable coastal communities in New Jersey and New York, resulting in billions of dollars in damages. In the meantime, sea-level rise, urban heat, and other factors impacted communities across the nation.

Congress, on the other hand, was still divided on the validity of the science of climate change. A report prepared at the request of Congress by the Congressional Research Service (CRS) in September 2013 tried to succinctly describe the current state of scientific evidence to support the concerns over the potential impacts of climate change.

CLIMATE CHANGE SCIENCE: KEY POINTS—REPORT SUMMARY

Though climate change science often is portrayed as controversial, broad scientific agreement exists on many points:

- The Earth's climate is warming and changing.
- Human-related emissions of greenhouse gases (GHG) and other pollutants have contributed to warming observed since the 1970s and, if continued, would tend to drive further warming, sea level rise, and other climate shifts.
- Volcanoes, the Earth's relationship to the Sun, solar cycles, and land cover change may be more influential on climate shifts than rising GHG concentrations on other time and geographic scales. Human-induced changes are super-imposed on and interact with natural climate variability.
- The largest uncertainties in climate projections surround feedbacks in the Earth system that augment or dampen the initial influence, or affect the pattern of changes. Feedback mechanisms are apparent in clouds, vegetation, oceans, and potential emissions from soils.
- There is a wide range of projections of future, human-induced climate change, all pointing toward warming and associated sea level rise, with wider uncertainties regarding the nature of precipitation, storms, and other important aspects of climate.
- Human societies and ecosystems are sensitive to climate. Some past climate changes benefited civilizations; others contributed to the demise of some societies. Small future changes may bring benefits for some and adverse effects to others. Large climate changes would be increasingly adverse for a widening scope of populations and ecosystems.

As is common and constructive in science, scientists debate finer points. For example, a large majority but not all scientists find compelling evidence that rising GHG have contributed the most influence on global warming since the 1970s, with solar radiation a smaller influence on that time scale. Most climate modelers project important impacts of unabated GHG emissions, with low likelihoods of catastrophic impacts over this century. Human influences on climate change would continue for centuries after atmospheric concentrations of GHG are stabilized, as the accumulated gases continue to exert effects and as the Earth's systems seek to equilibrate.

The US government and others have invested billions of dollars in research to improve understanding of the Earth's climate system, resulting in major improvements in understanding while major uncertainties remain. However, it is fundamental to the scientific method that science does not provide absolute proofs; all scientific theories are to some degree provisional and may be rejected or modified based on new evidence. Private and public decisions to act or not to act, to reduce the human contribution to climate change or to prepare for future changes, will be made in the context of accumulating evidence (or lack of evidence), accumulating GHG concentrations, ongoing debate about risks, and other considerations (e.g., economics and distributional effects).

Source: Leggett, Jane A. 2013. *Climate Change Science: Key Points*. Congressional Research Service, September 10, 2013. https://www.hsdl.org/?view&did=745266.

Excerpts from the report summary are presented in the box "Climate Change Science: Key Points."

Even prior to 2008 and the Obama Administration, many federal agencies and many communities were acknowledging the impacts of climate change on their programs and jurisdictions. Although not alone, the Environmental Protection Agency (EPA) was considered the lead federal agency because of its regulatory powers in this field, and at the state level California was the leader in climate change legislation.

But the Obama Administration made climate change a subtle but key component of its policy agenda although other critical issues such as health care and the unstable international environment almost always overwhelmed it. At the same time, state and local governments were addressing their climate change issues somewhat independently of the federal government but always looking to the federal government for leadership, and more importantly, funding.

Federal Government Efforts in Climate Change

The federal government has supported research on global climate change starting in the 1970s. This early funding concentrated on the science of climate change with funding being provided primarily to universities, national laboratories such as Oak Ridge, and some nongovernmental organizations. This funding also supported research and development (R&D) on the risks from future climate change. Other federal agencies, especially the EPA, used funding to produce research and analyses that supported regulatory actions taken to reduce the impacts of climate change.

As more potential and real impacts from climate change were being realized, funding began to expand into the areas of adaptation and eventually mitigation. This expansion of approach is documented in the Climate Action Plan announced by the Obama Administration in 2013.

The plan noted that "climate change is no longer a distant threat—we are already feeling its impacts across the country and the world. Last year was the warmest year ever in the contiguous United States and about one-third of all Americans experienced 10 days or more of 100-degree heat. The 12 hottest years on record have all come in the last 15 years. Asthma rates have doubled in the past 30 years and our children will suffer more asthma attacks as air pollution gets worse. And increasing floods, heat waves, and droughts have put farmers out of business, which is already raising food prices dramatically. These changes come with far-reaching consequences and real economic costs. Last year alone, there were 11 different weather and climate disaster events with estimated losses exceeding $1 billion each across the United States. Taken together, these 11 events resulted in over $110 billion in estimated damages, which would make it the second-costliest year on record" (Executive Office of the President 2013).

"In short, America stands at a critical juncture. Today, President Obama is putting forward a broad-based plan to cut the carbon pollution that causes climate change and affects public health. Cutting carbon pollution will help spark business innovation to modernize our power plants, resulting in cleaner forms of American-made energy that will create good jobs and cut our dependence on foreign oil. Combined with the Administration's other actions to increase the efficiency of our cars and household appliances, the president's plan will reduce the amount of energy consumed by American families, cutting down on their gas and utility bills" (Executive Office of the President 2013).

In this plan, three major components highlight the expanding emphasis on leadership, information sharing, building capacity beyond the federal government, and new technologies to mitigate the risks. As stated in the plan, the three components included (1) Cuts Carbon Pollution in America, (2) Prepares the United States for the Impacts of Climate Change, and (3) Lead International Efforts to Address Global Climate Change. The box "Components of the President's Climate Action Plan" presents descriptions of the types of actions called for in the Climate Action Plan in each of these three components.

COMPONENTS OF THE PRESIDENT'S CLIMATE ACTION PLAN

1. *Cuts Carbon Pollution in America*

   For example, the plan:

   - Directs the EPA to work closely with states, industry, and other stakeholders to establish carbon pollution standards for both new and existing power plants
   - Makes up to $8 billion in loan guarantee authority available for a wide array of advanced fossil energy and efficiency projects to support investments in innovative technologies
   - Directs Department of Interior (DOI) to permit enough renewables project—like wind and solar—on public lands by 2020 to power more than 6 million homes, designates the first-ever hydropower project for priority permitting, and sets a new goal to install 100 megawatts of renewables on federally assisted housing by 2020, while maintaining the commitment to deploy renewables on military installations;
   - Expands the President's Better Building Challenge, focusing on helping commercial, industrial, and multifamily buildings cut waste and become at least 20 percent more energy-efficient by 2020
   - Sets a goal to reduce carbon pollution by at least 3 billion metric tons cumulatively by 2030—more than half of the annual carbon pollution from the US energy sector—through efficiency standards set over the course of the Administration for appliances and federal buildings
   - Commits to partnering with industry and stakeholders to develop fuel economy standards for heavy-duty vehicles to save families money at the pump and further reduce reliance on foreign oil and fuel consumption post-2018
   - Leverages new opportunities to reduce pollution of highly potent GHGs known as hydro fluorocarbons, directs agencies to develop a comprehensive methane strategy, and commits to protect our forests and critical landscapes

2. *Prepares the United States for the impacts of climate change*

   Building on progress over the last four years, the plan:

   - Directs agencies to support local climate-resilient investment by removing barriers or counterproductive policies and modernizing programs, and establishes a short-term task force of state, local, and tribal officials to advise on key actions the federal government can take to help strengthen communities on the ground
   - Pilots innovative strategies in the Hurricane Sandy-affected region to strengthen communities against future extreme weather and other climate impacts, and building on a new, consistent flood-risk reduction standard established for the Sandy-affected region, agencies will update flood-risk reduction standards for all federally funded projects
   - Launches an effort to create sustainable and resilient hospitals in the face of climate change through a public-private partnership with the health care industry
   - Maintains agricultural productivity by delivering tailored, science-based knowledge to farmers, ranchers, and landowners, and helps communities prepare for drought and wildfire by launching a National Drought Resilience Partnership and by expanding and prioritizing forest- and rangeland-restoration efforts to make areas less vulnerable to catastrophic fire
   - Provides climate preparedness tools and information needed by state, local, and private-sector leaders through a centralized toolkit and a new Climate Data Initiative

3. *Lead International Efforts to Address Global Climate Change*

The plan:

- Commits to expand major new and existing international initiatives, including bilateral initiatives with China, India, and other major emitting countries
- Leads global sector public financing toward cleaner energy by calling for the end of US government support for public financing of new coal-fired powers plants overseas, except for the most efficient coal technology available in the world's poorest countries, or facilities deploying carbon capture and sequestration technologies
- Strengthens global resilience to climate change by expanding government and local community planning and response capacities

Source: Executive Office of the President. 2013. *The President's Climate Action Plan.* June 25, 2013. http://www.whitehouse.gov/sites/default/files/image/president27sclimateactionplan.pdf.

A report prepared for Congress on the funding of federal climate change activities noted that "Congressional debate on climate change funding takes place in the broader context of stark choices among competing fiscal demands. Direct federal funding to address global climate change totaled approximately $77 billion from FY2008 through FY2013. The large majority—more than 75%—has funded technology development and deployment, primarily through the Department of Energy (DOE)" (Leggett et al. 2013).

There were several other interesting facts in this report. The president requested $11.6 billion for federal climate change activities in FY2014. Despite making capacity building and adaptation a priority in The President's Climate Action Plan, only 1 percent of this funding is for adapting to climate change. The remainder of the requested funds provides 23 percent would be for science, 68 percent for energy technology development and deployment, and 8 percent for international assistance (Leggett et al. 2013) (see Figure 4.1).

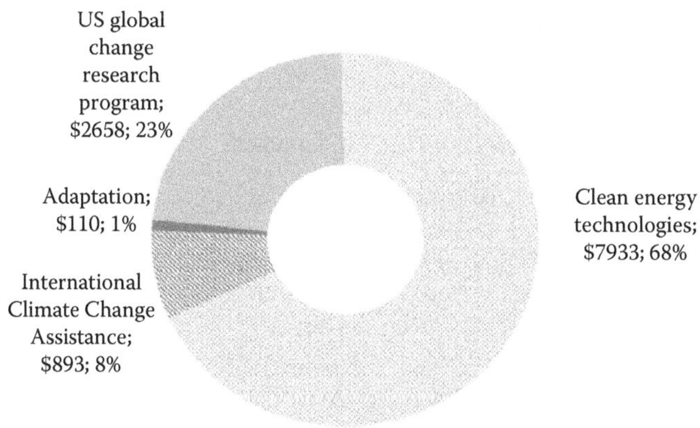

FIGURE 4.1 The president's request for climate change-related budget authority for FY2014. (From Leggett, Jane A., Richard K. Lattanzio and Emily Bruner. 2013. *Federal Climate Change Funding from FY2008 to FY2014.* Congressional Research Service. September 13, 2013. http://fas.org/sgp/crs/misc/R43227.pdf.)

Table 4.1 summarizes Federal Funding for Climate Change initiatives from FY2001–FY2013.

On November 1, 2013, the administration took another action to advance climate change adaptation. President Obama established by Executive Order the Task Force on Climate Preparedness and Resilience to advise the administration on how the federal government can respond to the needs of communities nationwide that are dealing with the impacts of climate change. The task force members "include state, local and tribal leaders from across the country who will use their first-hand experiences in building climate preparedness and resilience in their communities to inform their recommendations to the Administration" (The White House 2013a).

"The President signed an Executive Order that directs Federal agencies to take a series of steps to make it easier for American communities to strengthen their resilience to extreme weather and prepare for other impacts of climate change" (The White House 2013b).

President Obama has said that there is a moral obligation to our children and future generations to leave them a more protected and safer world. That is why in June 2013, the President announced a Climate Action Plan to cut carbon pollution, prepare communities for the impacts of climate change, and lead international efforts to address this global challenge. "The Climate Action Plan recognizes that even as we act to curb the carbon pollution that is driving climate change, we must also improve our ability to prepare for the climate impacts we are already seeing across the country. Across America, states, cities, and communities are taking steps to protect themselves from extreme weather and other climate impacts by updating building codes, adjusting the way they manage natural resources, investing in more resilient infrastructure, and planning for rapid recovery from damages that nonetheless occur" (The White House 2013a).

The plan defines the critical role the federal government has in supporting community-based preparedness and resilience efforts, making investments that promote preparedness, protecting critical infrastructure and public resources, and supporting the science and research needed to prepare for climate impacts.

State, Local, and Tribal Leaders Task Force on Climate Preparedness and Resilience

State, local, and tribal leaders across the country are already contending with more frequent or severe heat waves, droughts, wildfires, storms, floods, and other impacts of climate change. The task force will provide recommendations to the president on removing barriers to resilient investments, modernizing federal grant and loan programs to better support local efforts, and developing the information and tools they need to prepare. Task force members comprise governors, mayors, county officials, and tribal leaders representing a diverse range of communities. This project will be discussed later in the chapter.

Current Federal Agencies Actions on Climate Change

On Friday, October 31, 2014, the administration released a series of documents that outlined the vulnerabilities of the agencies to the impacts of climate change and the plans they have for addressing these vulnerabilities. Some examples of how the major agencies view their vulnerability include:

Department of Agriculture (USDA): Cited "the potential for a 100% increase in the number of acres burned annually by 2050" (Mooney 2014). The budget for fire suppression activities has increased from 13 percent in 1991 to 40 percent in 2014.

Department of Health and Human Services (HHS): "Outlines risks including more frequent or worse extreme heat events which is the leading weather related cause of deaths in the US" (Mooney 2014). Other things noted were changes in vector borne diseases such as West Nile and more asthma and allergic reactions in people.

TABLE 4.1
Federal Climate Change Programs: FY2001–FY2013 Enacted, and the President's FY2014 Request (in millions of nominal dollars, except where noted)

	Budget Authority (Fiscal Years, Enacted Except FY2013 and FY2014)														
Major Climate Change Program Areas	**2001**	**2002**	**2003**	**2004**	**2005**	**2006**	**2007**	**2008**	**ARRA**	**2009**	**2010**	**2011**	**2012**	**2013 Current[a]**	**2014 Request**
US Global Change Research Program (including the former Climate Change Science Program)	$1,728	$1,667	$2,078	$1,996	$1,864	$1,691	$1,825	$1,864	$641	$2,023	$2,195	$2,448	$2,506	$2,463	$2,658
Clean Energy Technologies (formerly Climate Change Technology Program)	$1,675	$1,637	$2,533	$2,870	$2,808	$2,789	$3,485	$6,394	$25,499	$5,216	$5,504	$5,646	$6,121	$5,783	$7,933
International Climate Change Assistance	$218	$224	$270	$252	$234	$249	$188	$202	$0	$323	$939	$819	$958	$797	$893
Adaptation	$0	$0	$0	$0	$0	$0	$0	$0	$0	$0	$69	$35a	$88	$90	$110
All Areas	$3,621	$3,528	$4,881	$5,118	$4,906	$4,729	$5,498	$8,460	$26,140	$7,562	$8,707	$8,948	$9,673	$9,133	$11,594
All Areas in Constant 2012 Dollars	$4,650	$4,471	$6,053	$6,189	$5,745	$5,361	$6,076	$9,027	$27,739	$8,025	$9,112	$9,166	$9,673	$8,955	$11,112

(Continued)

TABLE 4.1 (CONTINUED)
Federal Climate Change Programs: FY2001–FY2013 Enacted, and the President's FY2014 Request (in millions of nominal dollars, except where noted)

	Estimated Revenue Loss Effects of Energy Tax Provisions														
	2001	2002	2003	2004	2005	2006	2007	2008	ARRA	2009	2010	2011	2012	2013 Est.	2014 Est.
Energy Tax Provisions That May Reduce Greenhouse Gases	$0	$0	$580	$500	$369	$ 1,160	$1,520	$1,520	$0	$ 1,440	$4,140	N.A.	$10,132	$13,079	$9,839
Tax Expenditures in Constant 2012 Dollars	$0	$0	$719	$605	$432	$1,315	$1,680	$1,622	$0	$1,528	$4,332	$0	$10,132	$12,824	$9,430
Total Budgetary Impact[b]	$3,621	$3,522	$5,461	$5,618	$5,275	$5,889	$7,018	$9,980	$26,140	$9,002	$12,277	N.A.	$19,805	$22,212	$21,433
Total Budgetary Impact in Constant 2007 Dollars	$4,650	$4,471	$6,773	$6,793	$6,178	$6,676	$7,756	$10,648	$27,739	$9,553	$13,444	$9,166	$19,805	$21,779	$20,542

Source: Prepared by CRS based on data reported by OMB, including in Federal Climate Change Expenditures Report to Congress FY2014, August 2013; Federal Climate Change Expenditures Report to Congress, FY2008, May 2007, Table 8, p. 27; Federal Climate Change Expenditures Report to Congress, August 2003, Table 1; and Federal Climate Change Expenditures Report to Congress, July 2002, Table 1.

Notes: OMB adjusted the science and technology amounts for FY2003, FY2004, and FY2005 to reflect more recent accounting within these program areas. Remaining inconsistencies are likely across years due to changes in the scopes of what is considered for "climate change" and methods of allocating certain costs to programs. To convert from nominal to 2012 dollars, CRS used OMB, The Budget for Fiscal Year 2014, Historical Tables, Table 10.1-Gross Domestic Product and Deflators Used in the Historical Tables: 1940-2018. CRS used deflators for non-defense expenditures and adjusted index so FY2012=1.0. The total impact to the federal budget in each year as reported by OMB may differ somewhat from the sum of budget authority for each program area and the estimates of tax expenditures, due to exclusion by OMB of dollar amounts for certain activities to avoid double-counting.

[a] FY2013 Current Budget Authority, according to OMB, as of June 21, 2013. It reflects funds available calculated as the amount appropriated (FY2013 Enacted Budget Authority, data not shown), minus reductions triggered by the Budget Control Act of 2011 (P.L. 112-25) sequestration order issued March 1, 2013.

[b] Because the estimate for the adaptation efforts in this year is incomplete, this figure is likely an underestimate of the actual amount.

National Aeronautic and Space Administration (NASA): "Warns that climate change could actually interfere with its ability to get into space" (Mooney 2014). Also at risk are many of the NASA facilities located in vulnerable areas.

Department of Defense (DoD): Secretary Hagel described climate change as a "threat multiplier because it worsens many of the challenges that the military faces, including terrorism and disease" (Mooney 2014).

Department of the Interior (DOI): The DOI and especially the National Park Service (NPS) have already realized impacts in some of their facilities and in responding to the increasing wildfire threat (Mooney 2014).

Adaptation Programs at Federal Agencies

The Interagency Climate Change Adaptation Task Force chaired by the EPA recommended efforts to strengthen federal strategies and programs to prepare for and adapt to the effects of climate change at the national level. The task force acknowledged that many federal agencies already manage programs to support and inform adaptation at the regional, state, and local level. Examples of these actions are described in the box "Examples of Adaptation Programs at Federal Agencies" and also see Figure 4.2. FEMA's Climate Change Adaptation Policy Statement from 2011 is presented in Figure 4.3.

Federal Hazard Mitigation Programs

In the past there has been some controversy over the use of the word "adaptation" as it relates to learning to live with climate change, and the word "mitigation," which has been used by the risk and emergency management communities to define actions that can be taken to reduce the risk. This semantic argument has finally been resolved and now both concepts have been adopted in the efforts to limit the impacts of climate change. It is clear that the impacts of climate change, from sea-level rise, extreme weather events, and so on, have had a dramatic impact on the frequency and severity of natural disasters. In this context, it is important to understand the federal programs that support activities at the national, state, and local and private sector efforts to mitigate the impacts of natural disasters.

One of these programs, the NFIP administered by DHS/FEMA, continues to be controversial. Many opponents of this program believe that the availability of subsidized insurance for flood events has contributed to people moving into high-risk areas along the coasts and in high-risk riverine areas. FEMA is undertaking a review of the NFIP as part of its adaptation efforts.

On January 30, 2015, the White House released "a new flood risk management standard for new and rebuilt federally funded structures in and around floodplains. This new resilience standard will help ensure taxpayer dollars are well spent on infrastructure that can better withstand the impacts of flooding" (Podesta and Fugate 2015). The *White House Blog* entry noted the cost of rebuilding parts of New York and New Jersey impacted by Hurricane Sandy in 2011 and stated, "That's why when we invested billions to help communities rebuild from Sandy, we also committed to "build back better"—to rebuild infrastructure to a higher standard so it can withstand the increased risks posed by sea level rise and other climate impacts" (Podesta and Fugate 2015). A copy of the Fact Sheet released by the White House describing this new flood standard is presented in the box "Fact Sheet: Taking Action to Protect Communities and Reduce the Cost of Future Flood Disasters."

The following essay in this chapter, "Government Strategies for Natural Hazards Management and Mitigation," describes current and past federal hazard mitigation programs and the strategies and tactics employed by these programs to reduce the impacts of future disasters at the community level.

EXAMPLES OF ADAPTATION PROGRAMS AT FEDERAL AGENCIES

US EPA

- On October 31, 2014, the White House released the final versions of Climate Change Adaptation Plans produced by federal departments and agencies, including the EPA. EPA's final Climate Change Adaptation Plan (PDF, 64 pp, 1.7 MB) and the 17 Climate Change Adaptation Implementation Plans produced by the program and regional offices describe how the agency will integrate considerations of climate change into its programs, policies, rules, and operations to ensure they are effective even as the climate changes.
- On June 30, 2014, the EPA issued a new policy statement on climate change adaptation (PDF, 3 pp, 4.81 MB). This statement updates the initial policy statement (PDF, 3 pp, 595 KB) issued in June of 2011.
- Climate Ready Estuaries is a partnership between EPA and the National Estuary Program to assess climate change vulnerabilities in coastal areas, develop and implement adaptation strategies, engage and educate stakeholders, and share the lessons learned with other coastal managers.
- Climate Ready Water Utilities provides practical tools and resources for drinking water, wastewater, and stormwater utilities to better understand and adapt to climate change impacts.
- EPA's National Water Program Climate Change Strategy provides an overview of the likely effects of climate change on water resources and the nation's clean water and safe drinking water programs. It describes the goals and strategic actions being implemented by the National Water Program to adapt to those changes. Topics addressed include water infrastructure, water quality, watersheds and wetlands, coastal and ocean waters, and working with tribes.
- The EPA's Smart Growth Website under the interagency Partnership for Sustainable Communities offers a virtual map of projects around the country. View the map of Smart Growth projects to see what is happening near your community.
- The Office of Research and Development (ORD's) Global Change Impacts & Adaptation, as part of the ORD Global Change Research Program, assesses the potential vulnerability to climate change (and other global change stressors such as land-use change) of EPA's air, water, ecosystem, and human health protection efforts at the federal, regional, state, municipal, and tribal levels, as well as adaptation options to build resilience in the face of these vulnerabilities. The program carries out interdisciplinary syntheses across newly emerging scientific findings to identify potential impacts and characterize and communicate the uncertainty in the science to provide support for decision-makers and managers.
- The EPA's Tribal Science Council provides science information and resources gathered through tribal traditional knowledge, analytical science, and integrated approaches to science by tribes, the EPA, and other federal agencies and tribal organizations.
- The US Global Change Research Program (USGCRP) coordinates and integrates federal research on changes in the global environment and their implications for society. Thirteen departments and agencies, including the EPA, participate in the USGCRP. The USGCRP is currently developing an updated National Climate Assessment.

US Agency for International Development (USAID)

- USAID acknowledges the inextricable link between climate change and development (PDF). It has made adaptation one of its three pillars of climate change funding.

US Department of Agriculture (USDA)

- The US Forest Service (USFS) hosts a broad range of information and tools for land managers, including interactive climate maps, a climate change response framework for forest management, an adaptation guidebook, climate data, and examples of existing adaptation plans.

US DoD

- In the 2010 Quadrennial Defense Review, DoD establishes that climate change will affect DoD in two broad ways. First, climate change will shape the operating environment, roles, and missions that DoD undertakes. Additionally, DoD will need to adjust to the impacts of climate change on its facilities and military capabilities.
- In 2009, the Chief of Naval Operations formed Task Force Climate Change (PDF) to
  - Address implications of climate change for national security and naval operations
  - Answer the question "when" in terms of Navy decisions regarding climate change
  - Ensure the Navy is ready and capable to meet all mission requirements in the twenty-first century
- The US Army Corps of Engineers has updated guidance on how the agency's projects, systems and programs can respond to future changes in sea level.

US DOE

- DOE is developing and implementing strategies to mitigate climate change through changes in the department's technical programs, policies, and initiatives. While the majority of the DOE work focuses on reducing greenhouse gas emissions, reducing energy use can also have cobenefits as a climate adaptation strategy.

HHS

- The Administration for Children & Families contributed to the development of tribal adaptation plans.
- The Centers for Disease Control and Prevention (CDC) established the Climate and Health Program to lead efforts to prevent and adapt to the anticipated health impacts of climate change.
- CDC has launched the Climate-Ready State & Cities Initiative in eight states and two cities. Together, CDC and local health departments are developing ways to anticipate and respond to health effects by applying climate science and preparing adaptation strategies for the public health sector.
- The National Institute of Environment and Health Services (NIEHS) is researching the human health impacts of climate change, raising awareness, creating partnerships, and providing critical information to NIEHS stakeholders and the public.

US Department of Homeland Security (DHS)

- The Federal Emergency Management Agency (FEMA) is undertaking a critical review of the National Flood Insurance Program (NFIP)

US Department of Housing and Urban Development (HUD)

- HUD is supporting adaptation planning in communities (PDF)

DOI

- DOI manages US public lands, water, wildlife, and energy resources. In 2009 the Interior Secretary issued an order to develop a climate change response strategy. These strategy solutions include wetlands restoration, reforestation initiatives, and restoration of native species, improved water supply forecasting, and other efforts.
- The US Fish and Wildlife Service (USFWS) is integrating climate change into management of fish, wildlife, plants, and their habitats. The department published "Rising to the Urgent Challenge: Strategic Plan for Responding to Accelerating Climate Change" (PDF), which establishes a basic framework for the service and the broader conservation community that is based on adaptation, mitigation, and engagement.
- DOI manages 22 Landscape Climate Cooperatives—public-private partnerships seeking to create a holistic and collaborative adaptation approach grounded in science.
- The US Geological Survey (USGS) National Climate Change and Wildlife Science Center (NCCWSC) serves the scientific needs of managers of fish, wildlife, habitats, and ecosystems as they plan for climate change. NCCWSC manages the DOI's eight Climate Science Centers, which provide a full range of support for natural and cultural resources.
- The USGS Climate and Land Use Change Research and Development program contributes to understanding both the causes and consequences of climate change. Research efforts help identify potential vulnerability and understand the resilience of the earth's system to such changes.
- The US National Parks Service released a Climate Change Response Strategy (PDF) that integrates science, adaptation, mitigation, and communication to develop an ambitious coordinated strategy to respond to the impacts of rapid climate change on the national parks.

NASA

- NASA provides global information on changes in sea-level rise, temperature of the oceans, sea ice, and land ice. This helps provide other agencies with valuable information about global conditions.

US Department of Commerce

- The National Oceanic and Atmospheric Administration (NOAA) has numerous efforts underway related to coastal, marine fishery, water resource, and ecosystem adaptation. Learn more here.
- NOAA hosts climate.gov, a consolidated climate services website.
- NOAA is also conducting several studies of regional climates called Regional Integrated Sciences and Assessments.

US Department of Transportation (DOT)

- DOT has developed a Transportation and Climate Change Clearinghouse with information about the effects of transportation on climate change as well as the impacts of climate change on transportation. The "Adaptation Planning" section provides resources about approaches to planning for potential impacts caused by climate change.

Source: EPA.gov. 2015. "Federal and EPA Adaptation Programs." Accessed on January 4, 2015. http://www.epa.gov/climatechange/impacts-adaptation/fed-programs.html.

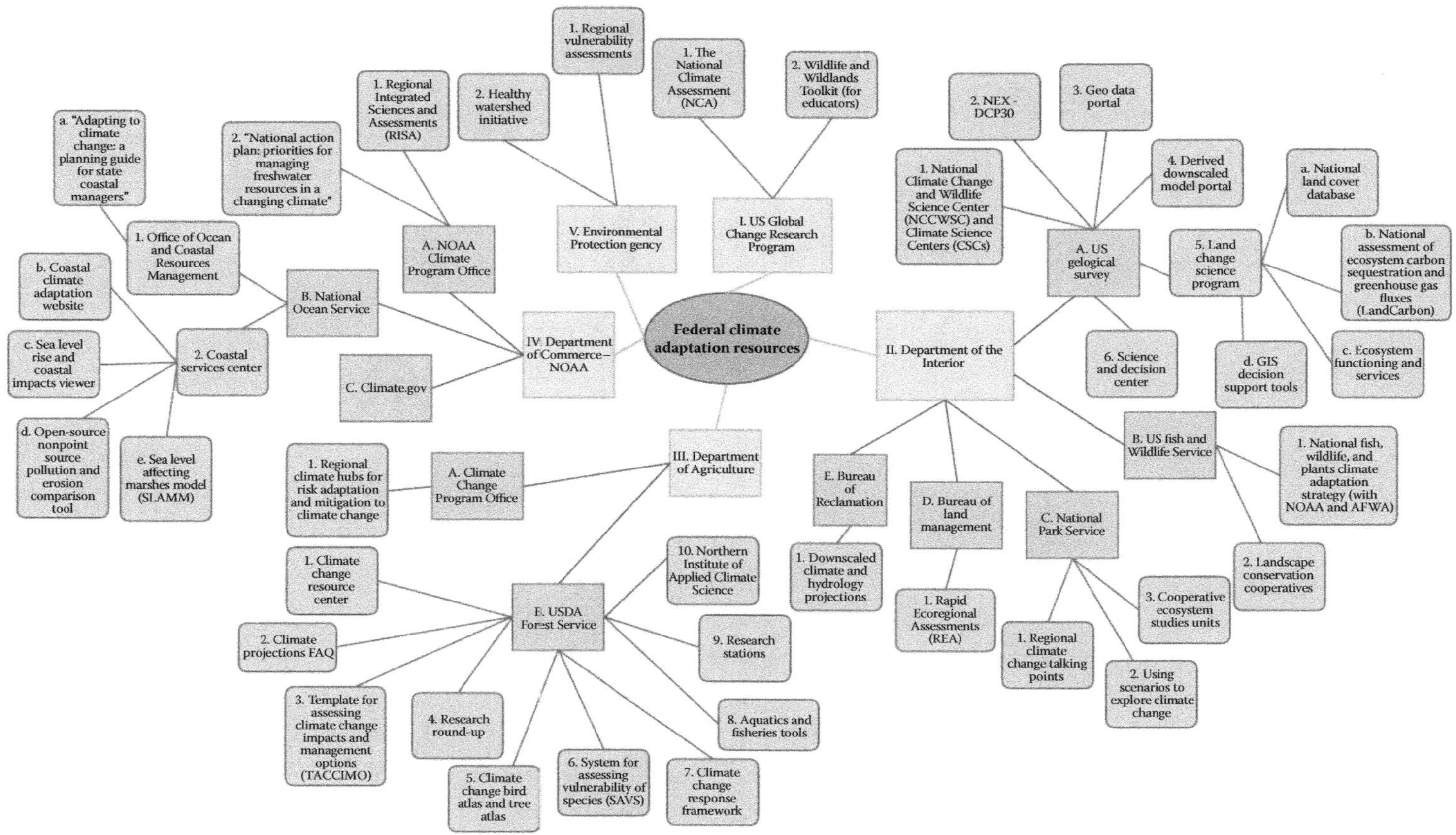

FIGURE 4.2 Federal climate adaptation programs. (From Resource Guide to Federal Climate Adaptation Programs for State Fish & Wildlife Agencies accessed at: http://www.floods.org/ace-files/documentlibrary/committees/NFB/Resource_Guide_to_Federal_Climate_Adaptation_Programs.pdf.)

FEMA ADMINISTRATOR POLICY

2011-OPPA-01

FEMA CLIMATE CHANGE ADAPTATION POLICY STATEMENT

I. **Purpose**

The purpose of this policy statement is to establish an Agency-wide directive to integrate climate change adaptation planning and actions into Agency programs, policies, and operations.

II. **Scope**

This directive applies to all Agency activities and is intended to guide FEMA personnel responsible for the oversight and implementation of organizational plans, policies, and procedures.

III. **Background**

While the scope, severity, and pace of future climate change impacts are difficult to predict, it is clear that potential changes could affect our Agency's ability to fulfill its mission. The challenges posed by climate change, such as more intense storms, frequent heavy precipitation, heat waves, drought, extreme flooding, and higher sea levels could significantly alter the types and magnitudes of hazards faced by communities and the emergency management professionals serving them. Some specific areas where climate change could influence our capabilities and the need for our services are:

- *Impacts on mitigation, preparedness, response, and recovery operations:* as coastal regions become increasingly populated and developed, more frequent or severe storms may increase the requirements for emergency services and response and recovery capacity.
- *Resiliency of critical infrastructure and various emergency assets:* continuity of operations, delivery of services, and emergency response efforts may be challenged and made increasingly complex by damages or disruptions to the interconnected energy and infrastructure networks.
- *Climate change could trigger indirect impacts that increase mission risks:* intensifying droughts, heat waves, and periods of heavy precipitation could create human and economic suffering that may lead to internal displacement, cross-border migration, and the spread of life-threatening diseases.

The need to address risks associated with future disaster-related events, including those that may be linked to climate change, is inherent to FEMA's long-term vision of promoting physical and economic loss reduction and life saving measures. Working within existing

FIGURE 4.3 FEMA Climate Change Adaptation policy statement. (From http://www.fema.gov/media-library-data/20130726-1919-25045-6267/signed_climate_change_policy_statement.pdf.)

(*Continued*)

statutes and authorities, FEMA will strive to be consistent in the Agency's incorporation of climate change adaptation actions and activities into on-going plans, policies, and procedures.

This policy statement identifies seven initial actions we will take to help integrate climate change adaptation considerations into our programs and operations. These actions also align with our vision of a Whole Community approach to emergency management, as it is expected that extensive collaboration with the public, all levels of government, the private sector, non-governmental organizations, and community organizations will be required.

IV. Policy and Procedures

A. In addition to the actions taken pursuant to Executive Order 13514, *Federal Leadership in Environmental, Energy, and Economic Performance,* with respect to federal facilities, FEMA will undertake the below high-level actions to integrate climate change adaptation considerations into the Agency's programs, policies, and operations. These actions directly support the White House Council for Environmental Quality's (CEQ) Implementing Instructions for Federal Agency Climate Change Adaptation Planning and are designed to enhance our climate change awareness, further evaluate the effects climate change may have on FEMA programs and operations, and identify potential areas for future exploration.

1. **To enhance climate research, monitoring, and adaptation capabilities, FEMA will continue to establish partnerships with other agencies and organizations that possess climate science and climate change adaptation expertise.** FEMA will continue to develop and maintain partnerships that enable the Agency to monitor the projected effects of climate change, and communicate climate science data and research needs related to emergency management and disaster resilience. FEMA will also collaborate with other Federal agencies, State, Local, Tribal and Territorial (SLTT) partners, intergovernmental organizations, nongovernmental organizations, the private sector, academia, and the international community to share lessons learned and develop best practices regarding climate adaptation.

2. **FEMA will continue to study the impacts of climate change on the National Flood Insurance Program (NFIP) and incorporate climate change considerations in the NFIP reform effort.** An initial 2-year study concluded that climate change is likely to have significant impacts on the NFIP; Special Flood Hazard Areas are projected to increase significantly across the nation, with impacts mounting over time as the number of policyholders are projected to double by 2100. In order to ensure the program serves the public most effectively, FEMA will continue efforts to understand the potential impacts of climate change on the NFIP and identify areas where future climate conditions can be included as part of the larger reform effort.

3. **FEMA will evaluate how climate change considerations can be incorporated into grant investment strategies with specific focus on infrastructure and evaluation methodologies or tools such as benefit/cost analysis.** FEMA will evaluate methods

2

FIGURE 4.3 (CONTINUED) FEMA Climate Change Adaptation policy statement. (From http://www.fema.gov/media-library-data/20130726-1919-25045-6267/signed_climate_change_policy_statement.pdf.)

(*Continued*)

for addressing future climate conditions through its grant programs to SLTT entities. FEMA will also study how to introduce long-term climate change risks into the benefit/cost analysis methods that guide the awarding of grants.

4. **FEMA will seek to understand how climate change will impact local communities and engage them in addressing those impacts.** FEMA will proactively engage and partner with SLTT communities to gain a greater understanding of their climate change adaptation challenges and activities, and look for ways FEMA can take action to support them in those efforts.

5. **FEMA will promote building standards and practices, both within FEMA programs and in general, that consider the future impacts of climate change.** FEMA currently promotes programmatic guidance and standards for use by SLTT partners to mitigate hazards through regulation of building and infrastructure construction. The current standards and guidance, based on today's climate, may not anticipate the risks structures will face as the climate changes. Therefore, it is important to review guidance and standards to determine the feasibility of incorporating future climate change considerations, and encourage the integration of adaptation measures into local planning and development practices.

6. **Through partnerships with the climate science community, FEMA will evaluate the potential impact climate change may have on existing risk data and the corresponding implications for Threat Hazard Identification Risk Assessment (THIRA) development and operational planning.** Changes in the climate will affect the accuracy and practice of using historical records to predict the magnitude, location, and frequency of future hazards—with significant challenges for important analytic processes and decisions. In response, FEMA will continue to work with the climate science and risk analysis community to evaluate the impacts of climate change on the viability of existing risk data.

7. **FEMA will continue to pursue a flexible, scalable, well equipped, and well trained workforce that is educated about the potential impacts of climate change.** Changes in the frequency and magnitude of severe weather events could potentially strain FEMA resources. FEMA will continue to assess and address its staffing and equipment needs to create a more flexible workforce by increasing employee readiness, cross-training staff, and increasing the pool of employees who are qualified and trained to respond to disasters or other events.

V. Responsibilities

Roles, responsibilities, and timelines for completing each of the above actions will be set forth in the follow-on FEMA Climate Change Adaptation Implementation Plan.

3

FIGURE 4.3 (CONTINUED) FEMA Climate Change Adaptation policy statement. (From http://www.fema.gov/media-library-data/20130726-1919-25045-6267/signed_climate_change_policy_statement.pdf.)

(*Continued*)

VI. Definitions

Climate Change: According to the Intergovernmental Panel on Climate Change (IPCC), climate change refers to "a statistically significant variation in either the mean state of the climate or in its variability, persisting for an extended period (typically decades or longer). Climate change may be due to natural internal processes or external forcings, or to persistent anthropogenic changes in the composition of the atmosphere or in land use."

Climate Change Adaptation: The IPCC defines climate change adaptation as "the adjustment in natural or human systems in response to actual or expected climatic stimuli or their effects, which moderates harm or exploits beneficial opportunities."

VII. Authorities

The Homeland Security Act of 2002, as amended (6 U.S.C. 101 et seq.), the Robert T. Stafford Disaster Relief and Emergency Assistance Act, as amended (42 U.S.C. 5121 et seq.), the President's Executive Order 13514 of October 2009, the 2010 Climate Change Adaptation Report drafted by ICCATF, and the Instructions for Implementing Climate Change Adaptation Planning issued by the CEQ.

VIII. Responsible Office

The Office of Policy and Program Analysis is responsible for the coordination and oversight of all aspects of this directive.

IX. Supersession

None

W. Craig Fugate Administrator FEMA	David J. Kaufman Director OPPA
Date: 1/23/2012	Date: 11/1/2011

4

FIGURE 4.3 (CONTINUED) FEMA Climate Change Adaptation policy statement. (From http://www.fema.gov/media-library-data/20130726-1919-25045-6267/signed_climate_change_policy_statement.pdf.)

FACT SHEET: TAKING ACTION TO PROTECT COMMUNITIES AND REDUCE THE COST OF FUTURE FLOOD DISASTERS

Across the country, extreme weather and other impacts of climate change are threatening the health, safety, and prosperity of our communities. This month, NOAA and NASA announced that 2014 was the hottest global year on record. And as the planet continues to warm, impacts like rising sea levels, intensified storms, and heavy downpours are contributing to an increased risk of flooding. President Obama is committed to ensuring that American communities thrive in the face of a changing climate. That is why, today, the President signed an Executive Order establishing a flood standard that will reduce the risk and cost of future flood disasters by requiring all Federal investments in and affecting floodplains to meet higher flood risk standards. By requiring that Federally funded buildings, roads and other infrastructure are constructed to better withstand the impacts of flooding, the President's action will support the thousands of communities that have strengthened their local floodplain management codes and standards, and will help ensure Federal projects last as long as intended.

This new Federal Flood Risk Management Standard, called for by the President's State, Local and Tribal Task Force on Climate Preparedness and Resilience, builds on the unprecedented actions President Obama has taken to support communities as they prepare for the impacts of climate change. Agencies will have flexibility in implementing the new Standard and will incorporate input from the public and stakeholders as they move forward, including through a series of public listening sessions across the country. This week, the Army Corps of Engineers released a comprehensive study that evaluates flood risks to the coastal areas affected by Hurricane Sandy and provides a framework to help communities address increasing flood risks. The study, which was called for by Congress, emphasizes the importance of improved planning, and notes that managing coastal storm risk is a shared responsibility by all levels of government. The Administration has made significant investments in resilient disaster recovery in the wake of devastating storms like Hurricane Sandy to ensure that infrastructure projects factor in climate impacts like rising sea levels, and to invest in making transit systems more resilient to flooding and extreme weather.

FLOOD IMPACTS ON COMMUNITIES

Floods leave behind big costs for communities and taxpayers. Between 1980 and 2013, the United States suffered more than $260 billion in flood-related damages. Recent examples include record flooding from excessive rainfall in central and northern Illinois in April 2013 that damaged homes and businesses and caused an estimated $1 billion in losses. And damages from Hurricane Sandy in 2012—when high wind and coastal storm surge devastated the Northeast—are estimated at $67 billion, with recovery efforts still ongoing.

More than 50 percent of Americans live in coastal counties, where key infrastructure and evacuation routes are increasingly vulnerable to impacts like higher sea levels, storm surges, and flooding. And according to the National Climate Assessment, more than $1 trillion of property and structures in the US are at risk of inundation from sea level rise of two feet above current sea level—an elevation that could be reached as early as 2050. That further jeopardizes the critical infrastructure Americans depend on every day for housing, transportation, energy, water supply, and more.

PROTECTING COMMUNITIES AND REDUCING THE COST OF FLOOD DISASTERS

In 2013, the President's Hurricane Sandy Rebuilding Task Force adopted a higher flood standard for the Sandy-affected region to ensure that Federally funded buildings, roads and other projects were rebuilt stronger to withstand future storms. The strengthened standard is similar to flood risk standards in place in the States of New York and New Jersey. The Sandy Task Force also recommended that the Federal Government create a national flood risk standard for Federally funded projects beyond the Sandy-affected region. In the President's Climate Action Plan, he followed through on this recommendation by directing agencies to update their flood-risk reduction standard to ensure that federally funded projects across the country last as long as they are intended. Over the past year, Federal agencies collaborated on this update. The new standard announced today gives agencies the flexibility to select one of three approaches for establishing the flood elevation and hazard area they use in siting, design, and construction. They can:

- Use data and methods informed by best-available, actionable climate science;
- Build two feet above the 100-year (1%-annual-chance) flood elevation for standard projects, and three feet above for critical buildings like hospitals and evacuation centers; or
- Build to the 500-year (0.2%-annual-chance) flood elevation.

To protect their homes, businesses and vital infrastructure from disasters, at least 350 communities across the country—ranging from Dallas and Nashville to Denver and Tulsa—and the States of Indiana, Montana, New York and Wisconsin have already adopted standards that either meet or exceed this new Federal standard. The Administration today is also releasing proposed guidelines—which will be available for 60 days of public comment—that when finalized will provide guidance to agencies on implementing the new standard. Once public input has been considered, including from a series of public listening sessions that will be held across the country, and the guidelines are finalized, agencies will implement the Standard through their own rulemaking or other procedures, which also will incorporate input from the public and stakeholders.

The new flood standard will apply when Federal funds are used to build, or significantly retrofit or repair, structures and facilities in and around floodplains to ensure that those structures are resilient, safer, and long-lasting. It will not affect the standards or rates of the National Flood Insurance Program. Each agency will carefully consider how to appropriately apply this standard, and consider robust public input before deciding how to implement it.

Source: The White House. 2015. "Fact Sheet: Taking Action to Protect Communities and Reduce the Cost of Future Flood Disasters." January 30, 2015. http://www.whitehouse.gov/administration/eop/ceq/Press_Releases/January_30_2015.

State and Local Actions on Climate Change Adaptation and Mitigation

State and local governments have been dealing with the impacts of climate change for quite some time and often under a different mandate brought on by a natural disaster, a zoning issue or revisions to state and local building codes. There were virtually no federal monies being provided to state and local governments to address the climate change issue. Many states and localities would take advantage of funding provided postdisaster, usually through FEMA, to add climate change to the list of risks they were trying to address in the disaster recovery phase. As a part of their focus on climate change, the Obama Administration recognized that support and collaboration with state, local, and tribal governments was critical to moving forward. So, on November 1, 2013, the administration took another action to advance climate change adaptation.

President Obama established by Executive Order 13653 Preparing the United States for the Impacts of Climate Change, the Task Force on Climate Preparedness and Resilience to advise the administration on how the federal government can respond to the needs of communities nationwide that are dealing with the impacts of climate change.

The Executive Order acknowledged that the federal government has an important role to play in supporting community-based preparedness and resilience efforts by establishing policies and

MEMBERS OF THE STATE, LOCAL, AND TRIBAL LEADERS TASK FORCE ON CLIMATE PREPAREDNESS AND RESILIENCE

Across America, states, cities, and communities are taking steps to protect themselves from extreme weather and other climate impacts. As part of his Climate Action Plan, President Obama signed an Executive Order on November 1, 2013, establishing the Task Force on Climate Preparedness and Resilience to advise the administration on how the federal government can respond to the needs of communities nationwide that are dealing with the impacts of climate change. The task force members, which include governors, mayors, county officials, and tribal leaders from across the country, met throughout the year and used their first-hand experiences in building climate preparedness and resilience in their communities to inform their recommendations to the administration.

State Officials

Governor Neil Abercrombie (Hawaii)
Governor Jerry Brown (California)
Governor Eddie Calvo (Guam)
Governor Jay Inslee (Washington)
Governor Jack Markell (Delaware)
Governor Martin O'Malley (Maryland)
Governor Pat Quinn (Illinois)
Governor Peter Shumlin (Vermont)

Local Officials

Mayor Ralph Becker (Salt Lake City, Utah)
Mayor James Brainard (Carmel, Indiana)
Commissioner Paula Brooks (Franklin County, Ohio)
Supervisor Salud Carbajal (Santa Barbara County, California)
Mayor Frank Cownie (Des Moines, Iowa)
Mayor Bob Dixson (Greensburg, Kansas)
Mayor Eric Garcetti (Los Angeles, California)
Mayor George Heartwell (Grand Rapids, Michigan)
Mayor Kristin Jacobs (Broward County, Florida)
Mayor Kevin Johnson (Sacramento, California)
Mayor Michael Nutter (Philadelphia, Pennsylvania)
Mayor Annise Parker (Houston, Texas)
Mayor Patsy Parker (Perdido Beach, Alabama)
Mayor Madeline Rogero (Knoxville, Tennessee)
Mayor Karen Weitkunat (Fort Collins, Colorado)
Mayor Dawn Zimmer (Hoboken, New Jersey)

Tribal Officials

Karen Diver, Chairwoman, Fond du Lac Band of Lake Superior Chippewa (Minnesota)
Reggie Joule, Mayor, Northwest Arctic Borough (Alaska)

Source: The White House. 2013a. "State, Local, and Tribal Leaders Task Force on Climate Preparedness and Resilience." November 1, 2013. http://www.whitehouse.gov/administration/eop/ceq/initiatives/resilience/taskforce.

prioritizing investments that promote preparedness, protecting critical infrastructure and public resources, supporting science and research needed to prepare for climate impacts, and ensuring that federal operations and facilities continue to protect and serve citizens in a changing climate.

The Executive Order established the State, Local, and Tribal Leaders Task Force on Climate Preparedness and Resilience comprised of state, local, and tribal officials (see box "Members of the State, Local, and Tribal Leaders Task Force on Climate Preparedness and Resilience").

On November 17, 2014, the White House released a report on the recommendations of the State, Local, and Tribal Leaders Task Force on Climate Preparedness and Resilience. The report included significant recommendations for federal action and changes to enhance and build the capability for climate change adaptation nationwide. Among the significant recommendations:

- Require or encourage strengthened building codes and floodplain ordinances
- Consider requiring the impacts of climate change to be incorporated into environmental review documents under the National Environmental Policy Act
- Improve floodplain maps to include information on sea-level rise, increased precipitation, and so on—events that would exacerbate flooding, erosion, and other hazards
- Reform policies inhibiting the deployment of Property Assessed Clean Energy (PACE) programs for resilient investments such as energy efficiency
- Reform cost-benefit analysis to ensure that federal investments account for the long-term benefits of adapting to climate change
- Improve the delivery of climate science, data, and tools to enhance application to lead to state and local action.
- Create a Climate Corps similar to AmeriCorps to build community capacity and develop the next generation of climate adapters
- Improve the delivery of disaster recovery funds to facilitate incorporation of climate and resiliency actions (The White House 2014b)

The task force also recommended that the administration develop implementation benchmarks and a means to track progress and report on that progress on a regular basis (The White House 2014b). A summary of the task force recommendations are presented in the Executive Summary of the report in the box "Presidential Task Force Recommendations."

State and Local Adaptation Plans

States and communities around the country have begun to prepare for the climate changes that are already underway. This planning process typically results in a document called an adaptation plan. As of the end of 2014, 14 states had completed their state adaptation planning process and had approved plans. Another nine states are engaged in the planning process. The box "Find Out Which US States Are Making Progress in Preparing for Climate Change" describes Georgetown Climate Center's online tool, the State Adaptation Progress Tracker, that tracks the development of state and local climate change adaptation plans across the country (Georgetown Climate Center 2014).

Local Innovations in Climate Change Adaptation and Mitigation

This section will focus on innovations in climate change adaptation and mitigation that are occurring at the local governmental level. In reality, local leaders and local communities have had to focus on the impacts of climate change on a more immediate basis when confronted with planning and zoning issues and/or in the aftermath of a disaster. To a significant degree their approaches to the problem have been creative and on the forefront of the climate change adaptation issue. We will highlight some of the actions different communities have taken, acknowledge some awarded communities, and provide case studies on significant projects.

PRESIDENTIAL TASK FORCE RECOMMENDATIONS

EXECUTIVE SUMMARY

As the Third National Climate Assessment makes clear, climate change is already affecting communities in every region of the country as well as key sectors of the economy. Recent events like Hurricane Sandy in the Northeast, flooding throughout the Midwest, and severe drought in the West have highlighted the vulnerability of many communities to the impacts of climate change. In 2012 alone, the cost of weather disasters exceeded $110 billion in the United States, and climate change will only increase the frequency and intensity of these events. That is why, even as efforts to reduce greenhouse gas emissions continue, communities must prepare for the impacts of climate change that can no longer be avoided.

At state, local, tribal, and territorial levels, leaders are making bold decisions on ways to invest in more resilient infrastructure, revise land use, update building codes, and adjust natural resource management and other practices to improve the resilience of their communities to climate impacts. The federal government has a critical role to play in supporting these efforts by ensuring that Federal policies and programs incorporate climate change, incentivize and remove barriers to community resilience, and provide the information and assistance communities need to understand and prepare for climate risks. The federal government also has a responsibility to protect its own investments, such as military installations and space launch facilities, and ensure that the lands and resources it holds in the public trust are managed for a changing climate.

In order to better support communities across the country as they prepare for the impacts of climate change, the task force proposes that the Administration advance actions across the federal government that align with the following overarching principles:

- Require consideration of climate-related risks and vulnerabilities in the design, revision, and implementation of all Federal policies, practices, investments, regulations, and other programs
- Maximize opportunities to take actions that have dual-benefits of increasing community resilience and reducing greenhouse gas emissions
- Strengthen coordination and partnerships among Federal agencies, and across Federal, state, local, tribal, and territorial jurisdictions as well as economic sectors
- Provide actionable data and information on climate change impacts and related tools and assistance to support decision-making at all levels
- Consult and cooperate with Tribes and indigenous communities on all aspects of Federal climate preparedness and resilience efforts, and encourage states and local communities to do the same

The diverse challenges posed by climate change will require a wide range of actions to ensure that communities across the country, large and small, are prepared. With coordination, thoughtful planning, and decisive action, Federal, State, and local governments, Tribes, and territories can ensure a safe and prosperous future.

Summary of Recommendations

1. Building Resilient Communities: Climate change will impact communities for years to come, and long-term efforts to build resilience will help communities thrive in the 21st century and beyond. By incorporating climate change considerations into its programs, the Federal Government can support communities as they rethink traditional approaches to land use and land management, building and infrastructure siting and design, and community planning.

2. Improving Resilience in the Nation's Infrastructure: Climate change poses a significant threat to the safety and reliability of critical infrastructure systems. Whether related to energy, transportation, freshwater management, coastal protection, or ecosystems, Federal action can improve the way climate impacts and greenhouse gas emissions are incorporated into public and private infrastructure investments, policies, and practices.
3. Ensuring Resilience of Natural Resources: Climate change puts America's vital natural resources and ecosystems at risk. By helping communities better protect and conserve the Nation's natural resources, the Federal Government can improve human and community resilience in cost-effective ways.
4. Preserving Human Health & Supporting Resilient Populations: Climate change presents a significant public health threat to individuals and communities, exacerbating illness and increasing the frequency and severity of dangerous extreme weather events. The Federal Government can support State, local, tribal, and territorial efforts to address the needs of populations most vulnerable to climate impacts, protect public health, and improve disaster preparedness.
5. Supporting Climate-Smart Hazard Mitigation and Disaster Preparedness and Recovery: Climate change will increase the frequency and severity of extreme weather events, which are often devastating to communities. Through more holistic hazard mitigation planning, improved data collection and mapping, partnership development, and program modernization, the Federal Government can improve efforts to prevent and mitigate the effects of extreme weather and other climate-related hazards.
6. Understanding and Acting on the Economics of Resilience: Climate change poses significant economic risk to all sectors and communities. Advancing measures to encourage more prudent investments in long-term resilience can better ensure a vibrant economic future as the climate continues to change.
7. Building Capacity for Resilience: To successfully prepare for climate change, communities must have the capacity to recognize, understand, and assess relevant climate-related hazards, risks, and impacts. The Federal Government can help communities build this capacity by continuing to shape or reshape programs, policies, information sources, and other forms of assistance that enable state, local, tribal, and territorial jurisdictions to prepare for climate change.

Additionally, the Federal Government should establish a process for tracking and reporting on progress made in the implementation of the recommendations, as well as specific benchmarks.

Source: The White House. 2014b. President's State, Local, and Tribal Leaders Task Force on Climate Preparedness and Resilience: Recommendations to the President. November 2014. http://www.whitehouse.gov/sites/default/files/docs/task_force_report_0.pdf.

In the fall of 2014, the White House announced the "Climate Action Champions competition, to identify and recognize local climate leaders and to provide targeted Federal support to help those communities further raise their ambitions" (The White House 2014a).

A key component of this program is that a coordinator will be appointed to each Champion program to mentor them, coordinate among the federal agencies and provide targeted support from a range of federal funding streams.

The 16 designated Climate Action Champions are listed in the box "Climate Action Champions" with a brief description of their climate change adaptation activities.

FIND OUT WHICH US STATES ARE MAKING PROGRESS IN PREPARING FOR CLIMATE CHANGE

OCTOBER 9, 2014

In recent years, a number of states have started taking action to prepare their communities for climate change. Many have even developed specific adaptation plans to guide their work. Until now, though, no one has ever been able to define how much progress states are actually making in implementing those plans.

The Georgetown Climate Center's online tool, the State Adaptation Progress Tracker, changes all of that. Now, anyone will be able to quickly determine how much progress their state is making and decision-makers will be able to learn from innovative examples of actions other states are taking.

"People want to know what their leaders are doing to prepare for climate change impacts, such as storms and rising seas," said Vicki Arroyo, executive director of the Georgetown Climate Center. "This tool provides transparency and enables individuals and communities to track state progress."

"This research shows that a number of states have started implementing changes that will actually make their communities more resilient. That's good news. Nearly half of all US states also have at least some planning underway to prepare for climate change. Unfortunately, the research also shows that many states are still not treating this issue with the urgency that is called for."

As of today, 14 states have finalized state-led adaptation plans (Alaska, California, Colorado, Connecticut, Florida, Massachusetts, Maryland, Maine, New Hampshire, New York, Oregon, Pennsylvania, Virginia, Washington). Another 9 have some level of planning underway (District of Columbia, Delaware, Hawaii, Michigan, Minnesota, New Jersey, Rhode Island, Vermont, Wisconsin).

A number of innovative efforts are underway in states across the country to minimize and avoid the impacts of climate change. For example:

- Massachusetts is investing $50 million to prepare its communities and infrastructure for climate change impacts. The Coordinated Climate Preparedness Initiative accelerated efforts to increase the resiliency of the state's infrastructure.
- In New York, Governor Andrew Cuomo signed into law the Community Risk and Resiliency Act (A06558/S06617-B), on September 22, 2014. The Act requires state agencies to consider climate impacts in many permitting and funding decisions. The new requirements will affect the siting of bulk storage and hazardous waste facilities and the development of oil and gas drilling permits, as well as other projects. The law calls for the adoption of official projections of sea-level rise for the state. State agencies have also been tasked with preparing model laws to help communities incorporate climate risks into local ordinances.
- California passed "cool pavements" legislation that will encourage the use of lighter colored paving and permeable paving materials that reduce the heat-island effect in urban areas. Extreme heat events, particularly in urban areas, can kill vulnerable populations (such as the very young, the very ill and those suffering from respiratory illnesses such as asthma).
- The Washington State Department of Transportation (WSDOT) issued guidance in March 2013 requiring the consideration of climate change impacts in environmental review documents. As a result, large state-funded infrastructure projects now consider climate change impacts over the project's lifespan. WSDOT has applied this guidance

to increase the resilience of infrastructure projects. For example, one recently approved project, the Mukilteo Multimodal Ferry Terminal, considered climate change impacts during the environmental review process. This project will ensure a resilient transportation connection between Whidbey Island [in Washington State] and the mainland by improving access, transit connections, and ferry operations. The project was designed to be resilient to future sea-level rise by using fill to elevate the site for the ferry terminal and locating related facilities, such as access roads, in upland areas.

- Maryland has incorporated climate change and sea-level rise considerations into its land conservation programs. The state prioritizes important coastal areas for conservation so that lands that will increase the state's long-term resilience are conserved first with state funds.

To view details about each state's progress and the specific goals that each state has completed, go to http://www.georgetownclimate.org/adaptation-progress.

Source: Georgetown Climate Center. 2014. "Find Out Which U.S. States Are Making Progress in Preparing for Climate Change." October 9, 2014. http://www.georgetownclimate.org/find-out-which-us-states-are-making-progress-in-preparing-for-climate-change.

A series of case studies concerning hazard risk reduction programs and climate change adaptation programs from communities across the country are presented in Chapter 6.

Conclusions

There has certainly been a lot of activity on climate change adaptation and mitigation in the last few years. Much of the innovation has occurred at the local level with financial support from the federal government coming usually after a presidential disaster has been declared. But communities have chosen not to wait and looked to innovative ways to fund adaptation, much of which has come through the foundations and other nonprofits.

The Obama Administration has made climate change an issue and has taken specific actions to make this a priority. They have sought to better coordinate the federal family through a revitalization of the Interagency Climate Change Adaptation Task Force as well as requiring federal agencies to produce climate change adaptation plans and having them report progress annually. They have convened the State, Local, and Tribal Leaders Task Force on Climate Preparedness and Resilience to better support communities. The recommendations from that Task Force have been provided to the president and implementation of those recommendations should be watched closely. They also initiated a competition to identify Climate Change Champions and are committed to providing additional federal support to these leading-edge communities.

However, a dichotomy does exist when a president's budget only provides 1 percent of funding to support climate change adaptation activities as opposed to programs in science, R&D, and other areas.

It's clear an influx of federal funding would be extremely welcomed to more climate change adaptation from the planning stage to the implementation of actions. Even if that does not materialize, should the administration decide to implement some of the recommendations in the State, Local, and Tribal Leaders report, significant strides could be made.

GOVERNMENT STRATEGIES FOR NATURAL HAZARDS MANAGEMENT

Brian Cowan

There was a house fire in my neighborhood last night, complete with fire engines and sirens. Our family went out to see what the commotion was about. It was not a neighbor close to our house but

CLIMATE ACTION CHAMPIONS

Blue Lake Rancheria Tribe (California): The Blue Lake Rancheria, a Federally recognized tribal government, began its strategic climate action plan in 2008 and is a regional leader in strategically planning and implementing both climate resiliency and greenhouse gas reduction measures. To date, the Tribe has reduced energy consumption by 35 percent and has committed to reduce greenhouse gas emissions 40 percent by 2018, utilizing a range of approaches including the use of biodiesel to power public buses and aggressive energy efficiency measures.

Boston (Massachusetts): The City of Boston has committed to reduce greenhouse gas emissions 10 percent by 2020 and 80 percent by 2050. The city is the first in the region to adopt Green Building Zoning, add climate resilience to the large new construction review process, and work with utilities on a regional micro grid. Additionally, Boston implemented a "Green Ribbon Commission," which represents businesses, non-profits, and community leaders from a variety of sectors working to develop shared strategies for fighting climate change in coordination with the city's Climate Action Plan.

Broward County (Florida): Broward County, a member of the Southeast Florida Regional Climate Change Compact, a regional partnership of four counties (Broward, Miami-Dade, Monroe, and Palm Beach) to advance climate mitigation and adaptation strategies, has committed to reduce greenhouse gas emissions 2 percent per year, reaching a 10 percent reduction target by 2020, and 80 percent by 2050.

Dubuque (Iowa): The City of Dubuque has adopted greenhouse gas emission reduction targets of 50 percent below 2003 levels by 2030. Alongside these aggressive targets, the flood prone community of Dubuque is focused on risk reduction and resilience, especially as it relates to development and redevelopment of community infrastructure.

Knoxville (Tennessee): The City of Knoxville has set a short-term greenhouse gas emission reduction target of 20 percent below 2005 levels by 2020. In order to execute on that target, the city developed a planning process that engages major stakeholders, including utilities and community grassroots organizations, and formalizes efforts to integrate energy provision, utilization, procurement, waste, and urban/agricultural use into the city's Energy and Sustainability Work Plan.

Metropolitan Washington Council of Governments (District of Columbia, Maryland, and Virginia): The Metropolitan Washington Council of Governments (COG) developed the National Capital Region Climate Change Report, demonstrating a fully integrated approach to climate change mitigation and outlining goals in the areas of greenhouse gas emissions, renewable energy, transportation and land use, sustainability and resilience, and infrastructure. The COG is committed to reducing greenhouse gas emissions 20 percent by 2020 and 80 percent by 2050. As a Climate Action Champion, the COG plans to establish an innovative Climate Champions Training Initiative to disseminate the Federal technical assistance and resources it receives to stakeholders across the region.

Mid-America Regional Council (Kansas and Missouri): The Mid-America Regional Council, a nonprofit association of city and county governments and the metropolitan planning organization for the bi-state Kansas City region, proposed the creation of a regional Resilience Working Group. Through leadership, planning, and action, the Mid-America Regional Council promotes regional cooperation and innovative solutions and seeks to reduce greenhouse gas emissions by 2 percent per year.

Minneapolis (Minnesota): The City of Minneapolis developed the Minneapolis Climate Action Plan, which includes greenhouse gas emission reduction targets of 15 percent by 2015, 30 percent by 2025, and 80 percent by 2050. The city has partnered with two utilities, Xcel Energy and Center Point Energy, to jointly plan, market, implement, and track strategies to meet climate and energy goals.

Montpelier (Vermont): Earlier this year, the City of Montpelier launched Net Zero Montpelier, a major initiative focused on making Montpelier the first effectively carbon-neutral capital city in the country by the year 2030. The city has demonstrated its leadership and innovation in climate mitigation and resilience by creating the first energy efficiency utility and the first standard offer program, and by making a commitment to eliminate fossil fuel use across all sectors.

Oberlin (Ohio): The City of Oberlin has developed a climate change mitigation and resilience plan for power production, solid waste, and transportation, utilizing an innovative and collaborative model that includes partnerships between the town, the University, the utility, industry, and international, national, and regional non-profits. The city established aggressive greenhouse gas reduction targets of 50 percent by 2015, 75 percent by 2030, and 100 percent by 2050.

Portland (Oregon): The City of Portland is a regional leader for greenhouse gas reduction and climate change mitigation. With support from 20 agency partners, Portland's 2015 Climate Action Plan is a strategy to put the city on a path to achieve an 80 percent reduction in greenhouse gas emissions compared to 1990 levels.

Salt Lake City (Utah): Salt Lake City developed a comprehensive and well-integrated portfolio of programs and policies, including renewable energy, transportation, code revisions, water systems, and building policies as priorities. The city established a joint resolution to reduce its carbon footprint by 20 percent below 2005 levels by 2020, and 50 percent below 2020 levels by 2040.

San Francisco (California): The city of San Francisco has established some of the most aggressive climate and sustainability targets in the nation, covering a broad range of sectors, including energy efficiency, renewable energy, transportation, water, green infrastructure, and waste. With robust goals to measure progress, San Francisco aims to reduce greenhouse gas emissions by 25 percent below 1990 levels by 2017, and 40 percent by 2025.

Sault Ste. Marie Tribe of Chippewa Indians (Michigan): The Sault Ste. Marie Tribe of Chippewa Indians demonstrates a holistic approach to climate action and preparedness through their energy strategy, emergency operations plan, integrated resource management plan, solid waste management plan, sustainable development code, and land use planning process, with ambitious goals including a net-zero energy goal. The tribe aims to reduce greenhouse gas emissions by 4 percent per year.

Seattle (Washington): Adopted in June 2013, Seattle's Climate Action Plan focuses on city actions that reduce greenhouse gas emissions and enhance resilience while also supporting vibrant neighborhoods, economic prosperity, and social equity. The plan aims to reduce greenhouse gas emissions by 25 percent below 1990 levels by 2017, and by 40 percent by 2025.

Sonoma County Regional Climate Protection Authority (California): The Sonoma County Regional Climate Protection Authority (RCPA) is the nation's first local government agency created specifically to address climate change. The RCPA has committed to pursue a 25 percent reduction in greenhouse gas emissions from 1990 levels through formal partnerships, pooling resources (financial and human), and working across silos.

It is anticipated that there will be additional funding made available through the following programs:

- Smart Grid Technologies: DOE announced that up to $3.5 million will be available through the Resilient Electricity Delivery Infrastructure (REDI) Initiative Funding Opportunity Announcement to help communities deploy pre-commercial and commercial smart grid technologies and tools that will help decision makers and resource managers to improve the recovery of electricity delivery services in their communities.
- Coastal Resilience Grant Program: The program will implement activities that build the resilience of forward-looking coastal communities and nearby ecosystems with the goal of mitigating the negative impacts of extreme weather events, climate hazards, and changing ocean conditions.

Other support to these Champion communities will include a variety of technical assistance. These include validated climate science, data, vulnerability assessments, and risk projection tools needed to make smart planning decisions in their communities and receive technical assistance in using the tools, participation in tabletop adaptation, preparedness, and resilience exercises, and technical expertise in renewable energy. There will be a special focus on providing technical expertise to those tribal communities that are chosen as champions.

Source: The White House. 2014. *Fact Sheet: 16 US Communities Recognized as Climate Action Champions for Leadership on Climate Change*. Office of the Press Secretary, White House. December 3, 2014. http://www.whitehouse.gov/the-press-office/2014/12/03/fact-sheet-16-us-communities-recognized-climate-action-champions-leaders.

one that is on the next block. Of course, had it been a next door neighbor, I would have been much more concerned that the fire department arrived quickly and put the fire out. Still, I was glad for their prompt response. The fire is a disaster for the family. There were no injuries—or worse—but their kitchen is not really usable, and the smell of smoke pervades everything. Naturally, we and other neighbors are doing what we can to help. But even with the ability to pay for the losses through insurance, it will be a long time before their household is back to normal.

As would happen with anyone, the event led me to think about my own house and its fire possibilities: are the smoke alarms working; shouldn't we have an escape plan so that we can get out in case of an emergency; does the electrical need to be looked at by a professional; should I remind my wife and children to be careful about the stove, and how often does it need to be said. Some of these are easy to do, and others may cost more than I think is necessary. And the longer I wait to do any of them, the harder it will be.

All of the thoughts and questions that ran through my mind are simply the same ones that anyone trying to manage their potential loss from hazards will ask themselves. Individuals, families, organizations, businesses, and governments all would go through similar decision-making processes: how high is the risk; what plans do we need; how much insurance is needed; how much money should be paid to avoid or lessen potential losses; what are the financial priorities; how do we make certain that our efforts are sustained over a long time.

As with decisions about one's health, a person can make most of his or her decisions relatively quickly. You can decide now to purchase or not to purchase health insurance because you know what you can afford to pay and you know what level of risk you are willing to accept. You can decide to lessen your health risks by regularly exercising—and you can do so through cost-free neighborhood walks or a costly gym membership. You can eat right or have junk food. You can see a doctor for regular checkups or not. You base your decisions on cost and on the level of threat—or at least your perception of the level of threat—to your health.

On the other hand, when an organization such as a business or a government determines that it will undertake hazard management decisions, the categories of decisions are the same, but arriving at those decisions becomes a more complex journey. Governments in particular need to determine who pays and through what mechanisms: Do their residents or businesses need to accept full responsibility for potential losses to their property (for example through insurance) or will the government provide some assistance; how much do they pay and when; what are the critical priorities (for example schools or bridges) and hence what are the financial priorities; who must be directly involved in the preparation and implementation of the hazard management plans and how will that involvement be achieved.

Federal Strategies for Risk Management

There have been many programs established by the federal government to address losses from natural hazards. The most well-known is its disaster assistance program. Its main purpose is to provide assistance after a disaster event. When losses from a natural hazard event are sufficiently large, the federal government helps to implement a recovery, and in some extreme events, a response to the event. (The term "response" here refers to the short period of time after an event when resources can save lives, affect emergency medical assistance, and provide immediately needed food or shelter. Recovery refers to the rebuilding effort in which resources are utilized to implement a return to communities' preevent status.) In order to provide examples of the principal components of risk and loss management strategies, we will briefly discuss here a small handful of the strategies the federal government has employed to address the potential impacts and losses from natural hazard events before those events occur. A discussion of the effectiveness of these strategies will also be provided.

Flooding

Flood hazard was initially addressed in the United States through public works—dams and levees—intended to eliminate the potential for flooding effects. The Tennessee River Valley, as an example, was a continuous source of flood losses, and the system of dams constructed there was intended to control the flood potential as well as using that control system for the generation of sorely needed electric power. The system of levees along the Mississippi River was also a major component of this man-made structural approach to flood hazard management. While there was clearly some success from this approach, it was not ultimately viewed as effective as it should be. In particular, the system of levees built along many of our rivers came to be less favorably viewed by some policy makers when flooding losses continued to occur. Flooding in the southern Mississippi River system from hurricanes in the late 1950s and early 1960s caused Congress to take a second look at the US approach to its management of the flood hazard.

The review resulted in one of the more innovative and well-balanced hazard management approaches, the National Flood Insurance Program. The skeletal structure of the program was this: a locality such as a city, town, or county would enact specified land-use and zoning measures that directed future development and construction away from flood hazard areas. The residents and businesses of the locality would then be eligible to purchase federal flood insurance for their properties. Premiums were collected by the federal government, either directly from the insured, or through an insurance agent. When first enacted by the US Congress, the program was voluntary. Participation was not very high, however, with only about 10 percent of US communities adopting the appropriate measures and joining the program. In the summer of 1972, another hurricane struck the east coast of the United States, again causing significant flood losses. As a result, Congress enacted in the following year a substantial revision to the law which stated that, if the US government would be backing the security of any mortgaged property in a flood risk area, then that property had to have flood insurance. Since virtually every mortgage at that time had federally backed guarantees, every flood-prone community in the United States would need to join the program.

The National Flood Insurance Program as enacted in 1973 included some notable aspects in its strategic framework:

- The federal government accepted the responsibility for providing each community with the definition, in map form, of its flood hazards. This acknowledged the prohibitive cost of such flood studies for communities, and also implicitly acknowledged the unique access of the federal government (for example, in the US Army Corps of Engineers) to hydrologic study experts.
- The development of these maps, and the flood studies to support them, would be developed with community participation and input. Communities would also have an opportunity to appeal study results (based on hydrologic data) before maps became effective and to request amendments or revisions of maps after they had become effective.
- Through its collection of flood insurance premiums, the program was essentially able to fund its priority requirements, including flood studies and mapping. Congress still needed to approve the program's annual operating budget, but its concern was the integrity of the flood insurance fund (i.e., the program's ability to pay flood insurance claims), not the annual government-wide budgeting process.
- Insurance rates for those structures already in existence in flood risk areas would be subsidized. However, all structures built after flood map effective dates would be charged actuarial insurance rates. This established a financial incentive for the implementation of flood loss reduction measures, internalizing the potential loss completely with the property owner.
- Disaster assistance would not be provided to homeowners or businesses that did not purchase insurance. Here the government was stating that it simply would not accept financial responsibility for losses once property owners had the opportunity to accept the responsibility themselves. In other words, it is not only trying to reduce flood losses through appropriate local land-use ordinances, it also wants to reduce its exposure to flood loss payouts from its disaster relief fund.

When the flood program was adopted, there was generally no flood insurance available in the private insurance market. Thus, in establishing federal flood insurance coverage, the federal sector was attempting to address a market failure. From a strategic point of view, however, it was also attempting to shift more of the burden of disaster flood losses to homeowners and businesses, and provide them with a reasonable and common means of paying for that loss over a long period of time. Strategically, the program was also forging a hazard management partnership between the federal and local sectors, and it enhanced that partnership by requesting state governments to join in the flood plain management arena if they were not already active in doing so, and it was basing its flood plain management advice and decisions on state-of-the-art technical studies. Lastly, the program was focused on the reduction of future flood losses through its specified flood plain management requirements for program participation (see Figure 4.4).

The National Flood Insurance Program has continued to be revised over the years. Amendments to the 1973 law have been numerous, including the removal of the universal flood insurance purchase requirement for all government-backed mortgages. But the systems and accomplishments of the 1973 Act endure; for example, approximately 20,000 local communities still participate in the program's flood plain management requirements, and banks typically will themselves undertake or require flood-risk review insurance when a home is located in the flood-risk area of a flood map. The National Flood Insurance Program can take credit for resulting in the avoidance of billions of dollars of flood losses.

FIGURE 4.4 "Adapting to Climate Change: Planning for an Uncertain and Expensive Future" was the topic of a day-long meeting, November 30, 2012, at the University of Rhode Island's Graduate School of Oceanography, Narragansett. (From https://www.fema.gov/media-library/assets/images/67962.)

Earthquake

The loss from major geologic hazards can be catastrophic and far-reaching. Large and major seismic events are devastating when they strike populated areas. California is often thought of first when US earthquake hazards are discussed, but the United States has numerous earthquake risk areas. These include the Pacific Northwest, Alaska, the Intermountain West, and the Central United States. Even the East Coast has its own earthquake threat. As recently as 2011, the Washington Monument in Washington, DC, sustained damage from a moderate earthquake in central Virginia.

Congress determined in 1977 that it would address the earthquake hazard in the United States and enacted the Earthquake Hazards Reduction Act. It created the National Earthquake Hazards Reduction Program (NEHRP). The program was based primarily on the conduct of research (both basic and applied) into the nature of earthquakes in order to better understand its causes and impacts. In the development of such understanding, the program would define quantitatively the earthquake hazard, enable prediction of its occurrences, and develop technologies for mitigating its impacts. Somewhat uniquely, the program was constructed on a partnership of four federal agencies—the National Science Foundation (NSF), USGS, the National Bureau of Standards (which later became the National Institute for Standards and Technology [NIST]), and FEMA—which would coordinate their efforts and their budgets for program purposes. FEMA was given a lead role and directed to support implementation activities by localities and states.

The National Earthquake Hazards Reduction Program has revised and altered its construct over time, but it fundamentally has reflected the following strategic components in its efforts to address the earthquake hazard:

- Increased knowledge and state-of-the-art technologies are continually sought. Use of the knowledge its research formulates and the implementation of developed technologies by local or state government is voluntary. The program therefore continually strives to develop the most easily understood and implemented technologies and guidance.
- To achieve implementation, the program's participating federal agencies rely on dissemination of information and technologies. Improving its technology transfer, information dissemination, and public awareness venues becomes a core effort.

- Since most of the program's budgetary resources are committed to research, there has been a low level of funds to directly support (through grants or other means) mitigation implementation.
- Effective partnering and collaboration with the state, local, academic, and private sectors is also a primary means for the program to reach its target audiences.
- As with most federal programs, the earthquake program confronted limited budgets for its efforts. Since the construct of the program was multiagency, its hazard-specific efforts compete for available funds within each agency as well as within the federal sector. Hence the identification of funding priorities becomes a major issue.

Comparing the NFIP and NEHRP

When comparing the flood and earthquake programs, a couple of interesting elements arise with respect to loss reduction strategies and their implementation. First is the element of financial incentive for implementation. There has generally not been affordable earthquake insurance available in the United States. While there have been attempts to enact such coverage, they have not been successful. But the attractiveness of financial incentive of some kind—insurance or other—to achieve loss reduction implementation on a broad level is clear. These incentives when properly constructed can lead toward an internalization of natural hazard risk by the property owner, a means to pay for loss over time, and a means to reduce the ultimate size of that loss.

A second aspect of the comparison of earthquake to flood programs is the attractiveness of multihazard approaches. Flood risks are more or less present in localities and the occurrence of floods is frequent. However, earthquakes are relatively rare, even in California. A multihazard approach allows any organization or governmental entity to plan for and address any hazard eventuality, based on comparative risks, losses, and priorities.

Notwithstanding any obstacles it may have to achieving earthquake risk reduction, the National Earthquake Hazard Reduction Program has effected significant successes using its strategy of partnering and state-of-the-art technology based on the best science. For example, it achieved the inclusion of earthquake-resistant building technologies in standard and recommended building codes and standards in large part because it collaborated closely with the academic and building sciences communities in the technologies' development and dissemination.

Project Impact

In the mid- to late 1990s, there was considerable effort placed toward the development of a strategy that would address natural hazard risks through the weaving together of observably successful components of strategies and programs centered on the actions that would manage natural hazards before their occurrence. Its development involved significant input and recommendations from across the spectrum: academia, scientists and engineers, federal agencies, state and local governments, business and industry, financial sector businesses including insurance companies, and public interest groups. It also included a review of strategies by an expert panel which then provided recommendations for programmatic construct. As can be easily seen, most of the specific strategic tactics that were recommended as a result of these efforts were not revolutionary. They have been used in other governmental and private sector efforts. However, the initiative that came to be known as Project Impact did have success because it mixed these tactics together, and its goal (to result in localities that would resist natural hazards losses and rebuild after them) was widely supported.

The components of this initiative included:

- A focus on risk loss reduction, as opposed to preparedness to respond to disaster events.
- Implementation at the local level—resources to support natural hazard risk reduction would be deployed and used by towns, cities, and counties.

- Each local-level project would have partnerships and these partners would commit the resources to work on the community's disaster resistance. The commitments did not have to be targeted at public buildings or needs (as an example, a business could commit to working toward its own risk mitigation for the good of the community. After all, the sooner a business is operating after a disastrous event, the quicker the community recovers). A publicly signed, informal agreement would articulate these commitments.
- The local jurisdiction would develop and formulate the priority needs of its infrastructure and public assets. This planning would be based on the understanding of its needs, its residents' and businesses' inputs, and vulnerability assessments where feasible.
- Public awareness would include celebrations of a community's achievement of its milestones. Public events with the participation of community partners were central to the initiative.
- Financial incentives would be fundamental to risk reduction implementation. The federal sector (principally FEMA) took the lead in this by providing public assistance grants. Other incentives (for example lower flood insurance rates) might be developed and deployed. In working with other federal agencies (such as the Small Business Administration), state government, and the private sector, additional financial incentives might be formulated.

The Project Impact initiative was viewed by its participants, especially the local governments, as successful. The opportunities it provided for success, and the public awareness of success, were important to them. Naturally, the presence of financial incentives was also a significant element. As with any governmental initiative, if the political support wanes, the initiative will not endure. Such was the case with Project Impact, and while other similar efforts in predisaster planning and preparation have been attempted, none to date have included all of the elements of Project Impact, and none have had similar success.

Three examples of Project Impact communities are presented in the box "Community-Based Hazard Mitigation at Work."

Elements of Success in Federal Strategies

Creating an environment or a culture that has the goal of managing and reducing the measurable impacts of natural disasters has no secret formula for success. The successful strategy adopted by any governmental entity to foster natural hazard risk management will be comprised of those tactics that are employed by anyone seeking to market a way of thinking, or put another way, seeking to have an element of thought present in individuals' and organizations' relevant major decisions. A successful strategy will rely on a suite of actions that support each other and advance the target toward the marketing goal.

The federal government programs outlined above demonstrate many of these tactics. These (and other hazard management programs, which have not been cited here) all use a range of these tactics to a greater or lesser degree. It can be said, however, that some of these tactics stand out in the successes that the programs have attained.

The NEHRP achieved a meaningful measure of success because it so fundamentally and centrally involved target disciplines in the development of the earthquake risk mitigation product—specifically, its development and promulgation of seismic safety building codes. There is an often-used saying in the earthquake risk reduction arena that earthquakes do not kill people, buildings do. Therefore the NEHRP has had as one of its major goals the construction and rehabilitation of buildings and infrastructure that resist earthquake forces. Involving architects, engineers, building code officials, building owners, government officials, and others in the development and dissemination of its guidelines for seismic building resistance was key to the successful integration of the guidance into standard codes.

The flood program, NFIP, made enormously successful use of two elements of successful risk reduction strategies: clear definition of hazard, and the presence of financial incentive/requirement. The NFIP has provided detailed mapping along with supporting hydraulic engineering

COMMUNITY-BASED HAZARD MITIGATION AT WORK

FREEPORT, NEW YORK

The Village of Freeport is located on the southern shore of Long Island in Nassau County, New York, approximately 13 miles east of John F. Kennedy Airport. From the start, Freeport relied on its waterfront location; it began as a fishing port and now is the recreational boating center of Long Island. Development in Freeport over the years resulted in frequent flooding, especially in the commercial district known as the Nautical Mile located in an area known as South Freeport.

In 1983, Freeport began to routinely elevate streets in South Freeport. Because of the cost, the time to complete the elevation of all streets at flood risk was estimated to be decades. To this point, the majority of the financing, between $1–2 million annually, came from the issuance of general obligation bonds. However, periodically, after 1983 Freeport has received financial assistance from both the state and federal Departments of Transportation. By the mid-1990s, many streets had been elevated, including Woodcleft Avenue, which is now a fishing and tourist attraction as well as the most significant commercial business district in Freeport. The Village of Freeport and private citizens raised $10 million to redevelop the Nautical Mile, including the installation of new bulkheads, replacement of overhead electric wires with underground wiring, and construction of new upscale restaurants.

Freeport used funding from six FEMA hazard mitigation grants since 1997 to elevate roads and 23 individual residences. Freeport used grant funds from FEMA's Project Impact to fund public awareness activities, replace and repair bulkheads, conduct a roadway grade raise and drainage improvement project, remove trees that threaten overhead power lines, and install hurricane resistant windows and doors in the Village's emergency operations center.

Source: National Institute of Building Sciences (NIBS), Natural Hazard Mitigation Saves: An Independent Study to Assess the Future Savings from Mitigation Activities, http://www.nibs.org/MMC/MitigationSavingsReport/natural_hazard_mitigation_saves.htm.

DEERFIELD BEACH, FLORIDA

Deerfield Beach, Florida, a coastal community of over 66,000 people, was the first Project Impact community to partner with FEMA. Deerfield Beach is well-acquainted with damages a natural disaster can invoke on a community: having been hit by seven major hurricanes in 75 years, residents knew more hurricanes were statistically almost a certainty. The community's determination to decrease damages sustained from future hurricanes increased after a particularly bad blow from Hurricane Andrew in 1992 followed by near misses of Hurricanes Erin and Opal in 1995.

With guidance from FEMA, Deerfield Beach identified and prioritized mitigation projects that would be most beneficial to the community. One of the first efforts undertaken was retrofitting Deerfield Beach High School, which also serves as a community shelter during emergencies. Hurricane straps were added to the cafeteria and auditorium and wind shutters were placed on all the school's windows. Additional projects included shuttering and disaster-resistant improvements to critical facilities, mentoring, shuttering for single-family residences for senior citizens and low income households, and a variety of public awareness activities.

Deerfield Beach worked very closely with a variety of business partners. The local Home Depot maintained a "Project Impact Aisle," offering products and informational materials on making buildings more disaster-resistant. During the initial 2 years of the program, the store also designated a senior manager as a Project Impact advocate, allowing him to spend

80 percent of his time in support of Project Impact activities. Solutia, Inc. donated hurricane-resistant glass to retrofit the Deerfield Beach Chamber of Commerce. Deerfield Builders Supply was a corporate sponsor of the annual Hurricane Awareness Week and member of the Local Mitigation Strategy working group and donated labor to install windows/doors in the Chamber of Commerce. Marina One Yacht Club built the first hurricane-resistant marine storage facility, designed to withstand 125-mph winds, offering 2,600,000 cubic feet of storage.

Source: FEMA, *Emergency and Risk Management Case Studies Textbook.* http://training.fema.gov/EMIWeb/edu/emoutline.asp.

WILMINGTON, NORTH CAROLINA

Wilmington is a coastal community in North Carolina that was impacted by six hurricanes between 1996–1999. Wilmington is one of the original seven Project Impact communities and the Port of Wilmington, the center of the community's economy, was an active partner in Project Impact.

As part of its commitment to reducing the damage from future hurricanes, the Port of Wilmington invested in a planning and mitigation effort that brought together federal, state, local, and private sector organizations. A risk analysis and four mitigation measures were implemented designed to eliminate or minimize the hurricane loses and ensure business continuity. These measures include securing of gantry cranes and mobile cranes to ensure they do not topple over, nonstructural measures to secure sensitive equipment, structural reinforcement of buildings including wind-resistant roofing, and structural reinforcement of electric power and telecommunications systems.[1]

A study of the benefits of Project Impact to Wilmington's labor market concluded, "The findings of this article are consistent with Project Impact's having a beneficial impact on the labor market of Wilmington. After the initiative, the equilibrium unemployment rate in Wilmington is significantly lower than before the policy intervention, controlling for the effects of other business cycle factors and trends. Additionally, the policy intervention is associated with a significant reduction in the long-run variance of the unemployment rate. The evidence is also consistent with the claim that the disturbance created by a hurricane is less after Project Impact was initiated than before. Taken together, these findings suggest that the activities and coordination efforts associated with Project Impact coincide with improvements in the Wilmington labor market characterized by a lower natural unemployment rate and a reduction of labor market risk. On one hand, these findings may be taken as evidence that Project Impact can improve the performance of a local economy. The results suggest that, at the very least, increased interaction between public and private sectors may be associated with improved labor market conditions."[2]

Source: [1]North Carolina Division of Emergency Management, Mitigation in North Carolina: Measuring Success (2002), Chapter 4: Sustainable Infrastructure and Critical Facilities, [2]Ewing, Bradley T. and Jamie Brown Kruse. The Impact of Project Impact on the Wilmington, North Carolina, Labor Market. *Public Finance Review* 2002, 30:296. http://pfr.sagepub.com/cgi/content/abstract/30/4/296.

information for thousands of US communities. This has provided those communities, insurance agents, bankers, home buyers, and others with a distinct and specific graphic understanding of the flood risks. By coupling the flood-risk identification with financial incentive, it has fostered a cultural recognition of flood risk in the business of home ownership and community housing development.

Project Impact made an intentional effort to bring together all of the tactics that had been shown to be necessary for successful programs. The initiative recognized that a broad, multi-hazard approach aimed at measurably lessening the impact of natural hazards in the United States would need the elements of a campaign: public awareness of risks, financial incentive, political support on multiple levels, a local community focus, recognition of achievements, public-private partnership, and enhancements of hazard definitions. These it attempted with notable success, albeit short-lived (as an initiative without specific legislated authority, it was subject to the priorities of changing political administrators). The program grew rapidly from a handful of pilot communities to hundreds of members over three fiscal periods. The federal government's role was acknowledged as merely one of the partners to a local jurisdiction's efforts, although perhaps one with a unique ability to supply knowledge, help create incentives, and bring a measure of accountability and visibility to the effort.

ASSESSING THE CLIMATE ADAPTATION RESOURCE AND SERVICE LANDSCAPE

Missy Stults
University of Michigan

John R. Nordgren
The Kresge Foundation

Sara Meerow
School of Natural Resources and Environment, University of Michigan

Melisa Ongun
University of Michigan

Rachel Jacobson
University of Michigan

Cory Hamilton
University of Michigan

INTRODUCTION

Local governments undertake a wide array of activities in the course of fulfilling their responsibilities to ensure the health, safety, and general welfare of their citizenry (Allan and Bryant 2012). With the growing realization that climate change is upon us, local and regional practitioners are increasingly looking for assistance with integrating climate into their existing range of responsibilities (Bassett and Shandas 2010; Friend et al. 2013; Smit and Wandel 2006). This often leads local governments to seek assistance and resources from external service providers. As such, it is not surprising to find that a rapidly growing number of information hubs, publications, tools, training materials, and consulting services have been developed to support local and regional efforts to prepare for the impacts of climate change (Bierbaum et al. 2013). To date, however, these technical and informational resources have not been cataloged and there is little understanding of how well these resources are helping local practitioners advance their adaptation efforts.

To help bring clarity to these issues, a research project was jointly undertaken between The Kresge Foundation and graduate students at the University of Michigan to (1) assess the current landscape of climate-adaptation resources and services available to support local and regional adaptation, (2) assess how well these resources align with the needs of local practitioners, and (3) identify strategic opportunities for moving the local adaptation field forward. This piece profiles the results from the first phase of this analysis, highlighting the types of adaptation-related resources and services provided by 85 primarily nonprofit and boundary organizations that support local or regional climate adaptation efforts. This research was conducted by a team of five researchers from the University of Michigan from October 2013–April 2014.

Methods

To undertake this analysis, a list of organizations that provide climate adaptation resources and services to local and regional practitioners was created. This list was developed by combining existing researcher knowledge about organizations providing adaptation-related services to local governments with information obtained through web-based research. Web-based research included scouring adaptation-related clearinghouses such as the Climate Adaptation Knowledge Exchange (CakeX), Georgetown Climate Center, ICLEI-Local Governments for Sustainability, USA's website, and a Google text search for the term "local climate adaptation resources." While identifying organizations, a focus was placed on nonprofit and boundary organizations, organizations that bridge the divide between science and policy-making (Dilling and Carmen Lemos 2011), as well as those organizations creating resources that are utilized by a wide array of local or regional practitioners. This meant that place-based organizations providing resources with little to no applicability or transferability to other local governments were excluded for the sample.

The list of potential organizations was shared with the project's advisory committee, which consisted of representatives from organizations actively working to advance local climate adaptation. The advisory committee reviewed the list, proposing additional organizations and helping prioritize which organizations to include in the analysis. In total, 85 organizations were profiled.

In tandem with the identification of organizations to profile, research was undertaken to determine the type of information to capture during analysis. Local government representatives from five communities across the United States were asked to help identify what type of information would be most useful to capture when reviewing adaptation-related resources and services. The decision to collaborate with local governments was pivotal in that it helped to ensure that the material captured during coding included information relevant to and structured in a way that is reflective of the ways in which local government staff are looking for resources or services. Through this process, information on the following variables was captured: (1) geographic area(s) served by the organization and the specific adaptation-related resource (based on the US National Climate Assessment regions), (2) sector(s) the resource is meant to support (a list of potential sectors was generated with local government representatives), (3) climate impact(s) (if any) on resource addresses (per the US National Climate Assessment), (4) phase(s) of planning process in which resource is used (based on the ICLEI-Local Governments for Sustainability Five Milestone process), and (5) the type of resource (a list of 19 resource types was generated in partnership with local government representatives) (see Table 4.2).

Based on the results of this engagement, a codebook was created to ensure that all researchers were as consistent as possible in their coding of available resource types. The codebook was tested through a process whereby each researcher profiled the adaptation-related resources and services from a single organization. Results were compared across researchers to determine the amount of consistency and the accuracy of the codebook. Based on these findings, minor edits were made to the codebook and a finalized version was given to each researcher. Each researcher was then assigned a number of organizations to review. Results from each researcher's analysis were stored in a centralized dropbox. After an organization's website was reviewed and all publicly available resources were identified, the researcher emailed an Excel file listing all the adaptation-related services or resources found for that organization and how they were coded to a representative from the organization (generally the identified adaptation contact or the organization's director). Each organization was then given an opportunity to review, edit, or add additional resources to this list. Any revisions from organizations were taken and integrated directly into the master research spreadsheet.

TABLE 4.2
Total Number and Percentage of Resources by Type

Type of Resource	Total Number Coded	Percentage
Best practice/case study	605	16.61%
Blog	39	1.07%
Clearinghouse	8	0.22%
Consulting	15	0.41%
Curriculum	20	0.55%
Fact sheet	639	17.55%
Grants	17	0.47%
Library	333	9.14%
Monitoring and evaluation	6	0.16%
Network	45	1.24%
Newsletter	51	1.40%
Planning guide	172	4.72%
Pledge/political activism	18	0.49%
Project in the field	232	6.37%
Scientific report/data source	579	15.90%
Talking points	216	5.93%
Tools	169	4.64%
Trainings/workshops	121	3.32%
Webinar	357	9.80%

Source: Leggett, Jane A., Richard K. Lattanzio and Emily Bruner. 2013. *Federal Climate Change Funding from FY2008 to FY2014*. Congressional Research Service. September 13, 2013. http://fas.org/sgp/crs/misc/R43227.pdf.

Weekly team meetings were held to answer any questions or concerns researchers might have experienced during their analysis. At the end of the analysis, all results were aggregated into a single Excel file and results were tabulated.

It should be noted that this analysis was not meant to be a comprehensive assessment of the climate adaptation resources and service landscape. Instead, this analysis presents an illustrative look at where key organizations working to provide climate adaptation related resources and services are currently focusing their efforts; a snapshot in time, if you will. The intent of this analysis, therefore, is to inform discussion and decision-making among adaptation service providers, philanthropic organizations financing the development of adaptation-related services and resources, and local governments using these resources.

Snapshot of the Adaptation Resource Landscape

For this analysis, 85 organizations, primarily nonprofits and boundary organizations, were coded and 3565 distinct adaptation resources or services were identified. As mentioned previously, for each resource, information was captured on: geographic area(s) served by the specific adaptation-related resource, sector(s) the resource is meant to support, climate impact(s) the resource helps address, phase(s) of the planning process in which the resource would be used, and the type of resource. The type of resource was based on 19 predetermined categories established in partnership with local government practitioners and the project advisory committee. Table 4.2 denotes the total number and percentage of resources available by each resource type coded in this analysis.

It should be noted that 77 resources we tagged as being two types of resources. As such, the total number of resources coded by type equals 3642. The most common type of resources found during analysis included fact sheets (17.55%), best practices/case studies (16.61%), and scientific reports and data sources (15.9%). The least common types of resources found during coding included monitoring and evaluation resources (0.16%), adaptation clearinghouses (0.22%), consulting services (0.41%), adaptation-related grants (0.47%), and adaptation-related pledges or political activism resources (0.49%).

Resources were also coded to determine the phase of the planning process they support. Resources that covered multiple phases of the planning process were coded for all phases to which the resource could contribute. If a resource was relevant to all phases and/or if the resource did not specifically apply to any phase, then the resource was coded as "generic." As Figure 4.5 indicates, the most common resources are those that support vulnerability assessments (1068 resources), adaptation planning (988 resources), and building community support (727 resources).

Resources were also coded based on the climate impact for which they were relevant and the local government sector(s) they support. Figures 4.6 and 4.7 show the total number of resources

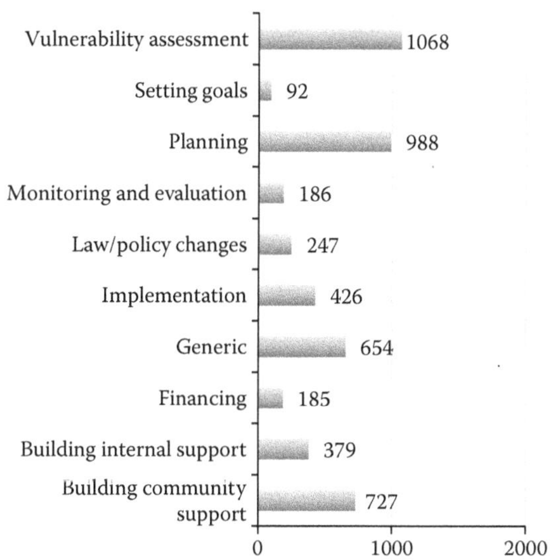

FIGURE 4.5 Total number of resources by phase of the planning process.

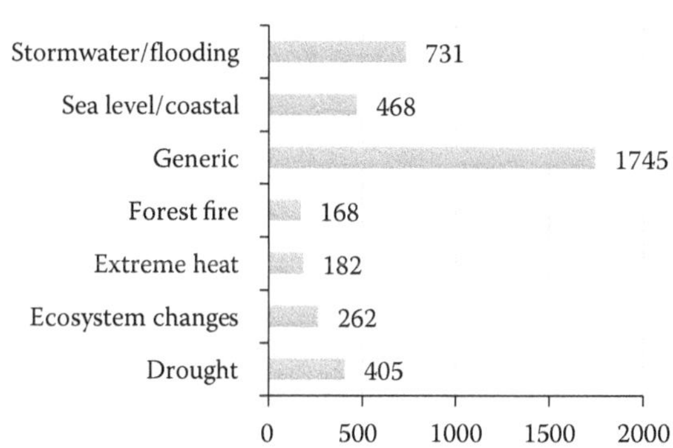

FIGURE 4.6 Number of resources by climate impact.

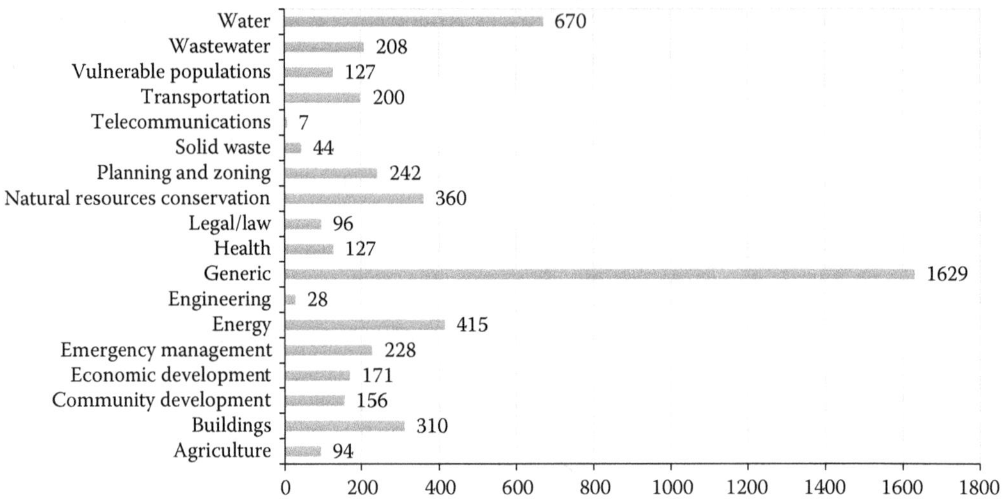

FIGURE 4.7 Number of resources by sector.

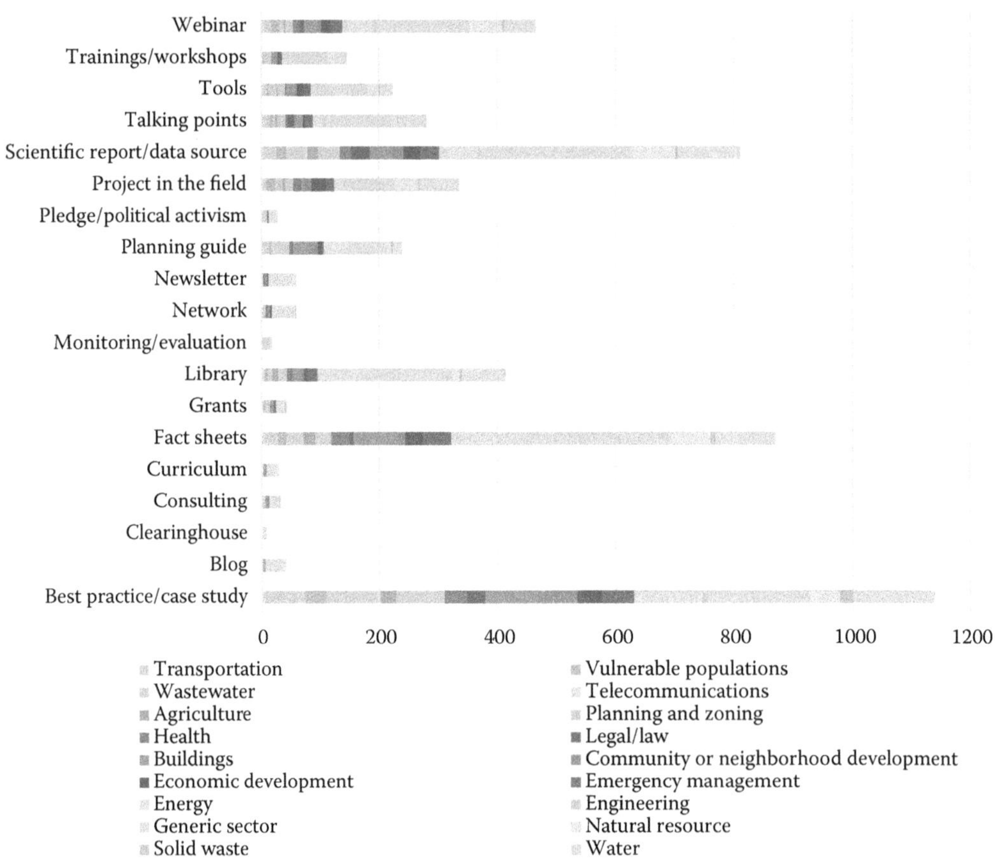

FIGURE 4.8 Type of resource grouped by local government sector that each resource is designed to assist.

coded in each of these categories. For climate impact, the most common entry was for generic* resources (1745 resources) that are applicable regardless of the changing climate condition. The least common climate impacts addressed in resources coded were forest fires (168 resources) and extreme heat (182 resources) (Figure 4.6).

The most common entry for a sector was "generic," meaning that a resource was applicable regardless of the sector (e.g., a resource was applicable across multiple sectors) (1629 resources were coded as generic). The second most common sector to which resources were applicable was water (670 resources). The sectors with the fewest resources or services of relevance include telecommunications (7 resources) and engineering (28 resources) (Figure 4.7).

In addition to counts of resources by type, sector, and climate impact, analysis was completed looking at how these variables combined. Figure 4.8 presents an example of this analysis, showing the results for resource type, grouped by the local government sector each resource is designed to support.

Results

The results of the coding analysis indicate that a significant number of resources and services are available to support local and regional climate adaptation efforts. The most common type of resources available include fact sheets, case studies/best practices, and scientific reports or data sources (Table 4.2). A large percentage of the resources that are available are generic in the sense that they are not specifically focused on a geographical region, a sector, or a phase of the adaptation planning process.

Specifically, these results demonstrate that a significant number of resources exist to support the early phases of the adaptation process, including undertaking risk and vulnerability assessments (1068 resources) and creating a climate adaptation plan (988 resources). However, resources focused on the later phases of the adaptation process, such as implementing strategies, making legal and policy changes, financing adaptation activities, and monitoring and evaluating progress, are the least common. These findings could be representative of the fact that the adaptation field is nascent, particularly compared to the climate mitigation field (Anguelovski and Carmin 2011). The results might also be indicative of the fact that the earlier phases of the adaptation planning process are difficult to undertake, meaning that an array of resources are needed to undertake things such as changes to and integration of highly fragmented governance norms and systems, methodologies for operating with high degrees of uncertainty, and techniques for establishing sufficient political and stakeholder support. Despite this, as local and regional practitioners begin moving from assessing and planning into action, more specific and targeted resources focused on helping them implement key adaptation actions will be needed.

Another interesting finding is the number of resources that are generic in nature. This may suggest that organizations are producing resources they think will be applicable across a wide range of climate impacts, sectors, or geographic regions. However, this could pose a challenge if the resources are too generic to be useful when facing a specific climate impact within a specific sector that necessitates immediate action. A fine line will need to be drawn between having a plethora of generic resources that fail to meet specific needs or having an abundance of specific resources that are only relevant in a small number of communities or circumstances.

Results from this analysis demonstrate that resources focused on addressing stormwater and flooding are fairly common. Resources focused on wildfire, extreme heat, and ecosystem changes are relatively infrequent. Given the serious public health ramifications of these changes, it is surprising that so few of the resources identified focus on these key sectors (O'Neill et al. 2009; Wardekker et al. 2012).

* The term "generic" was used during coding to indicate that a resource was not specifically focused on a geographical region, a sector, or a phase of the adaptation planning process.

Finally, findings based on the local government sector that adaptation-related resources or services are being designed to support found that water- and energy-related sectors are well resourced. However, telecommunications, engineering, and solid waste are underresourced sectors. As the adaptation field continues to grow and evolve, there may need to be increasing focus on providing resources and services to nontraditional sectors such as these, particularly if the goal is to mainstream climate considerations into existing operations (Biagini et al. 2014; Friend et al. 2013).

Conclusion

This preliminary research was undertaken to provide a snapshot of the current state of the local and regional climate adaptation resource and service landscape. From these results, it is clear that the adaptation field is growing at a rapid pace, producing a number of disjointed and duplicative resources. Despite this, major gaps in the types of resources and services being provided still exist. One notable example is the dearth of resources focused on financing climate adaptation activities, despite the knowledge that financial constraints are frequently cited as the largest barrier to moving forward with adaptation activities (Carmin et al. 2012; Eakin et al. 2011; Hunt and Watkiss 2011; Solecki et al. 2011).

Going forward, we hope the results of this research help to inform the future state of the climate adaptation service and resource landscape. In particular, we hope that resource and service providers will review this work critically, identifying where gaps and strategic opportunities exist to move the climate adaptation field forward. Additionally, we encourage more work to be done to understand the unique needs of local and regional practitioners and to ensure that the resources and services being provided are responsive to their needs and created in formats that are useful and understandable to key stakeholders.

This research aimed to profile the services and resources available to support local and regional climate adaptation. No effort was made to evaluate the utility of the services and resources that currently exist. Going forward, work should be done in tandem with local practitioners to determine how useful and usable existing resources and services are and what types of edits and changes would be warranted to make existing resources of higher utility. This work should not be undertaken by researchers or resource producers but should instead be undertaken in tandem with local government stakeholders since they are the desired end users of the resources and services profiled in this effort. Finally, the results outlined in this document should be updated and shared openly to ensure that the resources and services being created meet existing and future needs of local and regional adaptation practitioners.

REFERENCES

Allan, P., and Bryant, M. (2012). Resilience as a framework for urbanism and recovery resilience. *Journal of Landscape Architecture*, 6(2), 34–45.

Anguelovski, I., and Carmin, J. (2011). Something borrowed, everything new: Innovation and institutionalization in urban climate governance. *Current Opinion in Environmental Sustainability*, 3(3), 169–175. doi:10.1016/j.cosust.2010.12.017.

Bassett, E., and Shandas, V. (2010). Innovation and climate action planning. *Journal of the American Planning Association*, 76(4), 435–450. doi:10.1080/01944363.2010.509703.

Biagini, B., Bierbaum, R., Stults, M., and Dobardzic, S. (2014). A typology of adaptation actions: A global look at climate adaptation actions financed through the Global Environment Facility. *Global Environmental Change*, 25, 97–108.

Bierbaum, R., Smith, J. B., Lee, A., Blair, M., Carter, L., Chapin III, F. S., and Stults, M. (2013). A comprehensive review of climate adaptation in the United States: More than before, but less than needed. *Mitigation and Adaptation Strategies for Global Change*, 18(3), 361–406. doi:10.1007/s11027-012-9423-1.

Carmin, J., Nadkarni, N., and Rhie, C. (2012). *Progress and Challenges in Urban Climate Adaptation Planning: Results of a Global Survey* (p. 33). Massachusetts Institute of Technology, Cambridge, MA.

Dilling, L., and Carmen Lemos, M. (2011). Creating usable science: Opportunities and constraints for climate knowledge use and their implications for science policy. *Global Environmental Change*, 21(2), 680–689. doi:10.1016/j.gloenvcha.2010.11.006.

Eakin, H., Eriksen, S., Eikeland, P.-O., and Øyen, C. (2011). Public sector reform and governance for adaptation: Implications of new public management for adaptive capacity in Mexico and Norway. *Environmental Management*, 47(3), 338–351. doi:10.1007/s00267-010-9605-0.

EPA.gov. (2015). Federal and EPA Adaptation Programs. Accessed January 4, 2015. Available at http://www.epa.gov/climatechange/impacts-adaptation/fed-programs.html.

Executive Office of the President. (2013). The President's Climate Action Plan, June 25. Available at http://www.whitehouse.gov/sites/default/files/image/president27sclimateactionplan.pdf.

Friend, R., Jarvie, J., Reed, S. O., Sutarto, R., Thinphanga, P., and Toan, V. C. (2013). Mainstreaming urban climate resilience into policy and planning; reflections from Asia. *Urban Climate*, 1–14. doi:10.1016/j.uclim.2013.08.001.

Georgetown Climate Center. (2014). Find Out Which U.S. States Are Making Progress in Preparing for Climate Change, October 9. Available at http://www.georgetownclimate.org/find-out-which-us-states-are-making-progress-in-preparing-for-climate-change.

Hunt, A., and Watkiss, P. (2011). Climate change impacts and adaptation in cities: A review of the literature. *Climatic Change*, 104(1), 13–49. doi:http://dx.doi.org/10.1007/s10584-010-9975-6.

Leggett, J. A. (2013). Climate Change Science: Key Points. Congressional Research Service, September 10. Available at https://www.hsdl.org/?view&did=745266.

Leggett, J. A., Lattanzio, R. K., and Bruner, E. (2013). Federal Climate Change Funding from FY 2008 to FY 2014. Congressional Research Service, September 13. Available at http://fas.org/sgp/crs/misc/R43227.pdf.

Mooney, C. (2014). Green Goals for Federal Agencies. *Washington Post*, November 1.

O'Neill, M. S., Jackman, D. K., Wyman, M., Manarolla, X., Gronlund, C. J., Brown, D. G., Brines, S. J., Schwartz, J., and Diez-Roux, A. V. (2009). US local action on heat and health: Are we prepared for climate change? *International Journal of Public Health*, 8. doi:10.1007/s00038-009-0071-5.

Podesta, J., and Fugate, C. (2015). A New Flood Standard for Federal Investments Will Help Communities Weather Climate Change, Extreme Storms. *The White House Blog*, January 30. Available at https://www.fema.gov/blog/2015-01-30/new-flood-standard-federal-investments-will-help-communities-weather-climate-change.

Smit, B., and Wandel, J. (2006). Adaptation, adaptive capacity and vulnerability. *Global Environmental Change*, 16(3), 282–292. doi:10.1016/j.gloenvcha.2006.03.008.

Solecki, W., Leichenko, R., and O'Brien, K. (2011). Climate change adaptation strategies and disaster risk reduction in cities: Connections, contentions, and synergies. *Current Opinion in Environmental Sustainability*, 3(3), 135–141. doi:10.1016/j.cosust.2011.03.001.

Wardekker, J. A., de Jong, A., van Bree, L., Turkenburg, W. C., and van der Sluijs, J. P. (2012). Health risks of climate change: An assessment of uncertainties and its implications for adaptation policies. *Environmental Health: A Global Access Science Source*, 11(1), 67. doi:10.1186/1476-069X-11-67.

The White House. (2013a). State, Local, and Tribal Leaders Task Force on Climate Preparedness and Resilience, November 1. Available at http://www.whitehouse.gov/administration/eop/ceq/initiatives/resilience/taskforce.

The White House. (2013b). Executive Order—Preparing the United States for the Impacts of Climate Change, November 1. Available at http://www.whitehouse.gov/the-press-office/2013/11/01/executive-order-preparing-united-states-impacts-climate-change.

The White House. (2014a). Fact Sheet: 16 US Communities Recognized as Climate Action Champions for Leadership on Climate Change. Office of the Press Secretary, White House, December 3. Available at http://www.whitehouse.gov/the-press-office/2014/12/03/fact-sheet-16-us-communities-recognized-climate-action-champions-leaders.

The White House. (2014b). President's State, Local, and Tribal Leaders Task Force on Climate Preparedness and Resilience: Recommendations to the President, November. Available at http://www.whitehouse.gov/sites/default/files/docs/task_force_report_0.pdf.

The White House. (2015). Fact Sheet: Taking Action to Protect Communities and Reduce the Cost of Future Flood Disasters, January 30. Available at http://www.whitehouse.gov/administration/eop/ceq/Press_Releases/January_30_2015.

5 Elements of a Successful Adaptation Plan: Obstacles and Constraints

INTRODUCTION

This chapter presents two essays by two individuals, the first by Ron Brunner and John Nordgren and the second by Ron Brunner alone, who have been involved in community adaptation practices for many years. Ron Brunner is a consultant and Professor Emeritus at the University of Colorado, Boulder, and is a leading researcher in the field of climate change adaptation. John Nordgren is Senior Program Officer, Environment Program, at The Kresge Foundation and a major supporter of research on climate change adaptation including supporting the ongoing research project that produced three of the community case studies included in Chapter 6.

The first essay examines the results of a February 2012 workshop that brought together 80 practitioners from across the country involved in climate change adaptation at the community level. Brunner and Nordgren conclude that, "Looking back over the workshop, the individual and collective experiences of practitioners in climate adaptation can be characterized as an evolutionary process guided by a shared goal—*reducing the vulnerability of natural and human systems to climate change or increasing their resilience.* The community understands the importance of taking action despite an incomplete understanding of what success in meeting our goal might mean."

The second essay presents Ron Brunner's critique of the "Adaptation" chapter in the *Third National Climate Assessment* published in May 2014 and concludes that, "harvesting experience from more communities that have made substantial progress in reducing their losses and vulnerability to climate change, even if they have not taken projected climate changes explicitly into account; evaluating what works from the bottom-up perspectives of these communities; and evaluating systems for diffusing and adapting information about what works from each community to others. Such research is not the only resource involved in expediting progress, but it can help."

Also included throughout this chapter are four boxes that briefly examine the work done by the American Society of Adaptation Professionals (ASAP), profile the scope of the 100 Resilient Cities pioneered by the Rockefeller Foundation (100RC) program, summarize the findings of the 2014 Global Cities report developed by CDP, and describe the goals and objectives of the ongoing National Disaster Resilience Competition sponsored by the US Department of Housing and Urban Development (HUD) in collaboration with the Rockefeller Foundation.

All of the materials in this chapter could serve as valuable resources for community leaders as they develop and implement climate change adaptation programs.

CLIMATE ADAPTATION AS AN EVOLUTIONARY PROCESS: A WHITE PAPER

Ronald D. Brunner
John R. Nordgren* April 6, 2012
Kresge Foundation

Introduction

As recently as five years ago, climate change adaptation was regarded as taboo—akin to giving up on the climate crisis. It is fair to say that the nascent field of climate change adaptation has since gained widespread acceptance. Despite continuing debate in the US, public and professional acceptance of the need to address the impacts of climate change proactively has been increasing. As recently as three years ago, few cases of applied climate adaptation could be found. Now many efforts are working more or less independently to develop strategies and practices that define climate adaptation and contribute to successful outcomes. Much of what this emerging community of practitioners is doing is new, even if it is based on established practices. It has been likened to building a bike while riding it.

In this larger context, The Kresge Foundation over the past three years has deliberately seeded a number of promising projects in climate adaptation. The Kresge portfolio is intentionally diverse, representing a variety of approaches, topics, strategies, geographic locations, and scales of operations. To expedite progress, this is a good time to bring together representatives of these and other adaptation projects, to assess how they have progressed, and to make that assessment available to the larger community. Together with Tonya Graham of the Geos Institute (a Kresge grantee) and Peter Szabo of Bloomingdale Management Advisors, we planned and convened a workshop in Portland, Oregon, early in February, 2012. This white paper summarizes what we heard and learned at the workshop.

The 80 practitioners participating in the workshop represented organizations with more than 40 projects operating in nearly all 50 states.† The projects represented were as diverse as their geographic spread. For example, some focused on climate adaptation in small communities, others on larger regions, and still others offered information resources or other tools for climate adaptation practitioners. Some focused on helping natural systems, human systems, or both in adapting to climate change. Some addressed a single vulnerability, such as migration corridors or sea-level rise, while others took a more holistic approach to community resilience. Many but not all projects were grantees of The Kresge Foundation. The workshop was designed to (1) take stock of what has worked, what hasn't, and why; (2) compare future plans and needs; and (3) share some emerging tools and resources.‡

Four breakout sessions gave groups of 11 or 12 practitioners each an opportunity to share and compare what they had learned and to consider the future plans and needs of their projects. We employed scribes to take notes in these sessions. For the fifth breakout session, practitioners self-organized into groups according to their priority opportunities for networking or collaboration across projects. Four major tools and resources were presented and discussed in plenary sessions interspersed among breakout sessions. In the opening plenary address, Ron Sims shared his insights as County Executive for King County, WA, a recognized pioneer in climate adaptation. In the

* John Nordgren is Senior Program Officer, Environment Program, The Kresge Foundation. Ron Brunner is a consultant and Professor Emeritus, University of Colorado, Boulder. This white paper is based on the Kresge Grantees and Practitioners Workshop on Climate Change Adaptation that met February 7–9, 2012, at the Hotel Monaco in Portland, Oregon.

† Participants in the Portland workshop are listed at the end of this paper. The workshop and this paper would not have been possible without their open and dynamic engagement and many contributions.

‡ The four tools and resources featured were Arizona State University's Decision Theatre (http://dt.asu.edu/); the Climate Adaptation Knowledge Exchange or CAKE (http://www.cakex.org); the American Society of Adaptation Professionals (ASAP); and the Climate Access Network (http://www.climateaccess.org/). For more information on the workshop, including participant bios, their project summaries, and resources and agenda, see http://kresge.org/programs/environment/adaptation-climate-change/workshop2012.

closing plenary, a group of "listeners," as assigned, reported on their personal experience of the workshop. We concluded with comments from participants at large.

In the following pages, we attempt to distill the main findings and pull them together with examples. We have relied on our notes and the scribes' notes (sometimes quoted, often paraphrased), as well as our direct experience of the workshop.

Main Findings at a Glance

Looking back over the workshop, the individual and collective experiences of practitioners in climate adaptation can be characterized as an evolutionary process guided by a shared goal—*reducing the vulnerability of natural and human systems to climate change or increasing their resilience.* The community understands the importance of taking action despite an incomplete understanding of what success in meeting our goal might mean. As we move through this evolutionary process together, it is critical to learn from this workshop and similar opportunities.

Many of the projects and organizations at the workshop have moved beyond an initial emphasis on downscaling climate change projections and incorporating them into plans. Now they face the task of implementing the plans made. This brings to the fore many human factors, including political, social, and economic issues that compound more familiar uncertainties about the timing and severity of local or regional climate impacts. Because of these uncertainties, progress depends on trial-and-error learning. Collectively, practitioners have been learning the politics of finding common ground, which are necessary to mobilize resources, including the political will to implement plans, and thereby to sustain progress toward our shared goal.

Looking ahead, we need to persist in maintaining a diversity of trials, evaluating what works and what does not, and adapting successful models through persistence, networking, and collaboration.

Progress

The participants' discussion during the closing plenary session encapsulated the workshop's spirit. One participant urged us to "be bold" in taking action in the face of uncertainty. Even if it does not work out, she said, we can learn from it; and once the models of success are out there, other people can adapt them. Citing Ron Sims' inspiring opening address, another participant also urged us to be bold and to have confidence in ourselves. One of the seven listeners reported he was "blown away" by the consistency of what had turned up in taking stock of progress by the projects, especially the need to rely on personal engagement and relationships to advance adaptation. He concluded that "we know how to do this." Records from the workshop support these observations.

What worked? The short answer is that practitioners acted pragmatically. To move ahead, project leaders found it necessary to use whatever resources and flexibility were available to meet the needs of local or regional communities in different and changing circumstances. In particular, they learned to:

- Bring the right people to the table. They vary from place to place, but generally are people and groups who can make a difference in moving ahead. They include anyone who might help community members most vulnerable to climate change, often the disadvantaged, and local champions who have trust and connections in the community. They also include scientists, lawyers, and other experts; local, state, and federal officials; people in business; and youth.
- Engage them "where they are, not where we are," as one participant put it. They may be most concerned about taxpayer savings, fiscal conservatism, etc., and indifferent or hostile to climate change. If so, postpone discussion of climate change at the outset, and instead "appeal to their direct experience" of climate change impacts. "Help them figure out what they can actually do" to adapt while addressing their priority concerns at the same time.

- Rely on maps, photos, and other means of visualization in communicating vulnerabilities and possibilities. Anecdotes and stories based on case studies also work in making the abstract concrete. Back these up with sound science as circumstances permit. It is not enough to demonstrate vulnerabilities; catalyzing action depends on specific means to reduce vulnerabilities to species, landscapes, human life and limb, buildings, and anything else we value with the community.
- Encourage collaboration within and across groups in finding common ground sufficient for a community to act on implementation. Among the means that have worked are advisory boards, steering committees, workshops, various kinds of networks, and mainstreaming climate change into the existing, ongoing operations of communities or agencies. Collaboration often requires translating information across community groups, disciplines, agencies, and levels.
- Build personal relationships, credibility, and trust. This human capital is the foundation of effective communication and community action. It is not a given at the outset, but typically depends on a non-partisan, open, and transparent approach based on facts. Over time, it is the cumulative effect of incremental steps toward a long-term goal that succeed, as assessed by the partners involved. What may have been "we-they" at the outset becomes "we" partners.

The extent of convergence on these pragmatic practices is remarkable because the projects all began with different approaches at different times and places. Evidently, convergence on this pragmatic, sometimes opportunistic, engagement stems from a shared need to find common ground in each context to make progress. The long-term goal in each instance was often assumed and occasionally expressed, but not always as part of a climate adaptation strategy.

Nevertheless, judging from exceptions and qualifications, nothing works everywhere. For example, while an appeal to direct personal experience often works, there was a report of "a farmer [who] knows blueberry season starts earlier but [still believes] 'global warming is a government hoax.'" Evidently, "it is not a case of one size fits all." One breakout group reported "a tension between tailoring messages to local context and adhering to the meta-message" of climate science; they distinguished context-driven approaches from science-driven approaches. Another group reported that "there is no one best way to do it." For example, "when collaboration doesn't work, sometimes a lawsuit will work."

How do you know it worked when it did? Every project apparently did some evaluation, often informal, to explain its progress and justify new proposals for support if nothing else. While answers to the question of "what worked?" tended to converge, answers to "how do you know?" did not. Instead, this question turned up a variety of answers. Consider some examples paraphrased or quoted from the scribes' notes:

- From "our last big evaluation," a survey of information-resource users, one project concluded that "empirical data may never work to establish success metrics or define resiliency"; we may need to rely instead on what users used, qualitative observations and stories.
- According to another information-resource provider, "there are indicators that you can look at to see if you are on the right track, e.g., in the coral reef world, if corals aren't happy they turn white." Begin observing as actions are taken; don't wait 20 years and hope it is working.
- For another project, "one way to measure success is when people start attending meetings and workshops who were never there before." Similarly, "sometimes somebody picking up your phone call is a success."
- Another project found success in "getting political folks in the room with scientific folks for workshops, etc." For example, "out of a couple of large regional workshops, they identified some very valuable springs and seeps that are vulnerable to climate change."

- That project also found success "not so much in a huge adaptation plan" but in the way managers were "incorporating climate change in the back of the mind" and then into "those smaller [day-to-day] decisions that add up."
- In a project providing legal resources, a practitioner noted that "Success for my group is a change in law." Another project operating at the state level is hoping for success in local implementation of changes in state regulations that it helped put in place.
- Finally, success in Louisiana was halting large-scale federal maladaptive practices and projects; one project had a role in persuading the Army Corps of Engineers to de-authorize a planned levee. Similarly, "Success in Arizona is holding the line against moving backwards."

The first and last examples among others in the record indicate that the shared long-term goal is indeed reducing losses to climate change or increasing resilience. Taken together, the examples illustrate the variety of observations and criteria used to evaluate incremental steps toward the long-term goal in the absence of quantitative metrics. The criteria ranged from making contact with partners to blocking maladaptations that would have increased future vulnerabilities. The observations relevant to these "benchmark" or "milestone" criteria were consistently more qualitative than quantitative.

The examples shed more light on climate adaptation as an evolutionary process distributed among many projects that began with different approaches in different contexts. One participant described her information resource as "an evolving process; we need to keep making improvements and changes based on [user] needs and desires." Another described climate adaptation as "muddling through" and "coping and doing your best" because "no one is an expert." A third summed up the implications for evaluation: "Short-term, intermediate successes are valuable; with long-term goals, look at incremental steps along the way. When you achieve one, it's a success."

Why did it work when it did? The short answer is that projects succeeded in making incremental progress when and where they could adapt their limited resources to circumstances in a community, including the multiple interests and needs of community members. This did not always happen, as suggested by the challenges in the next section. Success also depended on circumstances beyond a project's resources and control. A number of practitioners agreed that "at the end of the day, [communities] need political will" to act.

Beyond a project team's own efforts, circumstances can often provide political will to act, most notably crises and emergencies from acute impact events. Hurricane Katrina and BP's oil spill in the Gulf were mentioned as examples of this reality. As one practitioner put it, "sometimes it takes the crisis: [something like] Katrina was known [in advance]; it was just a matter of time before that occurred." Another reported that the "oil spill created an opportunity to build a 'restoration economy'... More natural disasters are coming. We are not tapping into 'climate change' at all" to motivate our partners. Still another observed that a crisis is often the "tipping point" for making "practitioners willing to engage" in climate adaptation. In general, "extreme events in the next 10–20 years provide the opportunity to call [partners] to action." However, others were uncomfortable with a crises-driven or reactive approach to climate adaptation. In their view "adaptation by its nature is proactive."

Chronic impact events, which are characterized by less urgent circumstances than a sudden crisis, can motivate as well. For example, perennial water supply issues in the arid American West were described as a "lever" for adapting to climate change. Fear of legal liability motivated airport officials to address the effects of extreme weather events related to climate change. Similarly, federal and state mandates have helped motivate officials to consider climate change and collaboration on the ground in various places where projects are working. Some officials feel accountable for complying with mandates, and appreciate our help in doing so. However, "be careful what you wish for," cautioned one practitioner: "Mandates have led to massive resource expenditure, lawsuits, and potentially very little return on investment." Nevertheless, capitalizing on existing circumstances often allows a project to increase the return on investment of its own limited resources.

In addition to motivations, favorable circumstances may include latent resources waiting to be tapped, such as local knowledge. If tapped successfully, practitioners generally agreed, each incremental step ahead tends to generate more resources that enable more ambitious steps. In an early plenary session, Tonya Graham illustrated this point with the story of an aid worker in Vietnam. With only a small budget, he took on the massive problem of child malnutrition. In the beginning he hired a small group of village women to measure the height, weight, and other characteristics of poor children, and discovered a "bright spot": Some children were doing rather well, evidently because of a better way of feeding them through the course of a day. Showing others the better way, he eventually helped the Vietnamese improve the nutrition of more than two million children.* Tonya urged us to "find the bright spot" and "build on it."

Finally, one comment explained what works as *the creative integration of project efforts with specific local circumstances*: "Adaptation plans are best planned locally for what the specific community needs, not necessarily applying national or even regional plans to every community. We have to be incremental and creative and adaptive in our work to be successful." This comment reflects a general consensus among participants on what works.

Challenges

Practitioners offered few specific examples of what did *not* work in their past experience. However, they did identify an array of challenges they had encountered. The most formidable challenge was nationally-organized political opposition to climate adaptation. In particular, it was reported that "people are turning up at meetings and disrupting them" in a number of local communities. Apparently, participants did not consider disengaging from these communities. However, several tactics for dealing with disruption or opposition were tried or recommended:

- Rely on sequencing: "Build relationships with the right people, then take it public, rather than blowing the trumpets and beginning a public process right away. Find the local champions" first.
- Similarly, beware of the "big regional planning process" that can easily become "a target for disruptions" by organized opposition. It is politically prudent to let sleeping dogs lie until we feel strong enough to wake them up.
- Build an open, inclusive alliance that benefits community members and engages them on their own terms. In Montana, one such alliance stymied a property-rights group by including influential ranchers in the community.
- "Talk about what is relevant to [the opposition]; don't try to shove climate-change science down their throats. Interaction should be a two-way street." Among the things relevant to them are taxpayer savings and fiscal conservatism, particularly in distressed economies.

The general strategy was strength through inclusiveness. As one practitioner advised, seek "strength in numbers, in the diversity of people coming to the table" including "groups who have a vested interest" that might be served by climate adaptation; and "be more engaged with one another"... But no strategy can guarantee a win-win outcome based on a consensus of all major interest groups. Sometimes to move ahead it is necessary to settle for a majority, with a minority of self-assessed losers. Unanimity is unworkable, equivalent to giving every person and group a veto over taking action.

The projects also encountered less organized opposition from people who preferred old ways to new. Indigenous people in the Mississippi Delta, for example, preferred to have more levees built to protect their way of life. One project's pragmatic response was to stay engaged while looking for a

* *The Vietnam Story*, narrated by Jerry Sternin, develops the important concept of "positive deviance." See http://www.positivedeviance.org/about_pd/Monique%20VIET%20NAM%20CHAPTER%20Oct%2017.pdf.

local champion, but meanwhile to work with more open partners, the Army Corps of the Engineers and the Louisiana governor's office, to block a levee that would increase vulnerabilities to climate change. One participant suggested bypassing "hard places" in favor of communities that are ready: "Recognize where places are not ready for planning...have an informal conversation [but] recognize the limits" to what might be accomplished there. A different view was that "people matter more than the place" in some circumstances—people like "a trusted broker." Another comment implied waiting for a crisis to open up the situation: "People will sometimes willfully misunderstand... Maybe it's not correct to believe that we can educate our way out of this. Maybe people need to see their ideas fail."

In addition to opposition or indifference to climate adaptation, discussions in the breakout groups turned up other constraints on projects' ability to adapt to local needs and circumstances. The organizers took it for granted that limited funding is a pervasive and persistent constraint. (However, according to one workshop participant, too much funding after Hurricane Katrina exacerbated political divisions, and much of the funding was wasted without sound plans.) Our attempt to focus attention on other resources and other constraints turned up the following challenges to moving ahead.

- Established funding models were not always consistent with the nature of the problem. In some models, a successful proposal must promise measurable deliverables in a few years. But climate adaptation often requires more flexibility to respond to unforeseen problems and opportunities. And climate adaptation is a much longer-term process that engages many partners who share control over, and thus responsibility for, many interim outcomes along the way.
- Projects that overlap in a locale or region often operate as institutional "silos" despite similar goals. They tend to answer primarily to their separate sources of funding rather than to needs for progress in implementation on the ground, which typically include collaboration. Sometimes they speak with different voices, confusing or alienating the community. Sometimes they compete with each other for status, funding, and other resources.
- Communicating uncertainty about climate change risked "losing your audience." Also, it helped "opposition [that] has a clear message." Another view held that the "challenge is not uncertainty, but the public fixation with uncertainty, which leads to paralysis." We have "certainty that sea level is rising...there is actionable knowledge" to make the case for action to decision makers. Similarly, "scientific uncertainty is a misnomer. It's clear that sea level is rising and that temperature is increasing. We just don't know how much."
- Measuring progress in climate adaptation was a challenge. On the one hand, "operationalizing [costs and benefits] is so critical" to demonstrating progress. On the other, "it is hard to make the case for adaptation because it is hard to measure it; if it is successful, the metrics are not simple." The significance of water savings, for example, depends on the context. Standard measures are also problematic if "there is no one right way to adapt or to respond to climate change."
- Evaluating progress was often confused with measuring progress, which is not equivalent to it. As a process, climate adaptation is open-ended with no "end-point" to mark success or failure. Multi-valued interim outcomes are the result of multiple factors that interact. Thus it is difficult to attribute any one outcome to any one factor. Often success is "something that didn't happen"; it is not observable even if it can be estimated. For such reasons, "stories may mean more than the quantitative data." For practical purposes, "We know success when we see it."*

* For more on this challenge, see Ronald D. Brunner, Evaluating Progress in Adapting to Climate Change (November 11, 2011), a revision of a paper presented at *Practical Solutions for a Warming World: AMS Conference on Climate Adaptation*, July 18–20, 2011, Asheville, NC. Copies are available from the author.

- Networking was another challenge. For example, there was "Too much information and diffuse information and a lack of guaranteed quality of information," noted one practitioner. For another, "too much information on the Web" means "reinventing the wheel." Several practitioners had been *un*aware of most of the work represented in the workshop. Another used this workshop to make long-delayed personal contacts in four similar organizations.

Finally, as we face such challenges, we should not confuse labels with progress toward the long-term goal. One practitioner noted that "Sometimes adaptation can be in name only, equivalent to rebranding." Others in a number of projects substituted other labels for "climate adaptation" to describe their work and to get on with it. Federal funds are available for climate adaptation, but are not necessarily labeled as such.* And migration from areas vulnerable to climate change is a step toward adaptation whether or not it is enough or understood or labeled as such.

Looking Ahead

Much has been accomplished in the past few years. But looking ahead, much more remains to be done. This includes building on what has worked, dealing with the challenges we face, and exploring relatively new territory in climate adaptation. In the fourth breakout session, project representatives compared their current plans and needs in groups, and then self-organized into new groups to discuss the main opportunities for networking and collaboration across projects. We expect some of those discussions are ongoing and ultimately will bear fruit. Meanwhile, a number of the possibilities for working together were raised but not developed during the workshop. These might be organized and illustrated in three categories:

First, workshop participants were interested in organizing more workshops or conferences to mitigate institutional fragmentation, to make better use of resources represented by projects in this workshop and others like them, and to build a community providing mutual support. The underlying assumptions seemed to be that this would facilitate the flow of insights and information, supplement existing Web-based networks or stimulate new ones, and help to build a community of practitioners in climate adaptation. An annual workshop to follow-up this workshop was mentioned frequently, with The Kresge Foundation or the new American Society of Adaptation Professionals suggested as possible sponsors. Also mentioned were workshops to bring together projects that address climate impacts in the same region (e.g., the Pacific Northwest) and projects that work on a single climate impact (e.g., sea-level rise) in different locales and regions. Another suggestion was to bring together climate adaptation projects working on the ground with public officials and others working at the state and federal levels.

Second, workshop participants were interested in additional informational resources. These are typically impractical for an organization or community to compile on its own. Compilations of resources available in the following areas could assist many projects in improving and expanding their work on the ground:

- Economics, especially a compilation of methods to project the costs and benefits of action on a climate adaptation policy compared to inaction. Such projections are often necessary for local and state officials to buy into climate adaptation.
- Financial mechanisms, specifically a compilation of multiple sources for funding climate adaptation, including "private insurance, municipal bonds, state-level insurance, crop derivatives, flood insurance, FEMA, socially responsible investment" and business partners.

* For example, see Edward A. Thomas, Esq. et al., *The Patchwork Quilt: A Creative Strategy for Safe and Long Term Post-Disaster Rebuilding* (Natural Hazard Mitigation Association, June 2, 2011). http://xa.yimg.com/kq/groups/23535088/1306835394/name/Patchwork%20Quilt%20final%20for%20HUD-FEMA-SJ%20pdf%206-2-11.pdf. This updated guide to federal post-disaster resources was initiated in 1983.

- Insurance, especially "third-party research, consulting, and fact sheets on the insurance costs of climate change. This research should be amenable to place-based advocacy (local/regional statistics). Currently, the market is not valuing risk accurately."
- Legal assistance, especially on what *can* be done. "People don't know what they can do within legal constraints. They are afraid of being sued. It only takes one angry property owner. Federal government doesn't necessarily help, especially for regulatory decisions."
- Politics, especially a compilation of information to match the resources and erroneous rhetoric of nationally-organized opposition to local planning. These might include "templates/checklists for passing ordinances, fact sheets, talking points, [and other] resources" to support fact-based decision-making.
- "Mainstreaming," especially integrating climate-adaptation considerations into hazard mitigation plans and actions that already exist in many local governments. Such considerations could also be integrated into local business operations by including them as partners.

In addition, the use of stories was often advocated for a variety of reasons. One is to engage, motivate and inform people on the ground: "People remember stories." Others are to "magnify successes" and to "market the success of tools" for "scaling up and out successful projects" and for "bringing media attention to success stories" in climate adaptation. Project leaders are best prepared to write their own case studies for other practitioners, subject to verification by independent third-parties.

Third, workshop participants were interested in exploring a variety of larger topics with potentially important but less immediate payoffs for moving ahead. One of them came in the form of a question often raised but seldom answered: "How do we define success in climate adaptation?" Similarly, "What is our positive, long-term vision?" Such a vision might displace the "gloom and doom" scenarios that have held back progress in climate-change adaptation and mitigation. Perhaps for similar reasons, there was also interest in "the big picture" and our "theories of change." Openness to different perspectives on these and other issues might address one participant's concern about premature closure: "There is a tendency to work with the same types of people who work in the same way on the same things. Maybe that doesn't move the ball forward as much as we want. Maybe we need more collaboration with the private sector or people with different perspectives on these issues."

Conclusion

The closing plenary session underscored participants' needs for reassurance that we are all in this together, for confidence to act despite uncertainties, and for solidarity to persist in moving ahead toward the long-term goal. Thus what we feel may be just as important as what we know. *What we know* (or think we know) has been summarized in previous sections. The listeners, tasked to share summary observations, added highlights from their own experience of the workshop. Other participants then volunteered comments that emphasized *what we feel*, a theme that emerged in the later breakout sessions.

The first listener emphasized the importance of leadership, particularly leadership that reaches out to those parts of the private sector predisposed to advance climate adaptation. Another listener recommended a comprehensive view of our efforts in climate adaptation, one that includes our future as well as the past. She was impressed by the "trans-disciplinary sense of community" and "level of organization" we have achieved just in the last few years. For the future, another listener recommended recruiting more groups and additional voices to increase diversity in the climate-adaptation community. The final listener's report raised again that important question, "What constitutes success in climate adaptation?" She went on to suggest that we integrate our work in climate-change adaptation with climate-change mitigation.

Among those who volunteered comments, one practitioner reported that the workshop "validated" what he had been doing on the ground. He and others found "a support group" in other participants. Another observed that many workshop participants face "the same stressors" that she does. Still another affirmed that other workshop participants are struggling with "the same issues" he faces. The workshop raised more questions than he had asked before, and drew attention to the importance of working within the political structure. Another participant had been bogged down at home, and appreciated the workshop for opening up new possibilities. Another reported that she went from fatigue in the struggles at home to enthusiasm for the tasks ahead during the workshop. Several others commented on the wealth of additional resources represented in the room, with the implication that we have only just begun to tap it for our mutual benefit.

We affirm the need to work more strategically and intentionally to find common ground to make progress in a society with pluralistic interests and fragmented institutions, including many silos. The alternative is failure to integrate plans for climate adaptation with diverse community needs, leading to shelved plans, maladaptive actions or inaction, and future losses that might otherwise be avoided. Finding common ground is seldom easy, depending on the specializations we bring to climate adaptation. Some of us once felt secure in a single scientific discipline understood as rising above politics. Some of us once felt secure in a commitment to the environment as the single issue occupying high moral ground. Nevertheless, collectively we have been engaging potential partners where they are, encouraging collaboration across diverse groups, and building personal relationships, trust, and credibility.

In short, we have been learning about the tactics for finding common ground by practicing them. Learning often loosens bonds to older ways of thinking and doing, generating some discomfort in the process. Any discomfort can be alleviated by organizing ourselves to create new relationships and to accelerate progress in reducing losses or increasing resilience to climate change. This opportunity is based on what we have discovered more or less independently: a shared long-term goal, even if it is not always made explicit; convergent expectations about what works (and what doesn't) in pursuit of that goal; and many additional resources available to share in support of each other as we pursue that goal. By adapting to a changing climate *and* to the interests of multiple groups whose support is necessary to implement our plans, we are also making a difference in protecting things each of us values in our natural and human environments. There are many steps yet to be taken on many paths. Taking those steps and finding those paths depends in large part on organizing, communicating, and collaborating.

Kresge Grantees and Practitioners Workshop on Climate Change Adaptation February 7–9, 2012, Portland, Oregon Participants

Steve Adams	Institute for Sustainable Communities
Arun Agrawal	University of Michigan
Ben Alexander	Headwaters Economics
Susan Antenen	Conservation Biology Institute
Vicki Arroyo	Georgetown University
Dominique Bachelet	Conservation Biology Institute
Katherine Baer	American Rivers
Jessica Boehland	Kresge Foundation
Cindy Brown	The Nature Conservancy
Maria Brown	NOAA
Ron Brunner	Consultant to Kresge
Maxine Burkett	University of Hawaii
Molly Cross	Wildlife Conservation Society
Lois DeBacker	Kresge Foundation
Bob Doppelt	The Resource Innovation Group
Melanie Emerson	Sky Islands Alliance
Josh Foster	Oregon State University
Nancy Gilliam	Model Forest Policy Project
Bob Glazer	Florida Fish and Wildlife Commission
Tonya Graham	Geos Institute
Jessica Grannis	Georgetown University
Gwen Griffith	Cumberland River Compact
Marcus Griswold	Maryland Dept. of Natural Resources

(*Continued*)

(Continued) Kresge Grantees and Practitioners Workshop on Climate Change Adaptation February 7–9, 2012, Portland, Oregon Participants

John Hagan	Manomet Center for Conservation Sciences	Cara Pike	The Resource Innovation Group
Margy Hall	Model Forest Policy Project	Ray Rasker	Headwaters Economics
Kim Hall	The Nature Conservancy	Kara Reeve	National Wildlife Federation
Lara Hansen	EcoAdapt	Duke Reiter	Arizona State University
Nicola Hedge	The San Diego Foundation	Bruce Riordan	SF Bay Area Joint Policy Committee
Jennie Hoffman	EcoAdapt	Jackie Sartoris	Manomet Center for Conservation Sciences
Fran Hunt	Sierra Club	Chuck Savitt	Island Press
Alicia Hwang	Geos Institute	Nathaniel Seavy	PRBO Conservation Science
Brian Jackson	Environmental Defense Fund	Kristen Sheeran	EcoTrust
Mara Johnson	World Wildlife Fund	Ron Sims	(Fmr.) Deputy Secretary, US Dept. of HUD
Shawn Johnson	University of Montana	Minor Sinclair	Oxfam America
Livia Kent	Island Press	Ted Smith	University of MT/ Advisor to Kresge
Gayle Killam	River Network	Amy Snover	Climate Impacts Group
Marni Koopman	Geos Institute	Theo Spencer	Natural Resources Defense Council
Maria Carmen Lemos	University of Michigan	Bruce Stein	National Wildlife Federation
Frank Lowenstein	The Nature Conservancy	Peter Szabo	Bloomingdale Management Advisors
Telley Madina	Oxfam America	Gary Tabor	Center for Large Landscape Conservation
Adam Markham	Clean Air-Cool Planet	Mandy Valentine	Kresge Foundation
Kate Meis	Local Government Commission	Stacy Vynne	The Resource Innovation Group
Rob Melnick	Arizona State University	Eric Walberg	Manomet Center for Conservation Sciences
Eric Mielbrecht	EcoAdapt	Jen Watkins	Conservation Northwest
Louise Misztal	Sky Island Alliance	Wendy Wilson	River Network
Steve Nicholas	Institute for Sustainable Communities	Kimery Wiltshire	Carpe Diem West
Malia Nobrega	University of Hawaii	Brooks Yeager	Clean Air-Cool Planet
John Nordgren	Kresge Foundation	Emily Young	The San Diego Foundation
Reed Noss	University of Central Florida		
Julie O'Leary	Freshwater Future		
Jennifer Pagach	CT Dept. of Energy & Env. Protection		
Sascha Petersen	Adaptation International		
Steven Peyronnin	Coalition to Restore Coastal Louisiana		

Scribes:

Betsy Briar	Portland State University
Keith Henty	Geos Institute
Travis Leipzig	River Network
Sara Loreno	Portland State University
Nadia Jones	Portland State University
Ryan Tobias	Bus Project Foundation
Matt Wood	Portland State University

ASAP

Professionals across the United States are working to prepare for the impacts of climate change and make their communities, regions, states, and the country as a whole more resilient to those changes. These individuals realize that climate change is one of the most critical issues of our time and that it will affect many aspects of their jobs and our communities. They work not just in emergency management or hazard mitigation but other sectors focusing on climate science, public health, natural resources, governance, urban planning, transportation, forestry, water resources and more. They realize that while their individual efforts can make a difference, they cannot do it alone.

Starting in 2009, a core group of a couple of dozen of these practitioners began looking for a place to connect with each other, share information, and build on successes while moving away from approaches that weren't working. They needed a professional society that could help bridge the geographic and sectoral gaps that naturally develop, especially in the diverse, dynamic, and emerging field of climate adaptation. No such society existed and these individuals met annually at other conferences or through conference calls for a few years before securing funding to officially launch ASAP in 2013.

Thanks to the generous support of the Kresge Foundation, the MacArthur Foundation, and the Institute for Sustainable Communities, ASAP is now the only organization in the US that focuses solely on *connecting and supporting the people who are building climate resilience across the country.*

By the end of 2014, ASAP had over 800 members split across four large sectors.

- 35% are from the government sector representing local, state, federal, and tribal governments
- 30% work in nonprofit institutions
- 20% are affiliated with academic institutions
- 15% are from the private sector

ASAP continues to build out a portfolio of services focused on building resilience by supporting its membership. These services include support of regional and national adaptation conferences, bimonthly newsletters, webinars highlighting developments and hot topics in the field, interactive affinity groups on key topics such as policy, research, and communication, and a member-focused website that allows people to connect with each other. ASAP also leads some field building projects such as hosting the Prize for Progress in Adapting to Climate Change designed to highlight and share promising practices in the field.

Is ASAP the only way that professionals are connecting and collaborating to build resilience? Definitely not. It does, however, represent an important step in the evolution of an emerging field. The development of a professional society highlights how individuals are no longer working in isolation on this issue and are now working together to solidify the state of knowledge in the field, but more importantly they are working across sectors as well as within and across regions to better understand how climate change will impact our communities and what can be done to reduce those impacts, prepare for the future, and build resilience. This type of multisector collaboration and sharing of expertise is critical in developing effective responses to climate change.

Source: ASAP, https://adaptationprofessionals.org/about/.

100 RESILIENT CITIES—PIONEERED BY THE ROCKEFELLER FOUNDATION

100 Resilient Cities—Pioneered by the Rockefeller Foundation (100RC) is dedicated to helping cities around the world become more resilient to the physical, social, and economic challenges that are a growing part of the twenty-first century. 100RC supports the adoption and incorporation of a view of resilience that includes not just the shocks—earthquakes, fires, floods, and so on—but also the stresses that weaken the fabric of a city on a day-to-day or cyclical basis. Examples of these stresses include high unemployment, an overtaxed or inefficient public transportation system, endemic violence, or chronic food and water shortages. By addressing both the shocks and the stresses, a city becomes more able to respond to adverse events and is overall better able to deliver basic functions in both good times and bad to all populations.

Cities in the 100RC network are provided with the resources necessary to develop a roadmap to resilience along four main pathways:

- Financial and logistical guidance for establishing an innovative new position in city government, a chief resilience officer, who will lead the city's resilience efforts
- Expert support for development of a robust resilience strategy
- Access to solutions, service providers, and partners from the private, public, and non-governmental organization (NGO) sectors who can help them develop and implement their resilience strategies
- Membership of a global network of member cities who can learn from and help each other

Through these actions, 100RC aims not only to help individual cities become more resilient, but will facilitate the building of a global practice of resilience among governments, NGOs, the private sector, and individual citizens.

We began working with our first group of 32 cities in December 2013. In 2014, we received 330 applications from 94 countries for our second cohort. The second round of 35 cities joined the program in December 2014. A list of the current 67 cities participating in 100RC can be accessed at http://www.100resilientcities.org/cities#/-_Yz45MDE3NidpPTEocz5j/.

100 Resilient Cities—Pioneered by The Rockefeller Foundation is financially supported by The Rockefeller Foundation and managed as a sponsored project by Rockefeller Philanthropy Advisors (RPA), an independent 501(c)(3) nonprofit organization that provides governance and operational infrastructure to its sponsored projects.

Source: Rockefeller Foundation 100 Resilient Cities, http://www.100resilientcities.org/pages/about-us#/-_Yz45MDE3NidpPTEocz5j/.

MISSED OPPORTUNITIES: EVALUATING WHAT WORKS

Ronald D. Brunner

On the night of October 29, 2012, Superstorm Sandy struck New York City with high winds and a 14-foot storm surge at high tide. Many of us can recall vivid images from extensive coverage by national news centers in the city. These images include, for example, waterfront neighborhoods in Staten Island, Queens, and Brooklyn under water, with buildings, cars, and trees damaged or destroyed. By some estimates the city alone suffered $20 billion in economic losses and 44 deaths from Superstorm Sandy. According to Mayor Michael Bloomberg, "It was a wake-up call for just how vulnerable we are." Environmental activist Bill McKibben agreed: "If there was ever a wake-up call, this is it."* As one of the loudest in a long series of nature's wake-up calls, Superstorm Sandy appears to be a turning point in on-going efforts to put climate adaptation on policy agendas nationwide: news of it prompted many more communities, agencies, and organizations to consider their own vulnerabilities, perhaps enough to persuade other stragglers to follow as a matter of peer pressure if nothing else.

The ongoing efforts to put climate adaptation on policy agendas can be traced back to the late 1990s, when scientists, environmentalists, and other core experts began to realize that adapting to climate change was a complement, not an alternative, to mitigating greenhouse gas emissions (Parry et al. 1998; Pielke 1998). Concentrations of greenhouse gases in the atmosphere were already enough to intensify extreme weather events and their impacts. Taking the lead in adapting to climate change, these core experts relied on projected climate change impacts to promote planning and other proactive means to reduce vulnerabilities at local, state, and federal levels. Among other accomplishments, they influenced the establishment and work of the Interagency Climate Change Adaptation Task Force after the Obama Administration took office.† An assessment of the peer-reviewed literature and other sources on these efforts can be found in Chapter 28 of the *Third National Climate Assessment* published in May 2014.‡

National Climate Assessments implement a mandate in the Global Change Research Act of 1990 (Public Law 101-606, Sec. 106) to conduct periodic scientific assessments focused on integrating, evaluating, and interpreting the findings of climate science for policy makers. The perspective taken was, and to some extent still is, scientific and national in the global context: After an Overview (Chapter 1), the Third Assessment begins with the science of "Our Changing Climate" (Chapter 2), turns to the impacts of climate change on US Sectors (Chapters 3–15) and Regions (Chapters 16–25), and concludes with Response Strategies (Chapters 26–30). Chapter 28 on "Adaptation" as a response strategy "highlights efforts at the federal, regional, state, tribal, and local levels, as well as initiatives in the corporate and non-governmental sectors to build adaptive capacity and resilience in response to climate change."§ With few exceptions, Chapter 28 does not attempt to identify what works by evaluating these many efforts and initiatives according to its primary policy goal and criterion, "to build adaptive capacity and resilience."

Like the Third Assessment as a whole, Chapter 28 was underfunded. It is nevertheless is a valuable guide to conventional thinking among leading experts. It is also a snapshot of a highly complex, rapidly evolving situation taken from the top-down perspective characteristic of scientific assessments. As such, Chapter 28 is worth reviewing in search of missed opportunities for advancing the common interests of the various communities involved. From this somewhat different perspective, the primary policy goal and criterion is reducing the net losses and vulnerability of things valued by community members to extreme weather events and climate change.¶ Community members tend to value, for example, human life and limb, property and prosperity, and ecosystem

* Katherine Bagley and Maria Gallucci, *Bloomberg's Hidden Legacy: Climate Change and the Future of New York City* (Inside Climate News/ICN Books) Chapters 7 and 8. See also Alan Feuer, Building for the Next Big Storm: After Hurricane Sandy, New York Rebuilds for the Future, *New York Times* (October 26, 2014), MB1.

† See Progress Report of the Interagency Climate Change Adaptation Task Force: Recommended Actions in Support of a National Climate Change Adaptation Strategy (October 5, 2010).

‡ Rosina Bierbaum et al., Adaptation, Chapter 28, 670–706, in J. M. Melillo, Theresa (T.C.) Richmond, and G. W. Yohe, eds., *Climate Change Impacts in the United States: The Third National Climate Assessment* (US Global Change Research Program, 2014), accessible at http://nca2014.globalchange.gov/report/response-strategies/adaptation. An earlier version was distributed for public comment in January 2013. An even earlier version was published online as A Comprehensive Review of Climate Adaptation in the United States: More than Before, but Less than Needed, in *Mitigation and Adaptation Strategies for Global Change* (October 27, 2012), just two days before Superstorm Sandy struck New York City.

§ Bierbaum et al., Adaptation, 672, in the 2014 version.

¶ See The Common Interest, in Ronald D. Brunner and Amanda H. Lynch, *Adaptive Governance and Climate Change* (Boston: American Meteorological Society, 2010), 22–29, including a brief definition at 23: "In the simplest terms, the common interest is composed of interests widely shared by members of a community. It would benefit the community as a whole and be supported by most community members, *if they can find it*."

services. From the bottom-up perspectives of these people, there is little difference between adapting to extreme weather events and adapting to climate change, even if they can be distinguished. According to the Third Assessment, "Changes in extreme weather events are the primary way most people experience climate change."*

Harvesting Experience from More Communities

The focus of Chapter 28 provides a convenient point of departure: "While societal adaptation to *climate variability* is as old as civilization itself, the focus of this chapter is on preparing for unprecedented human-induced *climate change* through adaptation."† Accordingly, the authors featured cases of adaptation to climate change, the empirical foundation of their Key Messages. Consider three of the six Key Messages:

1. Substantial adaptation planning is occurring in the public and private sectors and at all levels of government; however, few measures have been implemented and those that have appear to be incremental changes.
2. Barriers to implementation of adaptation action include lack of funding, policy and legal impediments, and difficulty in anticipating climate-related changes at local scales....
6. The effectiveness of climate change adaptation has seldom been evaluated, because actions have only recently been initiated, and comprehensive evaluation metrics do not yet exist.‡

These Key Messages imply that we lack dependable knowledge of what works in climate adaptation: what little experience we have has seldom been evaluated. This postpones fulfillment of the Third Assessment's "goal of better informing public and private decision-making at all levels"§ insofar as that goal includes guidance on what works.

These Key Messages also overlook the huge base of experience in adapting to climate variability, the extreme weather events occurring within a climate. In recent decades, many US communities have responded to their losses from extreme weather events. A substantial but unknown fraction of them have overcome barriers to implementation and made significant progress—by their own evaluations, often without metrics—in reducing their net losses and vulnerability. By this criterion, they succeeded in adapting to climate change without taking projected climate changes into account.

The missed opportunity is harvesting experience from more communities like these to strengthen the empirical foundations of climate adaptation for different kinds of communities that have already heard nature's wake-up calls. They do not need more reasons for putting climate adaptation on their policy agendas. They need guidance on what has worked for similar communities elsewhere. Consider some models that have been largely overlooked in the peer-reviewed literature.

A leading example is Tulsa, Oklahoma, which had a long history of devastating floods, among the worst in the nation. The 2009 edition of this book included an account by Ann Patton, a participant-observer from the outset of Tulsa's continuing effort "to build a disaster-resistant, *sustainable* community" (Patton 2009). Prompted by a damaging flood in June 1974, citizens meeting in a flooded living room began to organize as Tulsans for a Better Community. They quickly acquired the technical expertise and advocacy skills necessary to devise a four-point program of flood control and pressure the city government to adopt it. Under continuing pressure after floods in 1976 and

* This is the second of 12 Report Findings in Melillo et al., *Climate Change Impacts in the United States: The Third National Climate Assessment*, 15.

† Bierbaum et al., Adaptation, 672, their emphasis. However, the "emerging areas of emphasis" in Chapter 28, 687, include "Adaptation to Extreme Events." This has been the focus of research in natural hazard mitigation for nearly four decades. The Natural Hazards Center was established as an international information clearinghouse at the University of Colorado, Boulder, in 1976.

‡ Quoted from Bierbaum et al., Adaptation, 671.

§ Melillo et al., *Climate Change Impacts in the United States: The Third National Climate Assessment*, 3.

1984, the city gradually strengthened its technical capabilities, organizational structure, and floodplain regulations, won voter approval of a bond issue for flood-control projects and a stormwater utility surcharge on water bills, and began buying out and removing from its floodplains buildings and mobile homes—more than 1000 of them by the mid-1990s. This earned Tulsans the lowest flood insurance rates in the nation under FEMA's new Community Rating System and has helped them avoid major flood damage since then. Meanwhile, late in 1998, FEMA's Project Impact named Tulsa the Most Outstanding Model Community for Disaster Mitigation, and awarded the city a 3-year, $500,000 grant to create public-private partnerships for mitigation of additional natural hazards including tornados, lightning, and extreme heat and drought. The effort continues under auspices of Tulsa Partners, Inc.

Tulsa is not alone in reducing its losses and vulnerability to climate change, and advancing other local community interests at the same time. Consider some additional models of possible interest to similar communities now looking for inspiration and guidance to mitigate their own flood losses:

- In 1955, a partnership including the Army Corps of Engineers sought to reduce flood losses in the Charles River Basin, Massachusetts. They succeeded by 1983, preventing development of all wetlands in the upper basin, thus allowing them to store excess precipitation and reduce peak flows (Hewes and Pitts 2009).
- After more flood damage in 1975, the village of Soldiers Grove, Wisconsin, sought to relocate its downtown to higher ground. Completed in 1983, relocation minimized damage from near-record floods in 2007 and 2008, stabilized the local population, and revived the local economy.*
- Following the Great Mississippi Flood of 1993, other small Midwestern communities relocated people and property out of floodplains with some success, including Grafton and Valmeyer, Illinois, and Pattonsburg, Missouri. Most of the documentation is from news reports.†
- Within four months of the Red River flood of 1997, Grand Forks, North Dakota, and East Grand Forks, Minnesota, removed 1100 homes from flooded areas. In the expanded floodplain they created a popular Greenway, some 2000 acres of open space for recreation.‡
- After another flood in 1995, a coalition emerged in Napa Valley, California, to devise a "Living River" flood management plan approved by ballot initiative. Since groundbreaking in 2000, "The project has increased the safety of Napa residents, spurred economic development, cleaned up acres of contaminated riverfront, and restored important habitats" for wildlife and recreation.§

Each of these communities has demonstrated what works in reducing losses and vulnerability to floods while advancing other community interests at the same time. Other communities have demonstrated what works in response to other extreme weather events. For example, Manhattan,

---

* William S. Becker, *Come Rain, Come Shine: A Case Study of a Floodplain Relocation Project at Soldiers Grove, Wisconsin* (Madison, WI: Bureau of Water Regulation and Zoning, Wisconsin Department of Natural Resources, 1983). See also Hewes and Pitts, *Natural Security*, 34–39.

† On Grafton, see Yu Xiao, Community Economic Adjustment, Post-Disaster Mitigation, and Long-Term Economic Recovery (2008), accessible at http://www.colorado.edu/hazards/workshop/08ps_abs.html, and Tim Bryant and Terry Hillig, With Levees Breached or Topped, the Mississippi River Is Reclaiming "Its Natural Flood Plain," *St. Louis Post-Dispatch* (June 22, 2008). On Valmeyer, see Patricia Leigh Brown, Higher and Drier, Illinois Town Is Reborn, *New York Times* (May 6, 1996), Marisol Bello and Peter Eisler, Town Relocated and Reborn; Ill. Community Finds Life Goes on after Flood's Destruction, *USA Today* (June 20, 2008), and Bryant and Hillig, op. cit. On Pattonsburg, see Margaret Stafford, Uprooted Town Has Remade Itself Since Midwest Deluge of '93, *Associated Press* (June 20, 1998) and James Lee Witt, *Stronger in Broken Places: Nine Lessons for Turning Crisis Into Triumph* (New York: Times Books, 2002).

‡ Hewes and Pitts, *Natural Security*, 36.

§ A local official quoted in Dave Dickson, Living River: The Napa Valley Flood Management Plan, in Bullock et al., *Global Warming, Natural Hazards, and Emergency Management*, 126–149, at 137.

Kansas, is a model for installing SafeRooms to protect life and limb from tornadoes. Miami-Dade County, Florida, is a model for protecting buildings from wind and water damage during hurricanes. Ruidoso, New Mexico, is a model for reducing vulnerability to wildfires. Properly evaluated, these are models that provide field-tested knowledge and information for decision-making: They reveal what has worked, how, in particular circumstances and therefore what might work in similar circumstances elsewhere.

Perhaps projections of climate change at the local level would have persuaded such communities to implement more or more ambitious actions on the expectation that losses from the most extreme weather events in the past will fall far short of future losses. But the fact remains that projections of local climate change impacts were not necessary for these communities to make significant progress in adapting to climate change, nor were such projections sufficient in communities that incurred major losses. In the four years prior to Hurricane Katrina in August 2005, people in New Orleans had access to a "steady stream of so-called doomsday treatises on the inevitability of a major hurricane laying waste to New Orleans…" These included articles in *Scientific American*, *National Geographic*, and "a landmark five-part series … published in June 2002 in the local *Times-Picayune*, titled 'Washing Away.'"* New York City was not ready for Superstorm Sandy, even though "Local scientists had predicted for at least 15 years that the changing climate would bring rising sea levels and more dangerous storms to the city's 520 miles of coastline." The Bloomberg administration "simply hadn't had the time or the focus or the money to do it all, to truly protect New York from the impacts of climate change."† Projections of local climate change are one factor in adapting to climate change, but progress depends on the integration of many more factors in each community.

The focus of leading experts on "unprecedented human-induced climate change" is understandable in view of their interest and expertise in climate adaptation and their long struggle to put it on policy agendas nationwide. The focus is also representative of a tendency of experts generally to attend to those parts of a large, complex problem for which they are prepared to make a contribution and to discount or overlook the other parts. For a climate scientist, the critical factor may be "Getting science down to the street" by downscaling global climate projections. For a planning expert, "Climate adaptation is just good planning." For a communications expert, "Messaging is everything." In the extreme, the means at their disposal may become ends in themselves. More experts of more kinds are offering their guidance to more communities seeking guidance on what works in climate adaptation. The effects of broader participation—or success in public engagement—are to dilute the relative influence of leading experts, and to underscore the need for integrating different kinds of expertise and other resources according to the multiple needs and circumstances of each community.

Evaluating What Works from the Bottom Up

The three additional Key Messages in Chapter 28 are consistent with experience in cases of adaptation to climate variability as well as adaptation to climate change:

3. There is no "one-size fits all" adaptation, but there are similarities in approaches across regions and sectors. Sharing best practices, learning by doing, and iterative and collaborative processes including stakeholder involvement, can help support progress.

* Douglas Brinkley, *The Great Deluge: Hurricane Katrina, New Orleans, and the Mississippi Gulf Coast* (New York: HarperCollins, 2006), 14. In retrospect, "Hurricane Katrina struck and played out in a manner that was eerily familiar" from an earlier projection of what would have happened if Hurricane Ivan in September 2004 had not missed New Orleans, according to Shirley Laska, Respect for Nature as Cornerstone to Community Resiliency: The View from Katrina "Ground Zero," *Natural Hazards Observer* XXXI (November 2006), 14–15.

† Bagley and Gallucci, *Bloomberg's Hidden Legacy*, Chapter 1.

4. Climate change adaptation actions often fulfill other societal goals, such as sustainable development, disaster risk reduction, or improvements in quality of life, and can therefore be incorporated into existing decision-making processes.
5. Vulnerability to climate change is exacerbated by other stresses such as pollution, habitat fragmentation, and poverty. Adaptation to multiple stresses requires assessment of the composite threats as well as trade-offs among costs, benefits, and risks of available options.*

These Key Messages imply that progress in adaptation depends on considering the particular context as a whole; the contexts include thousands of communities nationwide. This means attending to multiple goals, stresses, and other factors in each context and to differences (along with similarities) in approaches across contexts. Federal officials and their agencies lack the cognitive capacity, if nothing else, to manage the flood of information that would be necessary for them to play the leading role in many contexts, let alone every context under their jurisdictions. This calls into question the top-down perspective that is characteristic of scientific assessments in general and the Third National Climate Assessment in particular.

Consider some expressions of the top-down perspective in Chapter 28. The survey of "Adaptation Activities in the United States" begins with the federal government and the claim that "Federal leadership, guidance, information, and support are vital to planning for and implementing adaptation actions at all scales and in all affected sectors of society."† The account of Tulsa's efforts, one of 32 examples of "Adaptation Activities" in Figure 28.4, focuses selectively on three goals in the city's floodplain and stormwater program:

> Tulsa, Oklahoma, has a three-pronged approach to reducing flooding and managing stormwater: (1) prevent new problems by looking ahead and avoiding future downstream problems from new development (for example, requiring on-site stormwater detention), (2) correct existing problems and learn from disasters to reduce future disasters (for example, through watershed management and the acquisition and relocation of buildings in flood-prone areas), and (3) act to enhance the safety, environment, and quality of life of the community through public awareness, an increase in stormwater quality, and emergency management.‡

This snapshot, taken at an uncertain date, omits most of what was important to Patton in her bottom-up account of Tulsa's efforts. Notice also that the three-pronged approach is the lead; outcomes, actual reductions in losses and vulnerability, are overlooked apart from a reference to relocation. The few evaluations of success in Chapter 28 are based not on outcomes but on capacity building for which no comprehensive adaptation evaluation metrics exist. For example, the National Integrated Drought Information System (NIDIS) "has successfully brought together government partners and research organizations to advance a warning system for drought-sensitive areas."§ Whether NIDIS's warning system has helped any drought-sensitive area reduce its losses and vulnerability to droughts is not considered.

The missed opportunity is evaluating what works from the bottom up. Chapter 28 acknowledges that "Both 'bottom-up' community planning and 'top-down' national strategies may help regions deal with impacts such as increases in electrical brownouts, heat stress, floods, and wildfires."¶ It

* Quoted from Bierbaum et al., Adaptation, 671.

† Bierbaum et al., Adaptation, 672.

‡ Bierbaum et al., Adaptation, 685. This account is based on City of Tulsa, *Rooftop to River: The Tulsa Program*, 1–10, available at http://www.smartcommunities.ncat.org/articles/rooftop/program.shtml, the single source cited. In it there is no reference to "climate change," but there is one reference, at 7, to "climate" variability: "Capricious climate makes Tulsa vulnerable to weather emergencies, particularly tornadoes, violent thunderstorms, and floods."

§ Bierbaum et al., Adaptation, 687, and on evaluation metrics, 705–706.

¶ Bierbaum et al., Adaptation, 671.

also recognizes that "Most adaptation efforts to date have occurred at local and regional levels."* Nevertheless, perspectives from the bottom up have been largely overlooked.

From the bottom up, adaptation to climate variability or change is another complication in the ongoing process of community development. The point is to "mainstream" adaptation considerations into existing community operations, not to override them.

- Perspectives from the bottom up are typically focused on reducing past losses and future vulnerability in one's own community, not capacity building across many communities. In her "Tulsa Story," Patton reports that the community-based initiative sought to make Tulsa a disaster-resistant community by reducing losses to floods initially and other disasters later.
- Perspectives from the bottom up are typically whole community, not reductive to a three-pronged approach or the like. Patton's advice is to "Start with a small hub of your very best people.... Develop a shared vision. Then build a holistic, inclusive partnership around that heart.... The more comprehensive your program is, the larger your constituencies can be."†
- Perspectives and operations from the bottom up are typically contingent on the situation as it evolves, not a fixed master plan. For Patton, the initial politics of "conflict and confrontation" in the period 1974–1984 eventually led to the new politics of integration in the period 1990–1998 as "Tulsans began to pull together, united in a vision of a flood-free city."‡

In short, climate adaptation as a part of community development is an evolutionary process. Patton "describes how we made our way along, by trial and error, disaster by disaster, to reduce the risks that have plagued our lives since man moved to this locale." With her collaborators Patton learned to "Establish broad goals, specific objectives, and flexible strategies that can be adapted to avoid land mines, avert problems, and seize opportunities."§ Under these circumstances, effective leadership depends on community-based initiatives, such as Tulsans for a Better Community.

Somewhat different community-based initiatives emerged in the Charles River Basin, Soldiers Grove, and the Napa Valley in opposition to levees proposed by the US Army Corps of Engineers. These initiatives took the lead in developing alternatives that better served the common interest of their communities. (The Army Corps eventually became partners in them, except in Soldiers Grove.) William Becker, once publisher of the weekly newspaper in Solders Grove, described and justified the leading role of its community-based initiative: "Too often local plans are drawn by 'experts' from outside a community. Such 'top-down planning' often meets considerable local resistance." Soldiers Grove "not only got [local] people involved, it gave them control of direction setting and decision making as the community prepared to relocate." It was made clear to outside experts that they "would serve as resources in the community decision process, not directors of that process."¶ Local leaders effectively integrated expertise and other resources from local sources, units of the University of Wisconsin, private consulting firms, and multiple agencies of state and federal government. Soldiers Grove's relocation funding package, constructed piece by piece, included 14 contributions from eight state and local sources totaling $2.2 million, and 10 contributions from four federal agencies totaling $4.4 million.**

The leading role in adaptation as part of community development consists of integrating and adapting many resources to many needs and circumstances in each community, making each community unique. Community-based initiatives have played this leading role effectively in some

---

* Bierbaum et al., Adaptation, 678.

† Patton, A Tulsa Story, 108–109.

‡ Patton, A Tulsa Story, 90.

§ Patton, A Tulsa Story, 90, 109. See also Ron Brunner and John Nordgren, *Climate Adaptation as an Evolutionary Process: A White Paper* (April 6, 2012), 134–143 above in this chapter.

¶ Becker, *Come Rain, Come Shine*, 41.

** Becker, *Come Rain, Come Shine*, 36.

communities, and can do so elsewhere. It is difficult to imagine an agency of state or federal government with enough local knowledge and information, commitment, and persistence—not to mention a formal mandate and jurisdiction—to play more than a supporting role in adaptation. Insofar as funding is concerned, that supporting role is likely to decrease, if, as expected, state and federal budgets available to assist in adaptation do not keep up with increases in aggregate demand from more communities suffering losses from more and more intense extreme weather events.

FEMA's Project Impact is a model of the supporting role for federal agencies. Established in 1997, Project Impact was designed to help communities reduce their losses to natural hazards by building local partnerships. It also differed from the traditional federal program that "throws money at a specific and identifiable problem, with requisite rules and regulations attempting to control the outcome." Instead,

> The idea was to devolve hazard-mitigation responsibility and authority down to the lowest possible level. If money changed hands from the feds to the locals, it was only seed money for growing a process that would not end when the money ran out. In this model, the federal government became not an authoritarian parent, but instead a partner and facilitator to help inspire and empower locals to size up and solve their own problems—ideally before disasters struck. Local communities were helped to take ownership of their disaster problems and solutions.[*]

Project Impact worked in these ways in Tulsa, Miami-Dade County, Manhattan, Kansas, and many other local communities, although not in all of them. In any case, it was terminated by the Bush administration early in 2001. A major evaluation of FEMA's disaster mitigation grants from 1993 to 2003, including Project Impact, nevertheless found that "On average, across all grants, regions, and hazards studied, each dollar spent on mitigation saves society an average of $4 in avoided future losses."[†]

Diffusing and Adapting Information on What Works

Chapter 28 recognized that "the sharing of best practices and lessons learned can be pivotal means to advancing understanding and uptake of climate adaptation activity. The use of established information-sharing networks such as regional climate initiatives are illustrations of the types of networks that have supported stakeholder adaptation activity to-date."[‡] Such networks are among the channels of information, while best practices and lessons learned are conceptions of what works, the content of information. The assumption is that progress in climate adaptation nationwide can be accelerated by diffusing and adapting information about what works across communities and across levels in the federal system. There is no need to reinvent the wheels that work; the task is to improve upon them.

There is apparently little published research evaluating established means for diffusing and adapting information about what works in climate adaptation. Among the top-down means, it is not clear how or to what extent Chapter 28 or the Third Climate Assessment as a whole supported progress on the ground, where most of the action is.[§] It is not clear how or to what extent web-based national clearinghouses for information on climate adaptation have been used by people on

* Eric Holdeman and Ann Patton, Project Impact Initiative to Create Disaster-Resistant Communities Demonstrates Worth in Kansas Years Later, *Emergency Management* (December 12, 2008). Accessed at http://www.emergencymgmt.com/disaster/Project-Impact-Initiative-to.html. See also Brian Cowan, Project Impact: Building a Disaster-Resistant Community, 70–79, in the previous edition of this book.

† Philip T. Ganderton et al., Mitigation Generates Savings of Four to One and Enhances Community Resilience, *Natural Hazards Observer* XXX (March 2006), 1–3, at 1. See also Congressional Budget Office, *Potential Cost Savings from the Pre-Disaster Mitigation Program* (September 2007).

‡ Bierbaum et al., Adaptation, 682. See also an emerging area of research, A Compendium of Adaptation Practices, 687.

§ For observations on the local relevance of another scientific assessment, the Arctic Climate Impact Assessment, see Brunner and Lynch, *Adaptive Governance and Climate Change*, 18.

the ground. These clearinghouses tend to devolve the task of information filtering to the user* and provide search tools for that purpose—apparently on the assumption that the user understands what he or she needs to know and can evaluate the information available online. That assumption may be mistaken. In any case, the missed opportunity at the margin may be building on the relatively neglected alternatives for diffusing and adapting information on what works. These are the informal peer-to-peer and bottom-up interactions that have emerged more or less spontaneously as people facing an adaptation problem seek advice from others who have more experience coping with it.

When Tulsans were looking for guidance in the aftermath of floods in 1974, "one citizen, Bob Miller, even visited Rapid City, which had undertaken such a relocation project following a disastrous flash flood in 1972." Around the same time, "Ann Patton's research led her to the Natural Hazards Center and Gilbert White" at the University of Colorado in Boulder (Hinshaw 2006). Becker reported that in Soldiers Grove, "Shortly after relocation was proposed in 1975, a small group of villagers traveled to Niobrara, Nebraska, a small community relocating because of flooding problems caused by a dam downriver. The visit allowed villagers to see a relocation first-hand and to talk to its participants."† After the Great Mississippi Flood of 1993, Becker himself as a federal official drew on Soldiers Grove's experience to help with relocation in similar Midwestern communities, including Grafton and Valmeyer, Illinois, and Pattonsburg, Missouri. Also in the 1990s, FEMA factored these local experiences into the design of Project Impact. Dickson reports that "The concepts and lessons learned in the Napa project have already been transferred to consensus-based, environmental restoration projects across the United States and around the world…."‡ The number of visitors who continue to look for guidance in Tulsa, Soldiers Grove, the Napa Valley, and the like is an indicator of their status as models for other communities. The progress made by leading communities is an indicator of the *potential* for reducing losses and vulnerability in other local communities nationwide.

The content of information available from practitioners on the ground, and sought by others, can be inferred from the few relatively well-documented case studies, including those by Ann Patton, William Becker, and Dave Dickson, the sources for most of the scattered observations cited above. Unlike the brief account of Tulsa in Chapter 28, these case studies tend to be *comprehensive* because the authors understood that progress depended on integrating a range of factors not limited to downscaling or planning or messaging or program goals, for example. In one way or another they spontaneously integrated case-specific details relevant to the generalized adaptation process *as a whole*, as portrayed in Chapter 28: "Identifying risks and vulnerabilities ↔ Planning, assessing and selecting options ↔ Implementation ↔ Monitor and evaluate ↔ Review strategy and research; share lessons learned"§ and then iterate. Multiple iterations of this generalized model can be identified in Patton's account of nearly four decades of experience in Tulsa.

Well-documented case studies also tend to be *detailed* because progress depended on adapting to specific needs and circumstances as they arose in an evolutionary process. Funding, for example, is a requirement for all phases of the adaptation process, especially implementation, typically the most expensive. Local packages have integrated funds from various sources. Included among the local sources are a stormwater utility surcharge in Tulsa, a bond to purchase a site for relocation of Soldiers Grove's downtown, and a half-cent sales tax increase approved by ballot in the Napa Valley.

* Information filtering is the priority, according to Herbert A. Simon, *The Sciences of the Artificial*, 3rd ed. (Cambridge, MA: MIT Press, 1996), 144: "The real design problem is not to provide more information to people but to allocate the time they have available for receiving information so that they will get only the information that is most important and relevant to the decisions they will make. The task is not to design information-distributing systems but intelligent information-filtering systems."

† Becker, *Come Rain, Come Shine*, 44.

‡ Dickson, *Living River*, 144.

§ Bierbaum et al., Adaptation, Figure 28.3, 681, adds an important qualification: "This is not a stepwise or linear process; various stages can be occurring simultaneously, in a different order, or be omitted completely." The odds of making progress are severely diminished by a stage neglected or omitted.

Similarly, the politics of finding common ground are unavoidable in all phases of the adaptation process, especially in selecting the options to be implemented. Again, local leaders have relied on various means. Included among them are a grass-roots interest group to pressure city government for a decade in Tulsa, a relocation coordinator working quickly and closely with the Village Board and others at each step in Soldiers Grove, and a 30-month community coalition process to integrate community interests into a ballot measure approved by a two-thirds majority of voters in Napa Valley.

Taken together, comprehensive and detailed case studies can help people in other local communities understand the full range of general considerations on which their progress will depend, compare specific alternatives by which each general consideration has been satisfactorily addressed, and adapt what fits their own unique combination of needs and circumstances. Only such case studies can communicate the diversity and complexity of the evolutionary processes involved in making progress in climate adaptation. Moreover, if there is no "one-size-fits-all adaptation," as asserted in Key Finding 3, then one corollary is that there are no best practices—only better practices assessed according to particular needs and circumstances. Another corollary is that no lesson learned is applicable everywhere. Comprehensive and detailed case studies clarify the context of each practice and each lesson, and in doing so help each user assess what fits his or her needs and circumstances. There is always an assessment to be made.

Experts in climate adaptation can play a supporting role in these assessments of what works by improving the stock of comprehensive and detailed case studies. One opportunity is serving as third-party evaluators of case studies written by the practitioners involved. Experience in other policy areas suggests that practitioners sometimes exaggerate their claims of success, and practitioners looking for guidance sometimes accept such claims uncritically, especially if they are desperate. Independent evaluations reduce the risk that users of case studies will be misled by them. Another opportunity for experts is adding to the stock of comprehensive and detailed case studies by writing their own, based on collaboration with practitioners on the ground over an extended period of time. There are some precedents to build upon.* Additional case studies can be selected to add diversity to the stock, thereby increasing the odds that communities looking for guidance can find guidance from communities similar to their own.

Still another supporting role is evaluating systems involved in distributing and filtering information about what works in climate adaptation. For example, it is likely that local leaders just getting started in climate adaptation have the time—and the need—for only a few to several comprehensive and detailed case studies most relevant to their own needs and circumstances. They may need help finding those case studies in the flood of information available from national clearinghouses that may include hundreds of partial case studies. Perhaps more promising channels of information are the "regional climate initiatives" mentioned briefly in Chapter 28. Like the site visits mentioned in case studies, they rely primarily on peer-to-peer exchanges of information.

A leading example is the Climate Leadership Academy (CLA) of the Institute for Sustainable Communities. In September 2010, CLA brought together nearly 50 practitioners from 16 US cities for the first of a series of workshops. In conjunction with the workshops, CLA prepared a resource guide including recourse lists and case studies on "how various local government practitioners have made progress on climate adaptation planning" (Climate Leadership Academy n.d.) but on little more of the adaptation process. As workshop participants began to focus on sea-level rise, practitioners in cities in the southwest spun off to form the Western Adaptation Alliance focused on their own shared climate-related problems including drought and wildfires. In the Alliance's workshop

* See Thomas A. Schroeder, MD, Rashed Chowdhury, Mark A. Lander, Charles Chip Guard, Charlene Felkley, and Duncan Gifford, The Role of the Pacific ENSO Applications Climate Center in Reducing Vulnerability to Climate Hazards: Experience from the US-Affiliated Pacific Islands, *Bulletin of the American Meteorological Society* (July 2012), 1003–15; and Barrow as Microcosm, in Brunner and Lynch, *Adaptive Governance and Climate Change*, Chapter 3, 105–185.

in 2012, the information exchanged was mostly peer-to-peer and filtered by the specific needs and circumstances of practitioners on the ground.

By playing these supporting roles, experts can evaluate and correct their own theories of climate adaptation. They can also understand and appreciate what is involved in climate adaptation beyond their own specializations. That will facilitate applying their knowledge and skills to advancing the common interests of the communities they advise while advancing their own specializations at the same time. For reasons like these, the advice of one expert with experience on the ground is to engage community members "where they are, not where we are."*

As mentioned, systems for diffusing and adapting information on what works in climate adaptation have yet to be evaluated systematically. But evaluation is the most critical function in evolutionary processes guided by long-term goals: Innovations tend to arise spontaneously from the diversity of needs and circumstances; selection depends on evaluation of what has worked for communities individually and collectively; and what has worked tends to be adapted to somewhat different circumstances that encourage further innovation. Without evaluation, the status quo tends to prevail, and for many of us, the status quo in adapting to climate change is not satisfactory.

Conclusion

Experts specializing in adaptation can help expedite progress in adapting to climate change by focusing on the missed opportunities reviewed in this chapter: harvesting experience from more communities that have made substantial progress in reducing their losses and vulnerability to climate change even if they have not taken projected climate changes explicitly into account; evaluating what works from the bottom-up perspectives of these communities; and evaluating systems for diffusing and adapting information about what works from each community to others. Such research is not the only resource involved in expediting progress, but it can help.

Much depends on distinguishing the top-down perspective in scientific assessments from the bottom-up perspectives of practitioners in local communities, where most of the action is. Viewed from the bottom-up, adapting to climate change is not the main value outcome to be sought everywhere nationwide. Adapting to climate change is another consideration to be mainstreamed into the ongoing, open-ended process of community development involving many outcomes valued by community members. This is an evolutionary process in which each community takes the leading role. Agencies of state and federal government and nonprofit and for-profit organizations in the private sector have important supporting roles to play.

Both top-down and bottom-up perspectives contribute to expediting progress in adapting to climate change. But to the extent the latter have been discounted or overlooked in the peer-reviewed literature on which scientific assessments are based, there is more to be gained at the margin by adapting research to evaluating experience on the ground. Response strategies that have worked to reduce losses and vulnerability in any community are the appropriate point of departure for research, not past or projected climate change. Diffusing dependable information about what works for voluntary adaptation elsewhere is the destination.

Thanks to nature's wake-up calls, including Superstorm Sandy, the primary task is no longer putting climate adaptation on policy agendas nationwide. The primary task is to provide dependable guidance on what works.

* Quoted in Brunner and Nordgren, Climate Adaptation as an Evolutionary Process, 135.

GLOBAL CITIES REPORT 2014

CDP, C40 and AECOM are proud to present findings from an unprecedented number of cities disclosing their climate mitigation, adaptation and water management data. In 2014, 207 cities reported to CDP, an 88% increase since last year thanks to a groundbreaking grant from Bloomberg Philanthropies.

As a result, the data is clearer than ever before that cities are leading the way on climate change. In 2014, 108 cities reported their carbon emissions inventories. The cities of Denver (USA), London (UK), Madrid (Spain), Durban (South Africa), and Taipei (Taiwan) reduced their emissions by a total of 13.1 million tons CO2 equivalent since 2009, a 12% reduction. Cities are also increasing their resilience to the impacts of climate change. This year, cities reported 757 adaptation activities and 102 cities have climate adaptation plans.

There is significant opportunity for collaboration between city governments and business to improve climate resilience. CDP's data shows there is also increased action across the private sector. Last year, a record number of financial institutions, representing $92 trillion in assets, asked the companies they invest in to disclose their climate emissions, risks and actions, leading to greater management and accountability. In this report, we analyze data from cities and companies to understand what impacts cities expect businesses could face from climate change and how greater climate resilience makes cities more attractive to business.

Cities are reducing the climate risks faced by citizens and businesses through investment in infrastructure and services and by developing policies and incentives that influence action by others. These efforts to understand and reduce climate risks improve the cities' economic competitiveness. The city of Oslo, for example, reports, "[w]e estimate Oslo is relatively resilient compared with other Norwegian cities. This could then make Oslo more attractive for business settlement."

The benefits that business brings to cities, including jobs, tax revenue and services, are one of the drivers for cities to improve their climate resilience. Similarly, businesses are reliant on public infrastructure and environmental policies to support and guide their operations. This report shows how cooperation between cities and businesses leads to better resilience city-wide. Both sectors can benefit from a greater understanding of each other's climate change risks, and companies can help reduce city-wide risks by embedding local adaptation needs within their business operations.

We congratulate the 207 cities that disclosed their environmental data to CDP this year and the significant impacts they have already made in tackling climate change. While this report focuses on how cities' actions create more attractive, resilient places for business, it is not the responsibility of cities alone to mitigate and adapt to climate change. This report provides cities with unprecedented information and insights into the physical risks businesses face locally and how their actions help reduce those risks. We hope this will support cities in their mission to create attractive places to work and live.

Source: CDP, https://www.cdp.net/CDPResults/CDP-global-cities-report-2014.pdf.

NATIONAL DISASTER RESILIENCE COMPETITION

FACT SHEET SEPTEMBER 2014

The US Department of Housing and Urban Development (HUD), in collaboration with the Rockefeller Foundation (Rockefeller), is providing significant resources and support to communities to help them become more resilient. The National Disaster Resilience Competition (NDRC) is a two-phase process that will competitively award nearly $1 billion in HUD Disaster Recovery funds to eligible communities. Due to legislative program requirements, applicants will be limited to 67 states and local jurisdictions that were Presidentially-Declared Disaster Areas in 2011, 2012, and 2013 (effectively covering 48 of 50 states plus Puerto Rico and Washington, DC). See complete list at http://portal.hud.gov/hudportal/documents/huddoc?id=ElgbAppNatReslnceComp.pdfonly.

The competition will help communities recover from prior disasters and improve their ability to withstand and recover more quickly from future disasters, hazards, and shocks. To complement these funds, Rockefeller will provide technical assistance and training workshops to every eligible state and local government. This support will help applicants consider future risks and vulnerabilities in planning and decision-making, and assist them in applying for HUD funding. While distinct, these two programs will work together to help communities better understand their risks and identify ways in which they can protect the long-term well-being and safety of their residents.

OVERVIEW

Cities and towns face significant economic and social risks from extreme weather events including heat waves, drought, tropical storms, high winds, storm surges, and heavy downpours. In many places, these risks are projected to increase substantially due to climate change, sea level rise, and increased development in coastal areas and other vulnerable locations. Hurricane Sandy, recent droughts, and longer wildfire seasons illustrate the rise in natural disasters and the grave consequences of such events. In spite of advances in disaster preparedness, extreme weather is now affecting the safety, health, and economies of entire regions. American communities cannot effectively reduce their risks and vulnerabilities without considering future extreme events and the effects of climate change in their everyday planning and decision-making. Following a disaster, the recovery phase presents a valuable opportunity for communities to consider how to rebuild more resiliently. The National Disaster Resilience Competition (NDRC) will make nearly $1 billion available to communities that have been impacted by natural disasters in recent years. The competition will encourage communities to not consider how they can recover from a past disaster but also how to avoid future disaster losses. Applicants will need to link or "tie-back" their proposals to the disaster from which they are recovering, as well as demonstrate how they are reducing future risks and advancing broader community development goals within in their target geographic area(s). For example, a community that lost housing during a mudslide may consider constructing homes in a safer area for the survivors. It may also find a financing mechanism for affected downstream businesses to survive the effects of the event and prepare and plan for future hazards.

DEFINING RESILIENCE

A resilient community is able to resist and rapidly recover from disasters or other shocks with minimal outside assistance. Reducing current and future risk is essential to the long-term vitality, economic well-being, and security of all communities. By identifying future risk and vulnerabilities, resilient recovery planning can maximize preparedness, save lives, and bring benefits to a community long after recovery projects are complete.

This competition encourages American communities to consider not only the infrastructure needed to become resilient, but also the social and economic characteristics that allow communities to quickly bounce back after a disruption. For example, applicants need to consider how their projects will promote community development goals, ensure meaningful public engagement and participation, and build collaborations with neighboring jurisdictions and stakeholders who are critical partners in preventing, mitigating, and recovering from disasters.

Source: Department of Housing and Urban Development (HUD), Fact Sheet, http://portal.hud.gov/hudportal/documents/huddoc?id=NDRCFactSheetFINAL.pdf.

REFERENCES

Climate Leadership Academy, *Promising Practices in Adaptation & Resilience: A Resource Guide for Local Leaders*, Version 1.0 (Montpelier, VT: Institute for Sustainable Communities, n.d.), 5.

Hewes, W., and K. Pitts, *Natural Security: How Sustainable Water Strategies Are Preparing Communities for a Changing Climate* (Washington, DC: American Rivers, 2009), 40–46.

Hinshaw, R.E., Tulsa, Oklahoma, in *Living with Nature's Extremes: The Life of Gilbert Fowler White* (Boulder, CO: Johnson Books, 2006), 197–204, at 200.

Parry, M., N. Arnell, M. Hulme, R. Nicholls, and M. Livermore, Commentary: Adapting to the Inevitable, *Nature* 393 (1998), 741.

Patton, A., A Tulsa Story: Learning to Live in Harmony with Nature, in J.A. Bullock, G.D. Haddow, and K.S. Haddow, eds., *Global Warming, Natural Hazards, and Emergency Management* (Boca Raton, FL: CRC Press, 2009), 84–113, at 94, emphasis in the original.

Pielke, R.A. Jr., Rethinking the Role of Adaptation in Climate Policy, *Global Environmental Change* 8 (1998), 159–170.

6 Case Studies in Long-Term Disaster Risk Reduction

INTRODUCTION

The increased frequency and severity of disasters are two of the principal impacts of climate change. It is becoming increasingly clear that community leaders around the country must be prepared to deal with the fact that climate change will influence how often and how destructive disasters will impact their communities, their economies, and their environment. The lessons learned from communities that have dealt with chronic disaster risks over long periods should serve as a guide for communities seeking to adapt to a changing climate.

This chapter presents three case studies of communities from around the country that have taken an aggressive, community-based approach to reducing the impacts of their chronic hazard risks. This set of case studies concern disaster risk reduction actions taken by Tulsa, Oklahoma, Berkeley, California, and Napa, California.

For decades, Tulsa suffered chronic floods that resulted in increasing loss of life and injury and severe damage to private property, public infrastructure, and the community's economy and environment. Citizen action, community leadership, and a shared vision of a safer community drove the effort to make Tulsa flood-resistant.

The City of Berkeley is the home of the University of California Berkeley and sits atop the Hayward earthquake fault. City leaders, the university, and the community-at-large have worked together for years to create, fund, and implement a series of community-based hazard mitigation programs designed to reduce the devastation when the next earthquake strikes.

The case study concerning the Living River Flood Management project in the Napa (California) River Valley highlights several elements critical to the success of a regional approach to risk reduction such as a countywide planning process, involvement of the private sector, detailed knowledge of the risk and potential mitigation measures, and participation by the entire population of the county in making the plan a reality.

These three case studies were authored by individuals who have been deeply involved in these mitigation efforts and offer insights and ideas that community leaders across the country should consider in developing similar efforts to address the risks to their communities and to mitigate the negative impacts of climate change in the future.

A TULSA STORY: LEARNING TO LIVE IN HARMONY WITH NATURE

Ann Patton and Crystal Kline

INTRODUCTION

Stepping gingerly over muck-slicked floors, upturned appliances, soggy sofas, and sodden carpets, survivors in the Meadowbrook neighborhood gathered in Carol Williams's flooded living room. It was June 8, 1974. The latest in a series of Tulsa floods had just flashed down Mingo Creek, directly through their neighborhood—again. People had lost count of how many times the neighborhood flooded since it was built in the 1950s.

The water was down now, but it was dark and dangerous in Carol's living room, a haven for snakes and spiders, floors too slimy to walk, and nobody was sure about the wiring. The air was heavy with the stench of foul water. Carol recalled a woman running through the streets in the night, illuminated by lightning flashes, screaming, "My baby, my baby!" Carl Moose spoke quietly about wrenching his boat from the garage just in time to run his latest flood rescue, now becoming almost routine. Bob Miller said his family spent his daughter's ninth birthday stranded on their rooftop, watching their cat drown, with water lapping to their eaves—again.

Everybody agreed on one thing: We have to do something.

The '74 flood was neither the first nor the last on Mingo Creek. But the group that formed that day began a fight that would, in time, change the way Tulsa does business and would influence, to some degree, the nation's disaster programs, too.

This chapter describes some of what happened in Tulsa and what we learned about ways to build a community that is safe, secure, and sustainable. This chapter includes a bit about the place and characters; about death and disaster, about some of the programs and policies that helped move us forward.

Because this account must be abbreviated, it cannot properly acknowledge the many, many people who dedicated their time and talents to help improve our town. The Tulsa story must begin and end with thanks to these many partners, in our hometown and also from afar, who helped us learn from disaster and turn it into community progress.

Tulsa's Story

Some say a fair amount of human advancement arises in response to tragedy. So it has been in Tulsa.

This section describes how we made our way along, by trial and error, disaster by disaster, to reduce the risks that have plagued our lives since humans moved to this locale. It focuses on the years of significant change since the June 8, 1974, flood. Those years could be divided into a series of eras, and this writing follows that pattern:

1974–1984: Conflict and confrontation
1984–1990: Challenge and change
1990–1998: Integration
1998–2002: Collaboration and expansion
2002–2008: Sustainability

A Crossroads Place

Tulsa was born in northeastern Indian Territory, now Oklahoma, after the Trail of Tears, when Lockapoka Creeks camped on a high bank of the Arkansas River. We call their site Council Oak, after a venerable nearby tree.

This is a crossroads place. The town was built on rolling terrain, where the low, timbered Ozark hills meet the plains; at a weather junction where hot, dry air from the west collides with hot, humid southern air and cool northern fronts. We call this convergence zone Tornado Alley. It is prone to violent storms that can spawn tornadoes and flash floods that barrel down the many creeks that flow into the Arkansas River.

Its early tents, shacks, and dusty streets were peopled by pioneers, wildcatters, and Sooners, folk who made their own rules and lived by a frontier ethic: a man has a right to do what he wants with his land. In 1905, oil was discovered at Tulsa's doorstep, bringing a gush of wealth. The town boomed. Oil barons built a flourishing city with tree-lined boulevards and marble mansions. They established a tradition of fierce civic pride and generous donations to better their community. To this day, all current evidence to the contrary, Tulsans believe they live in the Oil Capital of the World.

The Arkansas River flooded pretty much every year, through the Roaring Twenties and into the Depression, with a possible exception of the Dust Bowl years. Major disasters produced changes. After the 1908 flood, Tulsa changed its form of government to the Galveston-disaster model, the

City Commission government. After the 1923 flood, Tulsans produced a landmark drinking-water system and preserved a 2800-acre open-space park in the Bird Creek bottoms. During World War II, after the 1943 flood, the US Army Corps of Engineers built Arkansas River levees around Tulsa's precious oil refineries. Floods in 1957 and 1959 produced the push that resulted in the Corps' Keystone Dam on the Arkansas River *upstream* from Tulsa. The Keystone Dam was closed in 1964, producing community euphoria. Tulsans believed they would never flood again, a fantasy that lasted for many years.

By and large, nature's extremes were viewed as something to endure. The Weather Service logged a tornado touchdown somewhere in Tulsa County, on the average, every year during the twentieth century; but Tulsans firmly believed an old Indian legend that no tornado would touch down in the city; something about hills to the west. The place also produced killing summer heat and winter cold, floods, and droughts; trouble was a way of life in Oklahoma.

Meanwhile, Tulsa was growing. Many early settlers had favored the high ground, perhaps because they were in close touch with nature or perhaps influenced by Native Americans who tended to honor natural mores. Now homes and businesses spilled over the highlands and down into the bottoms of tributary creeks with names such as Mingo, Joe, Fred, Dirty Butter, Bird, and Haikey—names that would become infamous, in time, as flood followed flood, over and over again: 1957, 1959, 1963, 1968, 1970, and more.

1974–1984: Conflict and Confrontation

By 1974, when Carol Williams convened that neighborhood meeting in her flooded living room, Tulsans had become numb to flooding. Mike McCool, now Tulsa's emergency manager but then a cop, cannot count the times he ripped off his gun belt and dived into a flood to rescue some hapless citizen. "It was just the way life was in Tulsa," he says.

After the Mother's Day flood in 1970, Tulsa joined the federal flood insurance program and promised to regulate floodplain land use—but the city neglected to adopt maps that would have made the regulations work. Flash floods came in rapid succession in 1971 and 1973, followed by four in 1974, dubbed "the year of the flood." The June 8 storm was the shocker: flash flooding and three tornadoes racked Tulsa, shredding the myth of invulnerability and leaving $18 million in damages. I was a newspaper reporter then, trying to make sense of it all, and I could not imagine a worse disaster.

Carol's group named itself Tulsans for a Better Community and began tireless agitation for flood control. They drew in supporters from across the city, including courageous maverick technical experts, such as fiery activist Ron Flanagan, a visionary planning consultant who dedicated his life to stopping Tulsa floods. They succeeded in creating a remarkable pool of expertise on the subject, luring in leading technical experts not only from Tulsa but also from across the country.

Locally, they perceived their enemy to be the Home Builders Association. Enmity reigned. It was the decade called Tulsa's Great Drainage War, as protestors played a clenched-teeth game with development interests. Generally, two steps forward toward stronger flood management were countered by a step or two backward when the next election favored pro-development interests, who dubbed the activists as "no growth freaks."

Tulsans for a Better Community matured into a skilled advocacy group, in part because members did their homework, tried to speak with facts, and knew when to attack and when to thank. Their advocacy program evolved into four major points:

- Stop new buildings that will flood or make anybody else flood worse
- Clear the most dangerous of the flood-prone buildings and turn the land into parks
- Carefully install remedial works, such as channels and detention ponds to hold and convey water, considering the off-site and future impacts of the works, watershed-wide
- Involve citizens at every point

Carol Williams epitomized the intense, diverse, and colorful group. Carol's specialty was using surprise, unorthodox techniques. She would identify a favorite dessert of a mayor or a department head and shamelessly curry favor by bringing it on her lobbying visits. It would not be a long stretch to say that she garnered a $150 million Corps' flood project on Mingo Creek with her fabled raisin pies for the congressman's aide. Carol could size up people quickly, usually by analyzing their shoes, and adjust her technique for the audience. She left one nonproductive meeting in disgust, saying, "What could you expect from an entire room of black wingtips?" When an embattled commissioner questioned why group members, mostly young mothers, brought their children to the endless string of flood meetings, Carol retorted: "We're training them to keep after you when we die."

The Memorial Day flood of 1976 struck in the middle of the night, a 3-hour, 10-inch deluge centered over the headwaters of Mingo, Joe, and Haikey creeks. The flood killed three and left some $40 million in damages to 3000 buildings. Enraged flood victims stormed City Hall, and newly elected commissioners, sympathetic, responded with a wave of actions. They imposed a temporary moratorium on floodplain building, hired the city's first hydrologist, Charles Hardt, assigned planner Stan Williams to develop a set of comprehensive policies, began master drainage planning, and gained public approval for the first flood control bond issue in many years. (Since 1977, Tulsa voters have not turned down a bond issue or sales-tax initiative for flood control, according to Tulsa Budget Director Pat Connelly.)

Within a couple of years, regulation was softened after a pro-development commission came into office, but the main body of the new program held. Although the battles were far from over, in large part Tulsa appeared to have stopped creating new problems. Over at least the next three decades, Tulsans could say proudly that there was no record of flooding in any new building that was constructed in accord with the 1977 regulations.

1984–1990: Challenge and Change

The 1984 election was another upset. Three of the five city commissioners were sympathetic to flood victims. In fact, the new mayor, Terry Young, had campaigned on a pledge to work on flood issues, and the new Street Commissioner (directly responsible for flood programs), J.D. Metcalfe, was a patrician industrialist who was a member of Tulsans for a Better Community. (I came into City Hall as J.D.'s aide, by the way.)

They had been in office 19 days when the worst flood hit on Memorial Day 1984, killing 14 and leaving $183 million in damages to 7000 homes and businesses. We huddled in the Emergency Operations Center, shell-shocked by reports of Tulsans drowning on lands that had flooded over and over before. Young and Metcalfe vowed right then that things would never be the same—whatever the political cost.

Within hours, we had mobilized a flood-hazard mitigation team. We proceeded with a great sense of urgency. We had learned over the years, disaster by disaster, what we needed to do to seize this moment and execute bold plans. Within days, we had assessed the damage, identified the areas of highest hazard, slapped on a rebuilding moratorium, and identified repeated flooded properties that were candidates for acquisition. One goal was to stop the flooding by clearing the most vulnerable buildings and moving their owners to dry sites. Within 15 days, when the Federal Emergency Management Agency (FEMA) came to town, we were able to meet them at the door with our plans in hand and ask for help to fund them.

It was a fight. FEMA didn't want to fund a buyout, then considered a radical, harebrained scheme. Political opponents charged that this was a "bailout" of people who should have known better than to live in a floodplain. We countered that many of the buildings we identified for buyouts were no longer viable; some had flooded to the ceiling five times in 6 years, and most would continue to be flood-prone even when all planned structural projects were completed someday in the distant future.

Ultimately, we were able to gain approval to purchase 300 single-family homes and 228 mobile home pads. Mayor Young won over FEMA, and the $17.6 budget included $1.8 in federal and $11.5 million in local funds, plus the proceeds of insurance claims for homes we purchased.

Within a year, we had established a Stormwater Management Department to centralize all flood functions, headed by planner/attorney Stan Williams, which was creating a unified local program to manage flood issues. Within 2 years, we had instituted a stormwater utility fee, a $2 monthly charge on everybody's water bill, for stable funding of maintenance, management, and planning. We conducted aggressive maintenance and public education programs. We held hundreds of public meetings to get citizens involved in master drainage plans for the entire city. Spurred by planners Sandra Downie and Ron Flanagan, we began including recreation facilities, including trails and soccer fields, in flood control channels and detention ponds, bringing in a new and positive constituency for stormwater management.

Mayor Terry Young lost the spring 1986 election, but Commissioner J.D. Metcalfe was reelected, and the program continued to evolve.

Another flood hit in October 1986, this time on the Arkansas River. The remnants of a hurricane dropped a 24-inch rain upstream of Keystone Dam, forcing the Corps to release upwards of 305,000 cubic feet per second downstream. It was a challenging time. Every major stream in northeast Oklahoma was at flood, including the Arkansas at Tulsa—despite Tulsans' fond belief that the Arkansas would never flood again.

At Tulsa, a private levee broke, flooding 64 buildings. Within days, Tulsa dispatched its hazard-mitigation team and cleared 13 destroyed dwellings, helping their owners move to dry sites.

Overall, the management team worked well, minimizing damages and dangers as much as possible. The new system had passed its first big test.

1990–1998: Integration

In the 1990s, Tulsans began to pull together, united in the vision of a flood-free city. Strong leaders successfully campaigned to change Tulsa's city government from the commission to mayor-council form. The change in 1990 meant that leaders such as Commissioner J.D. Metcalfe, who had championed the change, left City Hall. Action shifted into a new Public Works Department, headed by hydrologist Charles Hardt; and stormwater management slowly became institutionalized into city operations under Hardt's strong leadership. One of Hardt's skills is building bridges among warring groups, and he helped bring adversaries together to jointly build a safer city. Some former adversaries became strong advocates for floodplain management, and eventually it became a generally accepted element of the city's services (see Figures 6.1 and 6.2).

In 1987, researcher Claire Rubin had reported that Tulsa County had the most (to that time) federally declared flood disasters of any other community—nine in 15 years. Then, in 1992, FEMA ranked Tulsa's flood program tops in the nation in its new Community Rating System program. Tulsans generally understood that this community, which had one of the worst flooding problems in the nation, was becoming a national model, and they were proud.

Interest in floodplain management peaked again in 1993 when the Mississippi River flooded. With FEMA's new interest in mitigation, floodplain clearance became a respected tool. Tulsa

FIGURE 6.1 Sandbaggers fight Tulsa's 1986 flood. (*Tulsa Tribune* photo.)

FIGURE 6.2 Some charter members of the team that developed Tulsa's flood program. (Tulsa Partners photo.)

stepped up its ongoing floodplain clearance program. Capital packages routinely included modest funding for acquisition, which the city used as local match with FEMA funding for a continuing predisaster floodplain clearance program.

By the end of the decade, Tulsa had cleared more than 1000 of its most dangerous buildings from its floodplains, using the open lands for parks, trails, open space storage, and flood control works.

For the first time since at least statehood, the 1990s decade passed with no significant flooding in Tulsa.

1998–2002: Collaboration and Expansion

In 1997, FEMA director James Lee Witt launched a new initiative named Project Impact that was intended to empower local communities to reduce disaster losses. The idea was to scatter some FEMA money around the country, with few strings, and let locals come up with innovative ways to work out hazard-mitigation techniques to create disaster-resistant communities. The ultimate goal, Witt said, was to develop public-private partnerships to change the culture and to establish new cultures that value preparedness and mitigation. In late 1998, Tulsa was fortunate to receive a Project Impact grant for $500,000. I became director of the Project Impact program, named Tulsa Partners.

The Project Impact grant extended over 3 years and allowed us to expand our hazard-mitigation work beyond flooding into other hazards. We focused on windstorms and tornadoes, lightning, extreme heat and drought, winter storms, hazardous materials, and terrorism after the 9/11 attacks.

It also taught us the magic of working through public-private partnerships. Most first-responder organizations and major business leaders became enthusiastic participants in Tulsa Partners. In short order, we had a cadre of dedicated partners working on a very wide range of public education and demonstration projects. They ranged from our SafeRoom initiative to McReady (see box, "Example Projects") (see Figure 6.3).

These partners were and are amazing—able, selfless, altruistic, interested in working together without personal gain, and united by a common goal to build a disaster-resistant community. As James Lee Witt once said, there is something about the Project Impact process that reaches down into your community and brings out the best in the best of your citizens; he was right. As they worked together, they moved Tulsa into a new era of cooperation. (Wonder of wonders, Project Impact even brought us together with the Home Builders Association of Greater Tulsa, who became the best of partners for the SafeRoom initiative and the green-building initiative called the Millennium Center.)

It is really true that when it comes to disasters, we have outgrown most of the turf building and petty competitiveness in favor of collaboration and partnership.

Another important advance was long-needed predisaster multihazard mitigation planning, which got under way in earnest around the turn of the century. Shepherded by planner Ron Flanagan,

FIGURE 6.3 Tulsa Partners management team, 2000. (Photo by Ann Patton.)

Tulsa's plan was one of the first approved in the nation and laid out a roadmap for long-term work toward becoming a disaster-resistant community.

When the Project Impact grant expired after 3 years, the City of Tulsa and various other sponsors continued to fund the program for several years. In 2000 we had established a 501-C-3, Tulsa Partners Inc., now ably directed by Tim Lovell, which serves as a useful vehicle for mobilizing public and private donations and creating innovative programs.

2002–2008: Sustainability

Recent Tulsa Partners projects include continuity planning for nonprofits and small businesses, in conjunction with the Institute for Business & Home Safety, disaster safety for children and care providers, in partnership with Save the Children, and public education and planning for preparedness and mitigation.

At the turn of the new millennium, we expanded our scope again. We had long contended that hazard and environmental issues are two sides of the same coin. For example, disasters generate tremendous waste and losses, and environmental problems are, in essence, slow disasters. If a house blows away or washes away, it is not sustainable. The most recent expansion of our program includes a shift toward sustainability. Our upgraded goal is to build a disaster-resistant, *sustainable* community.

This goal is at the heart of one of Tulsa Partners' current projects. Still in the planning and fund-raising phase, it is named the Millennium Center. A dedicated group is working to build this demonstration house at the Tulsa Zoo to provide fun, family-friendly, hands-on education on how to live safely and in harmony with Mother Nature (see Figure 6.4).

FIGURE 6.4 Artist's rendering of the proposed Millennium Center at Tulsa Zoo, a demonstration project to show how to live safely in Tornado Alley in harmony with Mother Nature. (From: Tulsa Partners.)

EXAMPLE PROJECTS

Here are some examples of Tulsa-based initiatives that may offer lessons for people working to build disaster-resistant, sustainable communities and live in greater harmony with nature. Many of these projects were born in Tulsa Project Impact or share a similar philosophy.

TULSA PARTNERS

Tulsa Partners Inc. is a 501-C-3 nonprofit program that continues the work begun by FEMA's Project Impact: creating partnerships to build disaster-resistant, sustainable communities. It serves as a catalyst for collaboration in a broad range of programs generally related to grassroots disaster management and sustainability. Public-private partners collaborate to accomplish their mission: to advance community goals, enhance quality of life, and create a more livable, safe, and sustainable community, in harmony with each other and nature (see Figures 6.5 and 6.6).

Over the past decade, Tulsa Partners has fielded more than 300 partners and hundreds of volunteers. This program has received some dozen awards, including several national awards, as well as grants from national and local organizations.

FIGURE 6.5 Tulsa Partners is an open, inclusive group working for a safe and sustainable community. (Tulsa Partners photo.)

FIGURE 6.6 Volunteers paint murals in shopping centers, hold pancake breakfasts, sponsor special displays and events, and conduct other education and demonstration projects. (Photo by Ann Patton.)

Funded by grants and donations, Tulsa Partners operates through a governing board, advisory committee, and numerous project-specific committees. It specializes in incubating innovative projects, deriving lessons learned, then institutionalizing those projects with other groups and proceeding to explore new ideas. Some of the best programs have been started by partners within their own organizations, sometimes independently and other times in concert with Tulsa Partners. Several of these programs are described in the following paragraphs.

See also www.TulsaPartners.org.

STORMREADY

As members of Tulsa Project Impact, the National Weather Service Tulsa staff in 1999 created a new program named StormReady. StormReady established preparedness criteria communities should meet to help them survive weather emergencies. For examples a town would need to establish a 24-hour warning system and emergency operations center, develop a formal hazardous weather plan with trained spotters, and provide public readiness education. When the criteria are met, the Weather Service will declare the community is StormReady (see Figure 6.7).

The Weather Service quickly took the program nationwide. As of January 2008, 50 communities have been designated StormReady sites.

See also http://www.stormready.noaa.gov/.

SAFEROOMS

Tulsa lies in the heart of Tornado Alley, but houses generally have been built without basements or other shelters. In 1998, when Texas Tech University developed new technology for building tornado SafeRooms, Tulsa Partners seized the opportunity to popularize them.

SafeRooms are specially anchored and armored closets or similar small enclosures. They can be built in new or existing buildings, inside or outside, above or below ground, to provide safe shelter in even the most dangerous tornadoes (see Figure 6.8).

With $50,000 from FEMA, Tulsa Partners formed a partnership with the Home Builders Association of Greater Tulsa to create some high-profile demonstration SafeRooms, coupled with an aggressive public-education program. When the disastrous May 1999 tornado hit Oklahoma, President Clinton kicked off a FEMA-supported SafeRoom initiative, which was later replicated in some other states, too.

FIGURE 6.7 Tulsa Partners is an open, inclusive, and diverse group. (Tulsa Partners photo.)

FIGURE 6.8 Dr. Ernst Kiesling, Texas Tech inventor of the SafeRoom, checks a surviving SafeRoom after the Moore, Oklahoma, tornado in 2003. (Photo by Ann Patton.)

Within a few years, tens of thousands of SafeRooms were built across Oklahoma. They were used successfully in subsequent tornadoes such as the 2003 tornado in Moore, Oklahoma.

See also *Safe Rooms Save Lives*, http://www.fema.gov/media-library-data/20130726-1515-20490-6905/okcase studies.pdf.

McREADY

In 2003, Tulsa Partners teamed up with McDonald's to provide a month-long family-preparedness blitz in McDonald's restaurants. The program, dubbed McReady, became institutionalized statewide under the auspices of the Oklahoma Emergency Management Department.

FIGURE 6.9 Family preparedness is a central focus of Tulsa Partners. (Tulsa World photo. Used with permission.)

Spring is the worst season for Oklahoma severe weather. The McReady program links emergency managers across Oklahoma with partners such as the National Weather Service and the Oklahoma Gas and Electric Company. Every April, they set up educational kiosks in McDonald's restaurants—inexpensive grids stocked with family-preparedness guides and other storm-safety materials printed by partners. McDonald's stores print survival tips on tray liners and bags. The low-cost program lasts a month, offering information to the thousands of customers who frequent the state's 170 McDonald's stores each day (see Figure 6.9).

See also www.McReady.org.

TULSA HUMAN RESPONSE COALITION

Tulsa's nonprofit agencies have banded together to plan for and manage social services during emergencies. The Tulsa Human Response Coalition (THRC) includes 50 agencies and first-responder groups. THRC goals are to work together to foster collaboration and communication, share resources, and reduce duplication of effort during emergencies. The group facilitates the human-services aspects of planning, preparedness, mitigation, response, and recovery to ensure effective service delivery. Example projects include emergency mental-health services and life-saving intervention during extreme heat and winter storms or other crises (see Figure 6.10).

The following are among the noteworthy advances in the group:

1. A plan for managing spontaneous volunteers during a disaster
2. A communications and call center connecting callers with social services and other resources through the 211 helpline
3. A backlash-mitigation plan, an innovative plan to help the community handle a crisis that could result in retaliation against a specific group (such as Muslims after the 9/11 attacks)

See also www.CSCTulsa.org.

FIRST RESPONDERS

Tulsa first-responder organizations have a broad spectrum of programs, some inspired or encouraged by Tulsa Partners. Collaboration is the norm and the key to success. The Tulsa Area Emergency Management Agency serves as a central coordinator and supporter for many

FIGURE 6.10 McReady partners kick off their annual blitz of preparedness information, 2004. (Photo by Bob Patton.)

of these activities. First responders meet regularly in several venues, including a standing Homeland Security Task Force, to share information and ideas.

Activities include Community Emergency Response Teams, a police Disaster Response Team, and a Volunteers in Police Service cadre. Tulsa has a stellar constellation of partners and programs related to emergency medicine and public health, generally arising from the Emergency Medical Services Authority (ambulance services), Metropolitan Medical Response System, Medical Reserve Corps, Tulsa Health Department, and university medical programs. One current focus is planning for public-health emergencies and pandemic flu.

See also www.OKMRC.org.

DISASTER-RESISTANT BUSINESS COUNCIL

Tulsa is strongly committed to encouraging continuity of operations planning for businesses, nonprofits, and government agencies.

In 2007 Tulsa Partners volunteers formed the Disaster-Resistant Business Council (DRBC) to help spearhead continuity planning. The DRBC is chaired by State Farm Insurance executive Dave Hall. It is a national pilot for the Institute for Business & Home Safety's Open for Business® program. Members include the Tulsa Metro Chamber of Commerce, the Association of Contingency Planners, and the Oklahoma Insurance Department. The DRBC supports Open for Business planning through workshops, public events and education, direct training, and collaboration with other programs. Examples of recent events include disaster-planning workshops for long-term care, hospitals, businesses, and child-care providers.

See also www.IBHS.org.

SAVE THE CHILDREN/TULSA PARTNERS INITIATIVE

In 2007 Tulsa Partners joined with the international Save the Children group to establish a demonstration project. The general goal is to develop and document ways local coalitions can improve child safety in disasters.

With a wide variety of partners, the group is working to provide preparedness training for children and care providers; to provide continuity of care through Open for Business planning; and to mobilize neighborhoods and the general community in support of children and their child-care centers (see Figure 6.11).

FIGURE 6.11 Allison McKee is proud of her disaster safety kit provided by Save the Children. (Photo by Elaine Perkins.)

They also developed a model children's annex to the community's Emergency Operations Plan. Tulsa has formally designated child-care centers as critical facilities; that is, safe and secure child care must continue in place even in times of emergency if the community is to function well and recover.

The Save the Children/Tulsa Partners initiative will be documented in a guidebook to help other communities learn from Tulsa's demonstration program.

See also www.SavetheChildren.org.

PLANNING

Tulsa's commitment to hazards planning is perhaps best exemplified by Ron Flanagan, a planning consultant and activist whose dedication to Tulsa extends over more than 35 years. Notable plans include master drainage plans for all watersheds. Specialized plans guide floodplain management, prioritized capital and acquisition projects, protection of critical facilities, and hazard mitigation. Flanagan served as consultant and catalyst for many of those plans, including the City of Tulsa's hazard-mitigation plan; this prototype plan was adopted on November 25, 2002, one of the first in the nation. Mitigation plan updates include human-caused hazards. A plan for hazard mitigation in historic buildings is under way (see Figure 6.12).

Flanagan and others helped Tulsa expand the concept of disaster management to include broader issues and constituencies. For example, maintenance easements along drainage channels became the backbone of a community recreation trails system, now extending over more than 50 miles of trails with plans for more. Stormwater detention basins are used for open space and sports fields. Tulsa hazards planning tends toward integration into the larger community fabric.

See also www.rdflanagan.com/Tulsa/Tulsa_NHM_book.pdf.

ENVIRONMENTAL PROTECTION

Tulsa is greening up in recent years, with a growing commitment to the environment.

The Metropolitan Environmental Trust (M.e.t.) champions environmental causes in this area. The M.e.t. provides recycling education and services for Tulsa and its suburban communities. The M.e.t. has recycled 100 million pounds of newsprint in 15 years, for example.

FIGURE 6.12 Tulsa's Centennial Park stormwater detention pond in downtown Tulsa. (Photo by Ron Flanagan.)

Funded by local governments, grants, and private donations, the M.e.t. is considered an authority and program catalyst on many environmental issues. The M.e.t's secrets of success include working to build relationships, finding ways so everybody can win, and fostering collaboration, not competition.

Other programs focus on air and water quality. The poultry industry in northeastern Oklahoma and Arkansas has threatened the quality of Tulsa's drinking water; Tulsa has launched a vigorous team effort to protect its excellent drinking water, long a source of civic pride.

Another noteworthy environmental program is run by the private nonprofit Up with Trees, whose volunteers have planted more than 16,000 trees at 400 sites along Tulsa's streets and expressways since the program began in 1976. In 2007, Up with Trees joined with the city to plant trees along Tulsa flood-control channels and in detention basins, too.

See also www.MetRecycle.com and www.UpwithTrees.org.

THE MILLENNIUM HOUSE AND MILLENNIUM CENTER

Inspired by Project Impact, in the year 2000 Tulsa Partner Don McCarthy had a dream: to build a demonstration house that would show how to live safely, in harmony with the environment, at a modest price. He called it the Millennium House. Virtually without funding, except for $15,000 contributed by Tulsa Project Impact, McCarthy and volunteers nonetheless got his house built by 2004. It was open to the public for a year before it was turned over to a low-income family who can enjoy utility costs of little more than $100 a year.

McCarthy's Millennium House inspired an even larger idea. Tulsa Partners pledged to build a permanent demonstration house to show how to live safely in Tornado Alley while also protecting the environment. They named it the Millennium Center. After 2 years of team building, planning, and integrating hazard and environmental techniques, the planning group is currently raising funds to build and maintain the Millennium Center and its fun, family-friendly exhibits.

See also www.mctulsa.org.

A DISASTER-RESILIENT COMMUNITY

Tulsa's skill in managing disaster was tested most recently on December 9, 2007, when an ice storm destroyed tens of thousands of trees and threw 75 percent of Tulsans into darkness. The power outage was the largest in Oklahoma history, with more than 600,000 customer accounts without electricity for upward of a week or more (see Figure 6.13).

In the Emergency Operations Center, Tulsa Mayor Kathy Taylor and her team developed a three-part recovery program:

1. Removing and disposing of more than 2 million cubic yards of debris
2. Mobilizing volunteers, including church members and 96 electricians, to help with home repairs, restoration of electricity, and debris clearance
3. Restoring the city's shattered urban tree canopy, in a public-private wave of tree plantings by the city and Up with Trees

Meanwhile, across the city, neighbors helped neighbors. The Tulsa Community Foundation launched a campaign to raise funds for emergency human needs. The Tulsa Human Response Coalition established a one-stop center operated by the Tulsa Urban League to help low-income people with critical needs, and the electric company and community planners turned their thoughts to long-term mitigation measures, including burying power lines.

See also www.cityoftulsa.org/Storm.asp.

FIGURE 6.13 The December 9, 2007, ice storm left 75 percent of Tulsans without power. (City of Tulsa photo.)

These disaster-resilient programs evolved within days and weeks, born from a community habit—a culture, if you will—of collaborative hazard management developed over many years in Tulsa's search for ways to live safely in Tornado Alley in better harmony with nature.

TULSA CHRONOLOGY

1900–2000. National Weather Service records show tornado touchdowns somewhere in Tulsa County, on the average, nearly every year during the twentieth century.
1908, 1923, 1943, 1957, 1959. Major floods on the Arkansas River at Tulsa. Records show the Arkansas flooded more or less nearly every year from statehood (1907) until 1964, when Keystone Dam was closed upstream from Tulsa by the US Army Corps of Engineers.
1949, 1957, 1959, 1961, 1963, 1968. These years mark some of the early recorded floods on Mingo Creek and other tributary streams in Tulsa.
May 1970, Mother's Day. Floods on Mingo and Joe creeks cause $163,000 damages and prompt the city to join the flood-insurance program and adopt its first floodplain ordinance. But the city failed to adopt maps that would affect floodplain regulation.

US Army Corps of Engineers, Tulsa District, issues Haikey Creek flood report, the first in a series on Tulsa-area problem creeks.

June 8, 1974. Violent storms cause flooding on Mingo, Joe, Fry, and Haikey creeks. At least three tornadoes also ravage the city. Toll: more than $18 million in damages with more than 120 injured.
October 1975. Tulsa hires its first hydrologist, Charles Hardt, and begins developing comprehensive stormwater management policies.
December 5, 1975. F-3 tornado in Northeast Tulsa injures 38.
March 16, 1976. Tulsa storm sewer and runoff criteria adopted.
May 30, 1976. Memorial Day flood kills three and leaves $40 million in damages.

September 17, 1976. Tulsa imposes a building moratorium in floodplains with critical flooding problems until new maps and regulations can be devised. Moratorium lasted 2 years.
1977. Voters approve a bond issue for emergency flood-control projects, the first in many years.
1977–1979. Changes in maps and regulations are adopted, including requirements for stormwater detention and specific permits for floodplain or earth changes. Tulsa begins developing master drainage plans to coordinate changes within entire watersheds.
April 19, 1981. Easter tornado in East Tulsa leaves between $75 and $100 million in damage, mostly in an industrial area.
May 27, 1984. Memorial Day flood kills 14, injures 288, and causes $183 million in damages. Tulsa imposes a rebuilding moratorium in repeatedly damaged floodplains until an aggressive program buys and clears 300 homes and 228 mobile home pads. Regulations and standards are strengthened and extended to entire watersheds. Tulsa creates a new Stormwater Management Department to centralize and focus flood-control activities.
September 27, 1985. Tulsa establishes a stormwater utility charge on water bills, $2 a month per household. Proceeds are used for maintenance, management, and planning.
October 1986. Arkansas River flood causes $3 million in damages in Tulsa, $67 million in the region. Tulsa buys and clears 13 destroyed houses after that flood.
March 1987. Researcher Claire Rubin finds that Tulsa County leads the nation (to that time) in federally declared flood disasters, with nine in 15 years.
1990. After a change in the form of government to mayor/council, Tulsa integrates stormwater management into a new Public Works Department. Floodplain acquisition programs continue throughout the decade, with Tulsa purchasing a few properties every year, often with FEMA assistance. By the mid-1990s, Tulsa has cleared more than 1000 buildings from its floodplains.
1992. FEMA gives Tulsa's flood program its best rating in the new Community Rating System, giving Tulsa citizens the lowest flood insurance rates in the United States. Tulsa continued to lead the nation in CRS ratings for more than a decade.
April 24, 1993. East Tulsa/Catoosa F-4 tornado kills seven and injures 100.
April 19, 1995. The Oklahoma City bombing destroys the Alfred P. Murrah Federal Building, killing 168, injuring more than 800, and convincing Tulsans that we need to plan for human-caused hazards as well as natural ones.
1998. FEMA selects Tulsa for a $500,000 grant to start a new Project Impact program to create public-private community partnerships for multihazard mitigation. The goal was to create a disaster-resistant, sustainable community. Project Impact expanded Tulsa's floodplain vision to include other hazards, including tornadoes. A broad spectrum of educational and demonstration projects addresses reducing risk from tornadoes, extreme temperatures, lightning, floods, and other hazards. Demonstration projects include tornado SafeRooms, historic preservation, social services, fire and law enforcement, Citizen Corps, CERT, and terrorism protection. The grant extends over 3 years, during which time the group creates a 501-C-3 nonprofit named Tulsa Partners Inc. to continue multihazard mitigation programs.
May 3, 1999. West Tulsa tornado. Tulsa Partners expands its focus again, this time working to marry hazard mitigation with environmental protection. They are working to create a Millennium Center at the Tulsa Zoo, with fun, family-friendly exhibits showing how to live safely in Tornado Alley in harmony with Nature.

2004–2008. Tulsa celebrates anniversaries of its 1984 and 1986 floods. At this writing, in 2008, Tulsa can mark more than two decades without a major flood in the city—a dramatic change from the years when Tulsa experienced a flood on the average of every other year or more often. An ice storm on December 9, 2007, causes the state's largest power outage. Tulsa launches an aggressive program to restore power, curb losses, and restore tens of thousands of downed trees. These points of progress are tribute to the community's dedication to reducing risk and creating a disaster-resistant, sustainable community.

Lessons Learned

Here are a few of the lessons we've learned—almost all learned the hard way—on Tulsa's journey toward becoming a disaster-resistant, sustainable community:

- Start with a small hub of your very best people—kind, committed, selfless, and statesman-like. Develop a shared vision. Then build a holistic, inclusive partnership around that heart (see Figure 6.14).
- Engage a dedicated, able program champion.
- Establish broad goals, specific objectives, and flexible strategies that can be adapted to avoid land mines, avert problems, and seize opportunities.
- Think holistically. The more comprehensive your program is, the larger your constituencies can be.
- It is important to take a negative mission (such as regulating floodplain use) and convert it into a positive, synergistic mission (such as also providing community parks and open space).
- Find something that is working well and attach your program to it. It might be the Red Cross in one town or the United Way in another, or perhaps the churches or the library or the city council. Every town will have a good starting place.
- Partnerships should be mutually beneficial, and all strategies should be win/win. Learn to listen well to what your partners need, and find ways to deliver it—as long as it does not compromise your base principles.
- Marry opposites for a stronger program. As Dr. Mark Meo at the University of Oklahoma taught us, good public policy happens at the intersection of grassroots citizens and technical experts. It's true with many diverse populations. Engage academics and marry them with common-sense common folk, too, for another example.

FIGURE 6.14 Flood projects can be beautiful, as shown in the Centennial Park detention pond. (Photo by Ron Flanagan.)

- Never, ever underestimate the power of the news media. Find ways to inspire them to share your community vision. You are challenged to become a translator of technical jargon into memorable sound bites that motivate humankind.
- Plan to seize any postdisaster window of opportunity. It may be in your town, but you can also take advantage of disasters or trends (such as a jag of interest in green building) elsewhere that capture the public interest. Shamelessly take advantage of the *hazard du jour* and build on it.
- Once you are certain of your long-range goals and principles, dare to invite in your adversaries, listen sincerely, learn from them, and seek to convert them into supporters.
- Celebrate success. Always spin to the positive. There are no failures, only lessons learned.
- Find your best management style. We use a jazz-band system we learned from a Tulsa planner named Gerald Wilhite, with light central control and maximum freedom for innovation; shared vision holds it all together and keeps it working in harmony.
- Rejoice in independent successes. Perhaps the best measure of success occurs when people create independent programs that further your mission. The Tulsa motto (perhaps the secret to collaboration success) is "There is no end to what you can accomplish in this world if you don't care who gets the credit."

Tulsa Story Update 2015

Crystal Kline

It is now 40 years since Tulsans for a Better Community first came together in Carol Williams' flood-soaked home on Mingo Creek in 1974. In many respects, we have become an entirely different community. This city built on a river can now pretty much withstand a 1-percent flood, or a 100-year flood, as it used to be known. But we remain vulnerable to larger floods. Stated another way: We have reduced the *frequency* of flooding, but not all the potential for *less frequent but catastrophic* flooding, and disaster amnesia and complacency have become hazards of their own.

We have shifted to a multiple-hazards perspective in our planning. Thanks to Project Impact, lessons learned on how to work together with the community have been successfully applied to other hazards. As a result, other communities throughout the country have taken notice and adopted similar partnerships and measures. The working-level public and private sectors in Tulsa take disasters very seriously and refuse to participate in the kind of turf wars that can be such obstacles to progress.

We have also shifted from a single-purpose disaster program to one that includes the environment and sustainability. The climate and environment do have definitive connections to disaster preparedness. We were a bit ahead of our time when we first began to connect these dots, but the time is ripe now to engage in this significant and urgent conversation at both the local and national levels.

The Significance of Partnering with the Public and Private Sectors and with Nature

Tulsa did not succeed in overcoming nature; we learned to work with it. What we achieved in developing a highly successful, life and property-saving stormwater management program was only possible through learning to respect nature and through learning how to work together in a partnership that included all of us in both public and private sectors.

The stormwater mitigation projects have been supported with a combination of public and private money, including grants and stormwater fees included in each utility bill. There has also been a cost to the private sector in terms of increased standards, as well as vast private benefits in terms of expanded markets, increased safety, and greater livability.

The Gathering Place, a lavish privately financed park to be located where early settlers gathered at the water's edge, is a beautiful example of private support. Private philanthropists have joined forces, donating $350 million to expand Tulsa River Parks with the addition of a world-class destination park and recreation area. Included as part of the plan is Blair Pond, where park activity is centralized with water as the heart of the park, offering opportunities for boating, fishing, and wildlife habitat observation. The pond will be fed by both groundwater and runoff from Swan Creek, which is to the north of the park site, and a new wetland system will be used to filter out pollutants from stormwater runoff before the water is released into the pond.

It Takes a Disaster

Sadly, we also learned that unless a major disaster is fresh in the minds of a community, it is difficult to sustain public awareness about the need of preparedness and mitigation measures in anticipation of the next "someday" event. As new urgent local priorities emerge, if the past is any teacher, we know the public may not support the political courage, statesmanlike conduct, and long-range vision that disaster management demands. In the long term, as more time passes since the last major disaster and as the champions of change move on or die off, fewer people in the coming generations are aware of how past disasters necessitated the very delicate and balanced floodplain management ordinances, fees, and policies.

For so many years, Tulsa had experienced regular floods, and as we continued to develop—decreasing the amount of undeveloped drainage space where runoff waters could linger or flow (if there is no place for water to go, it will always go where it must, at its own time and pace)—each major flood became worse than the last. Thanks to the heroic and epic efforts of Tulsans for a Better Community, Mayor Terry Young, J.D. Metcalfe, Charles L. Hardt, Ron Flanagan, and many others too numerous to name, the Memorial Day Flood of 1984, which killed 14 people and wreaked damages that totaled more than $180 million, was the last of its kind.

We've had big rains—although not yet the monster rains that led to floods of such nightmarish proportions—but the stormwater management program and other mitigation measures that have been implemented since then have given the water a place to go, and thus prevented similar floods. Indeed, Tulsa has done such a fabulous job of learning to work with Mother Nature instead of against her that we have taken Tulsa from being known as the Worst Flooded City to ... well, a city that barely remembers we once had a flooding problem. In that, our greatest success may also be seen as our greatest hazard.

More than 1000 homes have been cleared from the floodplain; homes that once were subject to repetitive flooding. Homes like Carol Williams' home, and most others in her neighborhood. Pipes and channels have been converted into attractive community amenities. Today, the area of Mingo Creek that was once the site of such devastation is now a beautiful stormwater drainage area that doubles as a recreation area and includes soccer fields, running trails, and bicycle paths. Other such drainage/recreation areas exist throughout the city, with names such as Carol's Pond, J.D. Metcalfe Retention Area, and Flanagan's Way Trail. It is both a joyously beautiful and sadly dangerous thing that most people in the adjacent neighborhoods are unaware of the devastation and death that took place often just feet away from their own front doors only a few years before.

And with that forgetfulness, that sense of safety with no memory for how we got here, comes the peril of making the same mistakes, of undoing all the good we have achieved. With that collective amnesia and the resulting apathy, how do we hold the regulatory pieces together, and how do we hold the community support for the necessary funding? When it is so easy to use the stormwater management fund for capital improvements, and floodplain parks and open space for tax-generating buildings, how do we continue the public will to maintain the program when we haven't had any major floods in a generation?

We're Not Finished

One brightly positive development is the recent announcement that Tulsa was selected to be a part of the 100 Resilient Cities program. The Rockefeller Foundation created this significant program that funds cities from around the world to establish a new position in city government for a resiliency officer who would lead the city's continuity planning efforts in the face of a natural or man-made disaster. Tulsa takes its place in this program alongside major world cities that include Paris, Barcelona, and Athens. This designation is an honor that is a testament to those who worked and sacrificed for many years to manage Tulsa's flooding problem, and it serves as both a challenge to not allow ourselves to forget the past, as well as an opportunity to educate a new generation of Tulsans so as to continue the work we have begun in creating a disaster-resistant and sustainable community.

In addition to preserving the current successful stormwater management program, there are other challenges; residual issues that remain to be addressed if we are going to do justice to the work begun by Tulsans for a Better Community. The most urgent is the Arkansas River, which flows through Tulsa.

The Arkansas River has the potential for flooding at Tulsa on a catastrophic scale in ways we have not yet seen, especially if the city follows through with current plans to encourage river development. Danger points are river levees and dams designed to "protect" and foster development. A wise woman once told me, "Levees were built to fail." Today, the 70-year-old levees in Tulsa are in danger of failure due to advanced age. The 50-year-old Keystone dam has the capacity to pass nearly 1 million cubic feet of water per second, which would cause a flood of nearly biblical proportions downstream. Additionally, proposals to engineer the river, turn it into a series of lakes, and encourage river floodplain development could place even more Tulsans in jeopardy. Former City of Tulsa Public Works Director and Tulsa's first hydrologist, Charles Hardt, says, "We need to UN-develop the river… I think the proposed low-water dams are not an alternative that will work on this river, because of flooding issues. Anything that hangs up debris and interferes with the flow of water will be a negative and will be detrimental to life, property, and the ability of people to get medical attention and to get to safety…. The more we leave the river in a natural state, with open space and parks along the banks, the better off we will be."

The key to our future success in holding the commitment necessary in advance of disasters, and addressing these issues will be, as always, for the public and private sectors to continue to work with one another and to work with nature, not against her.

Charles Hardt said, "Every decision along the river needs to be weighed against the need to keep the river's right-of-way open, to minimize flood damage, debris, and death." We would add to this quote that every decision about Tulsa's growth and development needs to be weighed against the long-term environmental impact in terms of stormwater management and the need to prevent and/or minimize flood damage with its loss of life and property.

We have done some tremendous work in Tulsa and accomplished great things. We've saved lives and property, added to our city's economy and livability, and found creative ways to turn disaster into opportunity. But our work is far from finished. Tulsa's flood program has a proud past but an uncertain future. It is time to pass the baton on to a new generation of Tulsans, with the fond hope that they will preserve and improve on the decades of work to make Tulsa a truly disaster-resistant, sustainable community.

HAZARD MITIGATION IN BERKELEY, CALIFORNIA: PARTNERING FOR COMMUNITY ACTION

Arrietta Chakos

Introduction

Berkeley, California, on the San Francisco Bay's eastern edge, lies directly across the Golden Gate Bridge in a geologically active region. The precarious Hayward Fault slices through the city's hillside neighborhoods; other regional fault systems, including the San Andreas Fault, pose serious threats. Geologists forecast the region has a 70 percent probability that a major earthquake could strike in the next 25 years. The densely populated and developed region is susceptible to an array of natural hazards and has suffered most from wildland fires.

Potential damage from near-field effects of earthquakes, along with other natural hazards including urban/wildland fire, landslide, liquefaction, and creek flooding, is considerable.

Two regional disasters—the 1989 Loma Prieta earthquake and a disastrous 1991 urban wildland fire—moved local leaders to undertake efforts to better protect Berkeley's residents and its built environment. The 1989 earthquake was far enough away from the city to cause only minimal direct damage, but the significant regional damage, social disruption, and casualties coalesced into an alarming call. The subsequent 1991 East Bay Hills fire that hit both Berkeley and neighboring Oakland was more locally devastating—twenty-five people died in the East Bay Hills area and over 3200 homes were destroyed.

These disasters were reminders that local communities were often unaware of the consequences of development in hazard-prone areas, especially during periods without frequent natural hazard events (Chakos et al. 2002). The Berkeley area hadn't suffered a significant emergency for some time when the 1989 Loma Prieta earthquake struck, and officials responsible for safeguarding the city's well-being were largely unaware of the existing hazard vulnerability. There were, however, a few local leaders who recognized the disaster safety issues enough to know the need to accurately assess the community's risk and to develop a strategy to best reduce that risk given limited financial resources.

Community Leadership

Berkeley's City Council, at the behest of a council member, Alan Goldfarb, made the first steps toward community sustainability. An urban development expert, Goldfarb had long advocated for more involvement by city officials to respond to and recover from potential disasters. He worked with the regional council of governments, the Association of Bay Area Governments (Association of Bay Area Governments 1989), on a benchmark study about local government's responsibility and liability with respect to disaster readiness. Mr. Goldfarb convinced city council colleagues in July 1989 (a few months before the October 1989 Loma Prieta earthquake) to fund an Office of Emergency Services for Berkeley and pushed for an updated disaster-response plan. These actions prepared the way for the larger community changes that followed.

Once the 1989 and 1991 disasters hit, Berkeley residents saw they had to act on their own behalf to lessen the community's risk. Although the state and federal governments provided assistance for areas struck by disasters, this assistance was by no means enough to fully compensate people and their communities for direct (and indirect) financial and property losses. Berkeley's response in the postdisaster environment was an atypical focus on prevention. Common sense suggested that reducing risk before the next disaster through preparedness programs and strengthening structurally vulnerable buildings were more prudent approaches. Inspired by Goldfarb's leadership, local leaders made community safety and sustainability a common value and developed innovative, practical approaches to effectively reduce hazards' risk with long-term programs to support the viability of these efforts.

Strategic Approach to Survivability and Sustainability

The community took an innovative and multilevel approach that included

1. Developing disaster readiness networks in neighborhoods
2. Strengthening buildings and infrastructure
3. Acquiring state and federal partners to support local initiatives

Disaster Readiness Networks in Neighborhoods

Using disaster readiness as a community theme and collective value, Berkeley strived to become more disaster resilient. The City Council convened advisory commissions, the Disaster Council and the Fire Safety Commission, with appointees tasked to keep community readiness an important policy matter. The dedication and persistence of these groups ensure that hazard mitigation remains a high priority in municipal budget decisions.

Other initial efforts included

- Activating community residents to form neighborhood disaster preparedness groups. The city council funded preparedness classes taught by firefighters and medical professionals for those who lived and worked in Berkeley. The proviso was that block captains be assigned to the groups and that neighbors work together in trained teams to help one another in an emergency.
- Thousands of people have participated in the safety trainings and have been encouraged in this work by being awarded special disaster equipment and supplies funded by safety tax dollars. In Berkeley's 10 square miles, some 50 disaster-supply locations are established, including one at every public school.
- Annual community disaster drills use electoral precincts as the center of outreach efforts with the exercises similar to get-out-the-vote campaigns with precinct captains and well-organized community contact plans. Such efforts have successfully reached thousands of neighbors in a morning with only a few hundred volunteers, and are kept alive with the neighborhood networks essential to a disaster resilient community.

Strengthening Buildings and Infrastructure

The second aspect of the strategy was to strengthen the built environment to better withstand the kinds of expected regional disasters and to promote long-term sustainability. Local city council and school board members made community safety a priority by bringing attention to the need for local action to prevent future potential damage and sought voter approval of funding measures and developed several incentives programs, including those outlined below.

Between 1992 and 2002, Berkeley's city council took six special tax measures to its voters and got approval for over $362 million to retrofit every public school, every fire station, and essential municipal buildings. As well, a new emergency operations center and new schools were built, a pilot community warning system developed, and a project to install a backup water supply system to use in case a disaster cuts off the main existing supply begun. Berkeley voters approved all six special taxes for hazard mitigation with a supermajority vote. These taxes generated revenue for the seismic and fire-safety upgrades that took the last decade to complete.

The city government also developed programs without much fanfare to reduce risk in private sector buildings with new approaches to increasing safety that proved successful. The city crafted an innovative tax rebate, another Goldfarb initiative, to fund home-safety improvements. Owners who upgrade their homes are returned a percentage of the improvement costs as a rebate when they sell their house. This has been a successful program over the last 15 years, and Berkeley's retrofit rate (about 65 percent) is high in comparison with the rest of the region. The program was recently updated with better building standard requirements for more effective improvements. Using other fiscal, technical, and administrative incentives for private sector retrofit, many owners have retrofitted their buildings.

Other popular incentive programs have included a permit-fee waiver for seismic improvements in homes and some unreinforced brick buildings. Grant and loan programs were also aimed to assist eligible low-income seniors, disabled, and other low-income residents make safety improvements to their homes. The city's tool-lending library is a much-needed center for technical assistance where residents borrow tools needed to retrofit homes and confer with helpful staff members. Together, these programs spur community safety in privately owned buildings.

Acquiring State and Federal Partners to Support Local Initiatives

The third step of the strategy was to form partnerships with state and federal agencies to leverage local efforts. Cities are hard-pressed to implement predisaster safety programs without the technical and fiscal support of state and federal agencies; although contributions from these sources to local communities are decidedly scarce, they can be used strategically to supplement existing initiatives. Berkeley's officials saw that more intergovernmental affiliations would enhance their opportunity to secure outside technical resources and monies for locally funded projects. This realization became a crucial turning point for Berkeley, as it sought a working alliance with institutions and agencies that could help guide crucial hazard mitigation planning and decisions.

- Berkeley joined state coalitions of school districts and cities politically active in negotiating state-funded allocations for schools and local government buildings. With dedicated state legislators, the allied communities helped improve California laws on safety guidelines for schools and public buildings. With amendments, urban communities were able to compete for additional safety funding long out of arm's reach.
- FEMA grew to be a strong ally at the federal level (see box, "Berkeley and FEMA: Working Together").
- Berkeley's city government was in an optimal situation to benefit from a wide range of technical and policy resources in the San Francisco Bay Area. The University of California, Berkeley (UCB) campus provided expert consultation on questions with which city officials grappled. The city regularly worked with technical advisers from the campus on how best to evaluate existing buildings, how to define appropriate levels of risk, and how to make informed decisions about the efficacy of seismic retrofit proposals.

  The campus's role was quite important as a center with an array of interdisciplinary seismic safety experts—seismologists, geotechnical experts, structural engineers, architects, and public-policy experts. These same experts were typically Berkeley residents and actively worked to promote safety endeavors throughout the community, often giving their time at city council and school board meetings to brief elected representatives. Officials relied on UCB policy and research centers for continued policy and program direction as public sector construction projects proceeded.
- Other governmental agencies joined in—the Governor's Office of Emergency Services, Coastal Region (California OES), the US Geological Survey (USGS), the California Geological Survey (CGS), and the Association of Bay Area Governments (ABAG)—and steered the local governing bodies to consider policy and budget decisions that would be the most cost-effective and provide added measures of disaster protection.

The Bay Area Regional Earthquake Preparedness Project (BAREPP, a division of California OES) and the California Seismic Safety Commission led local presentations on risk, potential regional damage estimates, and mitigation, and educated the community on how it might address its seismic hazards.

Public discussion continued at many levels in the community, especially with community volunteers on municipal boards and commissions that directly advised the city council to act responsibly to limit the community's risk exposure. Berkeley's mitigation efforts progressed from its assertive seeking and use of the technical and policy resources at hand. Good ideas came from many sources, and they were scrutinized and shaped by a spirited public process.

BERKELEY AND FEMA: WORKING TOGETHER

HAZUS

In the mid-1990s, Berkeley officials were invited to serve on an oversight panel in the development of FEMA's Hazards US (HAZUS) loss estimation modeling application that over time became a baseline measurement for assessing community risk for many types of natural hazards. Because of this participation, Berkeley was an early HAZUS user and gained further local support for the ongoing hazards work using the findings of early HAZUS studies.

MITIGATION BENEFIT STUDY

Involvement in the HAZUS project led to more local government advisory support to FEMA from Berkeley for a benchmark study. Working with technical, policy, and practitioner experts, Director Witt's office prepared a report to Congress on the necessity for and benefits of predisaster mitigation. The report was instrumental in the establishment of FEMA's Project Impact, the Clinton administration's progressive policy response to the problem of natural disasters and their cascading effects. With seed funding for local governments, FEMA intended to knit together partnerships among cities, the private sector, and citizens to develop indigenous solutions to hazards risk in the nation's most risk-prone areas. This was no one-size-fits-all program. The ingenuity for local disaster problem solving was there with an engaged network of involved stakeholders working together. Berkeley and its activist community were ready for just such an experiment.

PROJECT IMPACT

FEMA officials knew well how to motivate municipal officials to get the best results from their common efforts. Project Impact's early days were marked by successful national conferences as the project was moved through the country. In December 1999, the annual workshop was held in Washington, DC, where the cities of Tulsa, Oklahoma, and Berkeley, California, were designated the Project Impact communities of the year.

The ripple effect of the agency award reinvigorated Berkeley's commitment to ongoing risk reduction. At the time, California cities had been hit by serious budget cuts and in such circumstances, disaster safety programs are often the first things to go in deference to the more immediate need for police, fire, and health services.

JOINING FEMA'S PROJECT IMPACT

FEMA and California's OES were mentors in Berkeley's early mitigation program implementation. Each agency advised the city through lengthy project planning for retrofit and new construction efforts with funding support to reconstruct schools, fire stations, and essential city buildings to modern building code standards for government buildings.

The designation as a Project Impact community gave Berkeley local and state funding leverage that kept critical programs funded for an additional 3 years.

The City Council allocated local monies, tripling the federal Project Impact funds for the Disaster Resistant Berkeley Program, an umbrella for the city's preparedness, response, and mitigation efforts. This investment signaled continued commitment to safety. Berkeley's interdisciplinary approach called for an involved staff team that leveraged internal technical and policy expertise. This integrative approach strengthened the city's capacity with its overall disaster response system and increased its recovery capacity.

Adding to the existing risk reduction and other incentive programs, Berkeley was spurred to deal with other technical issues with Project Impact state and federal partners, including

- Ongoing community organizing and safety programs. Community engagement efforts were further energized by Project Impact AmeriCorps participants made up of a 13-person team that worked for 3 months on neighborhood group and business preparedness. The team's ambitious schedule and outreach program involved 35 small businesses and 50 neighborhood block preparedness groups to renew safety plans and training. Project Impact's $335,000 seed funding was matched by three additional grants that supplemented by $60,000 grants from VISA USA and the State Department of Insurance for public information and home retrofit projects for low-income and senior homeowners.
- The community outreach activities revitalized the community with mitigation and preparedness work and built a more solid connection with business owners (see box "Public/Private Partnership"). This, in turn, strengthened voter support for a 2002 safety tax measure that funded more neighborhood network building through the emergency equipment and supply program. Without Project Impact support to fund compelling public information materials and outreach events, it would be difficult to know if continuing support would have been sustained.
- Hazards Mapping Project. The Hazards Mapping Project was completed with help from Project Impact connections with USGS and the CGS scientists with a compilation of the geologic hazards in Berkeley's immediate environs. A multiagency team worked for 2 years to assemble private and public-sector data and to map areas vulnerable to landslide, liquefaction, and ground-shaking hazards. The completed hazard maps delineate risk zones and were building blocks for more thorough loss-estimation scenarios that better refined local risk analysis.
- Land-use plan: with the federal imprimatur, the city mustered resources to take a more comprehensive approach to risk reduction and to solidify a cohesive program embedded in both the city's general and mitigation plans. Mitigation policy adoptions in these long-term land use planning documents would weave needed safety policies into the community's daily life. Berkeley's updated land-use master plan was published in 2003 after 52 public meetings over a three-year period. The final product included newly crafted disaster safety policies and mitigation goals in the plan's housing, transportation, and safety elements. City staff and the Planning Commission ensured that seismic safety was emphasized as a primary goal in the policy framework. Inclusion of these policies is a critical factor in the city's development and land-use planning, providing legal and regulatory authority for sustainable mitigation programs.
- Community Hazard Mitigation Plan. Publication of the general plan was a necessary step in the development of the city's comprehensive mitigation plan. This was followed by a community summit in December 2003 with 120 representatives from regional agencies, utility providers, and nonprofit agencies, along with the city, campus, and local school district. The summit focused community participants to come to consensus about how to affect disaster resilience and to identify future opportunities for working together. As a FEMA-designated Disaster Resistant University, UCB was also engaged in aggressive mitigation efforts with an ongoing $1 billion retrofit initiative. Berkeley's city council adopted California's first comprehensive Disaster Mitigation Plan in June 2004 (City of Berkeley 2004), and recently updated that initial plan. The Mitigation Plan signaled the community's commitment to safety, signifying the culmination of many years of investment in sustainability.

PUBLIC/PRIVATE PARTNERSHIP

The review of hazardous apartment buildings funded with FEMA monies was a unique collaboration among the city, FEMA, the Lawrence Berkeley National Laboratory (LBNL), and the Earthquake Engineering Research Institute (EERI). An inventory and risk analysis of most city buildings had been compiled some years before, but one class of buildings—soft-story, multiunit apartment buildings—had yet to be inventoried. In the 1995 Northridge earthquake, soft-story apartment buildings sustained serious damage, causing numerous deaths. City leaders wanted a realistic assessment of the threat posed by these buildings and did not want to ignore a potential residential-risk problem. In an unusual collaboration, senior EERI structural engineers and UC Berkeley engineering graduate students worked in teams to survey 150 buildings in fall 2001, gathering construction details on the most at-risk buildings. LBNL engineers, on a parallel project, did further research on this building type with local senior citizen residents, owners of a particularly vulnerable building. LBNL contributed to the development of sensing devices and conducted tests to help design effective retrofit options with additional grant funding from Hewlett Packard.

WHY PROJECT IMPACT WORKED IN THE BAY AREA

Communities face new challenges in the midst of daunting climate disruption threats. Given potentially insurmountable conditions, effective change, and strong community and government partnerships will be necessary elements of a successful response strategy. The Project Impact model worked because it was a healthy melding of local and federal interests. Communities were valued as genuine partners that brought resources, skills, and mitigation solutions to the table; FEMA's sponsorship without intrusion brought support that further encouraged local capacity.

These outcomes from the Project Impact partnerships demonstrate what a powerful model FEMA promulgated. It literally sowed the seeds of community change and empowerment in hundreds of communities. These cities and regions also formed useful networks that moved forward collective efforts through the sharing of best practices, formation of colleague alliances, and active advocacy for hazard mitigation at many levels.

The program embodies an adaptive leadership model (Heifetz 1994) as defined by Harvard University's Ronald Heifetz that defines how to mobilize communities to tackle the problems before them, activating the widest range of site-and problem-appropriate solutions. Project Impact served a crucial function as it used the best of its federal powers to enable widespread community transformation. The San Francisco Bay Area communities involved in the program—Oakland, San Leandro, Napa, Berkeley—modeled the community-protecting behaviors we need to see more of in our world. With neighborhood and business leaders as equal partners, the work to create disaster-resistant and sustainable cities was equitably and successfully begun.

Conclusion

After two regional disasters struck, community leaders and public officials in Berkeley, California, acted to reduce risk in the city. The investment in long-term sustainability came from many fronts and was a community-wide effort to keep disaster readiness and safety initiatives front-burner issues over an 18-year period. Though these efforts were initiated locally in response to severe hazard events, the city found an ally in FEMA and its Project Impact initiative. Project Impact became a regional umbrella in the San Francisco Bay Area that connected many independent risk-reduction efforts in the hazard-prone area.

Update: 2015

The City of Berkeley has continued in recent years to integrate planning and action to be a more resilient community. This comprehensive approach is reflected by efforts to reduce risk resulting from climate impacts, natural hazards, and energy system disruptions. The city's recent designation as a Rockefeller Foundation awardee in the 100 Resilient Cities (100RC) initiative further supports community safety implementation.

In November 2006, Berkeley voters approved ballot Measure G, which was a mandate to reduce community greenhouse gas (GHG) emissions by 80 percent by the year 2050. The measure called for development of a Climate Action Plan that mapped how to reach the measure's goals. The subsequent plan defined actions Berkeley could accomplish to reduce energy use in buildings and transportation and to reduce the community's solid waste.

The plan includes the city's GHG emissions sources and future emissions estimates, along with recommended actions. Berkeley's many programs included the nation's initial curbside recycling effort, use of biodiesel and car sharing, and requirements for energy-and water-saving measures when buildings are extensively renovated or sold. Most importantly, though, Berkeley was the first city to institute voter-approved climate protection.

The City of Berkeley is one of the first cities to work on local energy assurance planning. Because functioning energy systems are essential services for a community's daily life and in disaster response and recovery, cities must be ready and plan for energy system disruptions. Consequently, in 2012, the city addressed energy assurance through an assessment and planning effort that was linked to the Local Hazard Mitigation Plan update and outlines how to minimize impacts of power grid interruptions. City staff conducted a power supply and demand assessment, including energy use in government facilities and strategies to limit energy interruptions during emergencies. City staff reviewed the Climate Action Plan and related efforts to identify the city's energy profile information and linked planning through the Local Hazard Mitigation to develop a postdisaster contingency plan for up to 7 days.

In July 2014, Berkeley reached another milestone in its long-term resilience building with the launch of its 100RC program. The city's efforts "to prepare for, withstand and bounce back from catastrophic events and chronic stresses" was commemorated in a new partnership with the Rockefeller Foundation.

Berkeley's strategy will link across public and private sectors to strengthen resilience to physical hazards and technology disruptions and plan for adaptive responses to climate impacts such as sea-level rise and drought. The work will incorporate existing efforts, including work to seismically update buildings, improve neighborhood social capital and disaster preparedness, and renovate water infrastructure.

A chief resilience officer position is funded by 100RC, which selected Berkeley in its first network cohort. Neighboring Oakland and San Francisco were chosen and form the 100RC's only regional hub group in this global initiative.

"A community's overall resilience begins in neighborhoods. Berkeley residents and businesses, working alongside the city and other agencies, have made significant progress on efforts like increasing energy efficiency and solar installations, which reduce Berkeley's dependence on fossil fuels, and preparing for disasters through participation in Community Emergency Response Training. Many residents have also taken advantage of the City's Seismic Retrofit Refund Program to make their homes safer. Engagement and input from neighborhoods and civic groups will be crucial to Berkeley's resilience strategy" (City of Berkeley media release, dated July 16, 2014).

In the next 3 years, the Berkeley community will focus on how best to ensure that all its safety and adaptation planning are part of the city's daily life. Many programs are implemented so that disaster, climate, and social resilience are integral when local leaders and community members make land use and budget decisions. Even more prominent is the emphasis on community residents as active resilience partners in neighborhood planning and home safety programs. Berkeley is a city long respected for political action and social engagement; these traits have fueled the city's culture of resilience over the last 25 years.

LIVING RIVER: THE NAPA VALLEY FLOOD MANAGEMENT PLAN

David Dickson
MIG, Inc.

From all indications, we are entering into an era of natural disasters. Even though the causes of this change are global, the effects will be very local, affecting each of the communities we live in. In the community where I reside, California's Napa Valley, we have already had a preview of the devastation that climate change will bring. For our entire history, we have been overcome by major floods that destroy our community, taking away our lives, property, and peace of mind.

Unfortunately, flood-induced disasters like those that Napa faces will only become more frequent in California and elsewhere in the years ahead. Scientists who study weather patterns predict that the Bay Area in particular will be slammed with more extreme storms bringing more intense rainfall in the coming years. This is supported by research conducted by the National Oceanic and Atmospheric Administration (NOAA), the USGS, and other scientists at the recent California Climate Change Conference sponsored by the California Energy Commission and the California Environmental Protection Agency.

These same scientists predict that climate change in California will cause three troubling outcomes that will ultimately threaten the health and safety of every community. The first result is an increase in severe Pineapple Express storms from the Hawaii Island area. These storms carry intense amounts of warm rain that will lead to more flooding. The next major effect will be further loss of the Sierra snowpack as temperatures increase, leading to drinking water shortages. Finally, climate change will result in drier, warmer weather inland, leading to more wildfires.

How does this affect flood protection? Throughout California, levees, dams, flood-control channels, and bypass channels are being forced to manage water flows for which they were not designed. Even the most forward-thinking 1950s estimates of peak flood flows, such as those engineers designed for on San Lorenzo Creek, are now being shown to be at least 50 percent below what will now flow from the hills during each Pineapple Express storm.

This means that more and more communities will need to address the threat of flooding or risk the economic deterioration experienced by Napa over its history due to major, frequent floods. As we know, a community that is not economically healthy is not healthy. The community's heart is under attack, as economic problems cause social problems and put strain on almost every member of the community. This is the human aspect of what's at stake in flood protection.

Yet because of the huge cost of multiobjective flood protection mitigation, only a Napa-like planning and Community Coalition process is likely to result in the action needed to upgrade the flood-protection infrastructure of these communities.

California, in particular, has strict laws requiring voter approval of any new special taxes for flood control. A two-thirds supermajority is required. It used to be that communities relied on the US Army Corps of Engineers to pay for 80 percent or more of major infrastructure improvements, but those days are gone. The recent Water Resources Development Act authorized an additional 180 projects around the country to receive federal help. Yet, if recent history is any measure, federal appropriations for these projects will be a long time in coming, if they come at all. This will become more and more the case as climate change becomes a reality, forcing the federal government to transfer its limited dollars to the crisis of the year, such as Hurricane Katrina, leaving just enough funds to spread around to keep all of the other urgent projects going forward, if barely.

The Napa River project provides a case study of how a community has come to terms with its river and its flooding problem in a successful way. In this article, I will tell the story about the genesis of the Community Coalition planning process that secured the agreement and political support needed to pass a sales tax to raise the local share of what has turned out to be a $500 million investment in Living River flood protection throughout the Napa Valley.

The project has been under construction for 10 years now. What has been accomplished? What still needs to be done? The second section provides a project update, including the good, the bad, and the ugly of executing the largest public works project in the history of Napa. The final section concludes with some lessons learned, and outlines the elements that need to be in place for a disaster-prevention project of this size and complexity to be successful.

The Napa River is a thread that runs through the Napa Valley. Starting from its headwaters on top of Mt. St. Helena, the river levels out and meets up with the San Francisco Bay Estuary in the city of Napa, the major urban center of the Napa Valley. Given its position on the river, it is not so remarkable that the city sits where it does. The city is centered where the river meets Napa Creek and then turns back on itself in what locals call the Oxbow, making it the furthermost navigable point on the Napa River Estuary. The tides come in and out up to this point, about a third of the way up the 55-mile length of the river, which runs from Mt. St. Helena to the San Francisco Bay.

The Napa Valley community has had a love-hate relationship with its river since the area was settled in the mid-1800s. For decades, the river has provided fresh water for the Valley's many farms and vineyards, which still comprise its main industry to this day. Beautiful and idyllic, the river has also provided a home for fish and wildlife and a place for people to relax and play.

However, when it starts to rain, the river takes on a much more dangerous and threatening character. It floods over its banks, causing damage and loss wherever it flows. Unfortunately, this happens all too often. Napa is one of the most flood-prone communities in California, even though we have a total population of only 126,000 people. Since 1862, Napa Valley has endured 27 major floods. This can be devastating in the heart of downtown Napa, where the river can carry only 20,000 cubic feet per second.

In 1986, in the largest flood in Napa's recorded history, close to twice that volume overflowed the riverbanks. This 100-year flow also inundated the region in 1995 and 2005.

As with most rivers around the world, each time a serious flood happens, the community goes into crisis mode. It is best described by that knot in your stomach when you know people are being traumatized in your community, especially the elderly and more vulnerable, who invariably end up living in the floodplain because that is where the cheapest housing is. One is also thankful at these times for the emergency response system—the fire departments and human-service system, the shelters, the police, the water rescuers, the volunteers who bring food to the shelters, and the innkeepers who provide rooms to the evacuees. It is government at its best! Unfortunately, it was not that way in the 1986 flood, because there had not been a major flood in about 15 years, and people forget about floods quickly.

It is even worse in the small towns of Yountville and St. Helena, halfway up the Napa Valley, where one third of the town's housing stock is in the mobile home parks, which were built in the 1960s, before floodplain regulations. These mobile home parks have flooded regularly. The people in these parks are the ones that I think about during high water.

Historically, floods have not been the only problems connected to the Napa River. Fifty years ago, slaughterhouses, tanning factories, sanitation districts, and oil companies discharged their wastes directly into the Napa River. The river was diked and leveed, and industrial buildings and residences were built right on top of the natural floodplain terraces of the river.

But the tide has changed for the Napa, and today the river is arguably one of the most important waterways in the nation. A dedicated and diverse community of activists and agencies that fought to resurrect it has not only improved its water quality and secured thousands of acres of wildlife habitat along its banks but has created an important model that redefined America's approach to flood control.

Our community has tried to fashion a solution to the major flooding for its entire history. Since the 1960s, no fewer than four US Army Corps of Engineers proposals have been presented, voted on, and rejected. The projects proposed in the 1960s, the 1970s, and then again in 1995 just did not address the needs of the Napa community. They did not protect its migrating fish, riparian zone, and wetlands and did not protect it from floods or reduce the potential damage floods could cause to Napa's 7000 downtown structures, including its civic center, not to mention the lives of its citizens (see Figure 6.15).

Then a remarkable coming together occurred around flood control, which voter surveys said was the number one issue facing the community. Over a 30-month Community Coalition process, the community's business leaders, environmentalists, government officials, mobile home owners, neighborhoods, fishermen, canoers, Red Cross workers, gadflies, and others participated and coalesced around the concept of a Living River flood-protection and restoration plan for the Napa River. On March 3, 1998, the voters weighed in with the required two-thirds majority to raise taxes in Napa County in order to implement the Living River Flood Protection and Estuary Restoration Plan.

I had the fortunate opportunity and privilege to manage the planning process and build the community-based structure needed to bring about compromises and achieve this community consensus. I was the process architect and manager. It helps that I am a self-confessed consensus junkie. I have lived and worked in the Napa Valley community for over 30 years, and I had the networks, relationships, and understanding of the parochial and esoteric political sand traps that exist here.

The first thing our community demanded was that the US Army Corps of Engineers change its relationship with the community. The community wanted to take control of their county's government to make it work for the community. The Corps needed to agree to come out of its offices and mix it up with the community. It needed to hear and listen to us. The San Francisco Bay environmental community had to be embraced and accommodated, because the 425-square-mile Napa River watershed is the last undammed tributary flowing into the San Francisco Bay Estuary. It is also a critical salmon and steelhead habitat and home to special status species, including California freshwater shrimp, the salt marsh harvest mouse, and California Clapper rails. The regulatory agencies made it clear early in the process that they would not permit a typical Corps approach of encasing the river in concrete.

With the help of Senator Barbara Boxer, we got the Corps to the table, agreeing to use the congressional planning appropriation of $1 million to focus on the local Community Coalition process. The Corps needed to trust the process, but it was a new experience for them in many ways.

A Coalition Steering Committee was formed, composed of local elected officials and the presidents of the Friends of the Napa River, the Napa Valley Economic Development Corporation, and the Wine Institute.

FIGURE 6.15 Napa, California, February 14, 2006: This California home was raised 10 feet to prevent, or mitigate, flooding. (Photo by Adam DuBrowa.)

The first thing the committee did was develop a set of goals:

1. Protection from the 100-year flood
2. A living, vital Napa River
3. Economic revitalization
4. A cost that the citizens could support
5. Retaining our valuable federal project authorization (50 percent funding)
6. Watershed-wide planning and a solutions-integrated system

In essence, they wanted it all. In order to achieve a two-thirds vote on a tax increase, every influential sector of the community had to be satisfied—in fact, excited—about transforming a floodplain. So the goals were presented to a coalition of 27 local stakeholder organizations to see if they would commit to a process to develop a flood plan addressing all of the goals. If, in the end, they could not commit, well, at least we had given one last concerted community effort.

Everyone warily agreed to sit at the table and assist in resourcing the effort: the Corps as well as the 27 government agencies with jurisdiction over the Napa River and any development within its sphere of influence. Over 24 months, there were eight town-hall-type meetings involving 200 to 250 of Napa's finest minds, who actively participated to conceive a plan, check its constructability and science, and determine its financial feasibility. These meetings became a celebration of progress. Over the first 6 months, the theme of the Living River became a rallying point, a point of guiding light against which any idea would be tested to determine if it contributed to it or threatened the achievement of that goal.

The hard technical work took place in a continuous process to support the coalition's direction. The coalition hammered out financing plans, urban-design concepts and standards, and definitions of the Living River based in science. The community learned about things like dissolved oxygen levels, continuous fish and wildlife riparian corridors, geomorphically stable channels, and a river system's natural width-to-depth ratio. We were told how we fit in the big picture by the likes of Luna Leopold, the son of Aldo Leopold, the great environmentalist. Luna was in his late seventies at the time and is considered the father of modern river geomorphology.

The old timers of Napa have always believed that no flood-control solution was possible because of the tidal action in the Napa River. The scientists sat with them and talked these things out. The scientists had to demonstrate with computer models how the tides interact with the flood flows, and how the Living River Plan accommodated both flows to protect the city from flood damage. We learned right away, of course, that you cannot control floods. You plan for living with them. The lessons of 1993 on the Mississippi River and the Galloway Report were vital to the coalition.

The four technical committees were organized according to different focus areas: Living River, Up-Valley watershed management, urban design, and finance and regulatory issues. Each was made up of a cross section of paid staff, government staff, hired consultants, the Corps of Engineers, and local citizens with special capabilities such as landscape architecture, natural-resource management, and political organizing. The committees met in the same auditorium each Friday for 6 months, preparing details to present to the larger Community Coalition.

The community held a celebration of achievement at Chardonnay Hall at the fairgrounds with over 200 coalition participants in June 1996, when the concept was developed enough to pronounce it a plan. Then began a 1-year period of verification to see if the details supported a plan that could actually be implemented.

After 2 years of relentless and intense research and negotiations, the Corps, 27 other governmental agencies, and 25 local nongovernmental organizations hammered out a revolutionary Living River plan. Where the Corps had proposed floodwalls and levees, the Coalition proposed terraced marshes and broad wetlands. Where the Corps had proposed dredging the river deeper to allow it to carry more water faster, the Coalition proposed making it wider, by returning much of its floodplain. The plan had stiff requirements. We wanted to reconnect the river to its natural floodplain and maintain the natural depth-to-width ratio of the river. We wanted to restore historical tidal wetlands and implement watershed-management practices to maintain the natural riparian corridors along the river and tributaries. We needed to clean up contaminated river-adjacent properties, replace eight bridges that now act as dams during high flows, relocate, purchase, or elevate 150 homes, businesses, and mobile homes that were in the floodplain, and purchase over 900 acres of river adjacent agricultural lands.

Original estimates for the plan totaled $250 million. About $100 million was to come from the federal government and state environmental restoration grants and highway bridge funding. Then, $150 million was to come from local taxpayers and the tourists who visit Napa Valley. A half-percent increase in the local sales tax taps the tourists, who pay about one-third of the local sales tax. This was a very appealing feature of the finance plan to the citizens. Other tax-increase proposals were soundly rejected in community surveys conducted under the direction of the Community Coalition. The Coalition said the tax must expire after 20 years, and two citizen oversight committees were required in the tax measure to scrutinize expenditures and oversee the technical aspects of project implementation.

Professional public opinion surveys were conducted. By March 1997, the plan was verified to a point that we knew the voters would support it, the Corps of Engineers could participate in an environmentally restorative program of flood management instead of flood control, and the environmentalists would compromise and ultimately support the tax increase and actively campaign for the effort.

The community coalition process itself became the campaign. All 27 organizations at the table either supported or were silent during the campaign. A well-financed public-issue campaign was bankrolled by individual contributions, investments by several large wineries who wanted to bring the city of Napa into a more intimate relationship with the wine industry and Up-Valley ambience, and by the environmental and business communities. Groups that are usually at odds came together around the flood problems of Napa.

All five cities of the Napa Valley and the county agreed in a Joint Powers written agreement on how the tax proceeds would be equitably shared to address flood protection on a watershed-wide basis.

On March 3, 1998, 23,000 Napa County voters turned out in a special election to vote on Measure A, the flood-control measure. Only one issue was on the ballet. It was a very high turnout for an election like this. At the end of the evening, the community celebrated victory, with a 300-vote margin. At the end, every participant in the process felt that his or her efforts had made a difference. It was a very sweet victory.

Project Update: A Community Lives through It

After 10 years, the Napa River Living River Flood Protection and Estuary Restoration Project is about 75 percent complete. It has ushered in a new era for the city of Napa and a major transformation of the city's southern entrance and downtown. Old levees have been removed or breached, creating more than 1000 acres of new wetlands. Five new bridges that used to act as dams during flood flows have been reconstructed. The city has managed to survive the major community disruption that is the result of such a massive undertaking. Costs have almost doubled over the original estimates for the project, but fortunately higher-than-expected proceeds from the half-cent sales tax and State of California bonds for flood control have managed to keep the local expenditure side of the equation in balance. The Federal Corps of Engineers' funding, however, has lagged, therefore postponing flood protection.

South Wetlands Opportunity Area

At the entrance to Napa in the southern reaches of the project's 7-mile span, the first phase of the project, known as the South Wetlands Opportunity Area, was completed in 2001. Levees were removed and breached to allow the tides to restore a marsh plain of about 1000 acres, which floods twice daily during high tides. A floodplain that is at a slightly higher elevation and is inundated once every 2 years on average was also created, essentially giving the river back its bank-side sponges. The marsh and floodplain has combined with other terracing and grading along the river to help lower downtown water surface elevations by several feet during flood events.

The Corps of Engineers and the California Department of Fish and Game, who have been monitoring fish in the restored floodplain for the past 6 years, are finding that native fish seem to be drawn to the new marsh and floodplain. Shorebirds can be seen in abundance probing in the new mudflats while ducks fly overhead. At a recent Bay Area science conference, the Corps said that the floodplain areas have increased rearing habitat for fish.

Heather Stanton, the project manager at the Napa County Flood Control District, emphasizes what a unique opportunity the community had. "The Project is very land-intensive in terms of restoration. The whole theory is to return the river to its floodplain where possible. That would be a constraint for some cities."

Oil Company Road

Another physically intensive phase of the project involved purchasing, demolishing, and relocating 33 buildings and warehouses, including nine residential units and 53 mobile homes, as well as relocating the Napa Valley Wine Train tracks and concessionaire building. According to Stanton, while residents were given relocation assistance and most moved willingly, some of the industries had a difficult time finding new sites. Pulling back the riverbanks—where industries on Oil Company Road were located before tankers started bringing heating oil and gasoline to Napa in the 1850s—also meant surprises. The city inherited 11 properties contaminated with petroleum, and despite the regional water board's promise to remediate the sites, the Flood Control District ended up paying for the removal of 170,000 cubic yards of contaminated soil. Ultimately the oil companies who polluted the land were forced to help bear the cost.

This phase and expensive land acquisitions in the north, along with the discovery of unforeseen underground utility, sewer, and water lines in the excavation area, were the primary causes of the 100-percent increase in the local share of the project budget.

Urban Riverfront

Across the river from the formerly contaminated sites—which are now new mudflats and floodplain—Army Corps flood walls have been completed and architecturally upgraded with local city funds. The walls double as protection for the historic Hatt Building and the 1884 granary, now called the Napa Mill Complex, designed and financed by local developer Harry Price to include an upscale inn, general store, several restaurants, and a pie shop. On top of the flood walls is a pedestrian promenade where walkers, bicyclists, and tourists can gaze out over the river and watch the bird life in the mudflats and marsh.

The more traditional, urban downtown stretch of the project was a key area of compromise during the Community Coalition process. The environmentalists wanted natural river on both riverbanks but agreed to allow fortification of the bank on the city side through the urban downtown reach.

The City of Napa, along with consultants, is guiding the project design, including the reconstruction of four new downtown bridges in accordance with a new River Parkway Master Plan. The main idea is to turn adjacent development to face the river where possible and to treat the entire river corridor as open space. The city has made a major effort to ensure that the design for the river edge adds value to adjacent properties.

Four bridges have been built and were designed to reflect the aesthetic of the floodwall promenade.

Oxbow Bypass

The wet-dry bypass for the large, horseshoe-shaped Oxbow turn in the river is a key feature of the flood-protection strategy. The bypass is a 900-foot-long, 280-foot-wide channel that will divert flood flows away from the Oxbow during large storm events and help speed the water downstream. Until the bypass is constructed, the level of flood protection remains less than needed and flooding can occur.

It is here, at the Oxbow Bypass, where the Community Coalition and Army Corps design process has gotten bogged down.

The latest incarnation of the bypass, designed with input from an advisory panel, is a seasonally dry channel that will act as a metaphor for the entire Napa Valley. The sides will be planted with native trees and grasses to resemble the Valley hills, while the bottom will replicate the Valley floor with a grid system of boulders, a lawn (so that flood flows are not a concern), and finally, a small low-flow channel, which is a sort of analogue to the Napa River.

How many of the trees and other design features proposed by the community the Corps will accept remains a question. According to city staff, Friends of the Napa River members, and other Community Coalition members, when it comes to the Corps and community-based design, the devil is in the details. It is hard to get these big federal entities to get down to the level of detail that the community is interested in. Some members of the original Community Coalition design team worry that the softer approach promised by the Corps seems to be getting harder. They are worried that areas where banks were to be stabilized predominately with plants will now be covered by more rock than was originally planned.

For issues like this, the Coalition can count on the Technical Advisory Panel established in the sales tax ordinance to act as a bulwark against too much mission creep on the part of the Army Corps.

To the north of the Oxbow is one of the Coalition's most prized achievements: 12 acres of riparian forest tucked into the Oxbow, acquired with $3 million from the State of California's urban Streams Restoration program. The Coalition has implemented a plan to preserve the Oxbow's native trees while removing invasive plants, adding some wetlands, and restoring old sloughs.

The Response of Local Elected Officials

Overall, local officials in Napa are pleased with the results of the project for the City of Napa, although they are frustrated by the slow pace of funding from the Army Corps of Engineers. Maintaining political support and enthusiasm is critical to satisfactory completion of a long-term public works project.

One especially enthusiastic supporter of the project is the Mayor of the City of Napa, Jill Techel:

> This project looked good on paper but as it has evolved it is even better. The first results were the reclaimed wetlands. The wildlife that has returned and grown is amazing. It really is something to see the birds and especially the new chicks running along the shoreline and getting ready to fly during springtime. The new trails are so appreciated by the local community. We had placed our back to the river and now we are opening it up and it is every bit the treasure we thought it would be. The trade-offs are key. We had multiple goals; to meet them there was and continues to be give-and-take. We had historic structure issues, resource agency issues, safety issues, economic future issues. However, as long as everyone takes a step back and considers what is good for the City in the long-term, compromises can be made and the project can move forward.

The Napa County Flood Control and Water Conservation District has also been a key, local player and observer of the project. Bill Dodd, who is Chair of the District, summed up the benefits of the project succinctly:

The project has increased the safety of Napa residents, spurred economic development, cleaned up acres of contaminated riverfront, and restored important habitats. Since groundbreaking in the summer of 2000, the project has reached multiple major milestones, including restoration of over 1000 acres of historic wetland in the Napa River estuary, construction of five bridge replacements, and cleanup of 11 acres of riverbank contaminated by petroleum spills. The severity of flooding has already been reduced by the partially completed project. This was clearly shown during the flood of December 31, 2005. The restored wetlands, mudflats and tidal terraces have stimulated huge growth in the bird count and expansion of riparian and riverine habitat. Partially complete river trails are highly popular in the community.

Put to the Test: Half-Completed Flood Project Survives Napa's Third 100-Year Flood in 30 Years

As Bill Dodd observed, even at partial completion, the project has mitigated the severity of flood events dramatically. Nowhere was this shown more effectively than on New Year's Eve 2005, when the partially completed project was inundated with the third 100-year flood since 1986. A big storm hit the region that night. Nearly 10 inches of rain fell within a 24-hour period, causing the river to rise to 23 feet and overflow its banks, flooding the entire region again, including downtown Napa. The question for everyone involved in the project was: How will we fare?

The *Vellejo Times Herald* measured the situation a few weeks later, in its January 23, 2006 issue: "Despite a thorough drenching that flooded Napa's downtown and caused an estimated $70 million damages to the city and $115 million worth of losses countywide, the half-completed Napa River flood control project still proved its worth." According to Napa County's emergency services manager, Neal O'Haire, the project worked in part to draw the water south, away from the Napa Valley and into the marsh and wetlands (reported in an article in the February 1, 2006, issue of www.bohemian.com). For Jill Techel, Mayor of the City of Napa, the partial success of the project was clear: "The good news, if there is good news when you flood, is that because of the work that has been done the water receded in less than 24 hours. In the past it has stayed for three days."

Put to the test, the project proved its worth from a flood-protection perspective even before completion. Unfortunately, though, a flood-control project is sort of like re-roofing your house: even with 75 percent completed, you still get wet.

Funding: Good News and Bad News

The most successful aspects of the project are the transformation of the City of Napa and the surrounding environment, flood protection, and economic development that has and is occurring. The key objectives of the project are being achieved. A less successful aspect has been the rising costs and cost overruns required to achieve implementation.

Fortunately, several additional sources of local and government funding have allowed the project to proceed with cost overruns of almost 100 percent in the cost of land acquisition, relocation, and utility relocations. It has been somewhat disappointing that the spirit of the Community Coalition has not carried over to property negotiations, where some opportunism has occurred. The City of Napa portion of the project cost is now about $350 million and another $175 million has been spent or is scheduled to be spent in the upper Napa Valley communities.

Funding for the local share items of the project has kept pace with cost increases through the achievement of layered funding from multiple federal, state, and local funding that results from the advantages of a multiobjective project that includes elements beyond flood control. These sources are shown in Table 6.1.

Unfortunately, the Corps of Engineers funding has not kept up with the local financing. Under the cost-sharing agreement with the Army Corps, the local community is responsible for the cost

TABLE 6.1
Multiple Benefits Equals Multiple Funding Sources

US Army Corps of Engineers	$130 million
Grants/Loans: Environmental Restoration	
California Coastal Conservancy	$2 million
State Lands Commission	$2 million
CALFED	$5 million
DWR Urban Streams Restoration	$1 million
California River Parkways Grant	$4 million
Clean Water Act: SRF, 2 percent loan	($65 million)
FEMA Hazard Mitigation Grant Program	
Napa Creek Home Relocations	$1 million
Mobile Home Park Protection	$7.5 million
Countywide Home Elevations	$4.5 million
Floodway Buyout	$5 million
Yountville Floodwall/St. Helena Planning	$8 million
Federal/State Highway Funds	
3rd Street Bridge	$9 million
1st Street Bridge	$2 million
Maxwell Bridge	$22 million
State Road Subventions	$75 million
Local Sales Tax Increase	$240 million
Total Project Cost (includes Maintenance Trust Fund)	**$520 million**

of lands, easements, rights-of-way, relocation of utilities, structures, and railroads (except railroad bridges). Almost 100 percent of the local elements are completed. The Corps' portions—including all excavation, setback levees, terrace grading, in-channel work, and railroad bridges—are in various states of completion due to funding shortfalls.

The reasons for the shortfalls are complicated, but part of it is due to the long process of securing money from the federal budget every year, as reported in the *Vallejo Times Herald* on January 23, 2006, a few weeks after the 2005 New Year's Eve flood: "Every year the Army Corps, the President and Congress go through a complex bartering system to reach an agreement on funding for the Federal share of the Napa project.

In 2005, the Army Corps of Engineers estimated the project needed $34 million. It received $12 million. The next year was not much better. The Army Corps of Engineers estimated that it needed $24 million for the project in 2006. The president only budgeted $6 million. Ultimately, the project received $14 million, according to the July 18, 2007, issue of the *Napa Valley Register*, which is still significantly less than what the Corps needed to complete its project tasks.

Given the cost of flooding in terms of property, cleanup, and lives, the federal government may be acting penny-wise and pound-foolish. As reported in the January 23, 2006, issue of the *Vallejo Times Herald*, using a slightly different metaphor, Barry Martin, spokesperson for the Napa County Flood Control and Water Conversation District observed, "The government has probably paid more than the cost of the whole project in just damage claims and cleanup over the course of the last three floods. Rather than treating the symptoms, you're treating the disease when you finish the whole project."

Supporting Martin's observation is the fact that since the project's inception in 2000, the region has saved an estimated $1 billion in flood-induced damages, indicating that an ounce of prevention really is worth a pound of flood-induced damage claims.

Economic Development

The story of the economic benefits of the project for the region is perhaps best told by someone who was on the ground at the time, Steve Kokotas, former executive director of the Napa Economic Development Corporation:

> The project has fostered a major economic boom in downtown Napa and throughout the Napa Valley. By simultaneously mitigating the risk of flood and improving the natural beauty of the riverfront, the city was able to attract developers and other investors interested in transforming downtown Napa into the world-class destination it deserves to be. The last six years have seen the construction of new hotels, spas, resorts, arts centers, wine-tasting rooms, and other commercial, arts, and entertainment development along the riverfront. More than a dozen vineyard owners opened up downtown tasting facilities that increase their products' visibility while reducing the negative impact of heavy traffic and tourism in the Napa countryside. All of these new amenities, along with the new natural and recreational opportunities provided by the restored river, entice visitors to stay longer and contribute more to the local economy. They have also created many job opportunities. In short, by protecting the downtown from major flooding, the project turned the town's biggest liability, the river, into its greatest asset.

The project has significantly improved the urban landscape of Napa. The most noticeable aspect is the influx of private investment and development in the region, surpassing $560 million since 1999, according to the Napa Community Redevelopment Agency. The new $70 million COPIA Center for Wine, Food, and the Arts was an important early anchor of the downtown riverfront revitalization, representing the first time the city has been able to enjoy the benefits of the Napa Valley wine industry. New hotels, spas, wine-tasting rooms, and resorts cater to a recent spike in tourism.

According to the December 10, 2006, issue of the *Napa Valley Register*, city officials reported that the downtown could easily have more than one million square feet in new construction within a decade. This adds up to a more than 50 percent increase in the downtown's current nonresidential square footage. "All of a sudden ... [there are] plenty of buyers who believe in downtown," says Bill Kampton, a commercial broker who is leasing space at Napa Square, a new office and retail complex in the downtown.

Real estate is also booming. The Napa Community Redevelopment Agency reports that real estate transactions since 1999 have totaled an extra $209 million over and above the $560 million in private investments. Craig Smith, executive director of the Napa Downtown Association, attributes this to the sense of security the flood-control project engendered in investors. "When the flood plan passed, that's when outside folks started buying property and investing in the future" (*San Francisco Business Times*, Vol. 22, No. 9, October 5–11, 2007). As of now, these investments appear to have borne fruit. Since the project's start, property values have increased by 20 percent, bucking the trend of the dot-com bust that hit Northern California. Sweetening the deal for property owners is a corresponding 20 percent decrease in flood-insurance rates.

The city's cultural and arts communities have also revived along the riverfront, supporting in addition to COPIA, the construction of the new Oxbow School of the Arts, and restoration of historic buildings such as the 1914 Opera House. Equally noticeable is the number of people who have returned to the riverfront for recreational purposes, the result of cleaning up contaminated sites and installing attractive new bridges, pedestrian promenades, and trails that afford good views of the river's restored nature and wildlife. Ecotourism is also on the rise, as outdoor enthusiasts take advantage of new opportunities for fishing, kayaking, boating, and hiking along the restored river.

The result is shaping up to be the biggest transformation of downtown Napa since the city's establishment. Recent development is responsible for what the *Napa Valley Register* called in their December 10, 2006, issue, "the greatest construction surge in downtown's history." A headline in the same issue proclaimed, "Downtown Napa is getting its mojo back." Perhaps more subtly, by mitigating the threat of recurrent, potentially devastating damage from floods, the project has finally allowed the city of Napa to benefit from the economic boom experienced by the rest of Napa

Valley. This has had a profound effect on the community's sense of itself. Residents in the city no longer feel like the Napa Valley's poor, ugly stepchild. "I think downtown is now perceived as part of the Napa Valley," says Harry Price, owner of the retrofitted Napa Mill Complex and a partner in Napa Square, the new office and retail complex, as reported in the December 10, 2006, issue of the *Napa Valley Register.*

Of course, as with all economic booms, not everyone is necessarily benefiting in equal part. There is some threat that gentrification of parts of Napa is occurring due to the economic benefits of the project, displacing some lower socioeconomic groups. The community will need to keep a careful eye on the development efforts to make sure they are inclusive and sustainable.

Happily, the dialogue among diverse groups in the community that was initiated during the development of the Flood Management Plan has continued, forming a democratic and consensus-driven basis for decision making in the region on a variety of topics including housing, tourism, land-use policies, and vineyard runoff, to name a few. All of these developments have resulted in a marked increase in the quality of life of Napa residents, who, just a few years ago, were demoralized by the periodic flooding and the bleak economic outlook it caused.

Perhaps the best aspect of the economic boom in Napa Valley is that—unlike in so many other places that have experienced sudden, exponential growth—it did not occur at the expense of the environment. In fact, restoring and protecting the environment is what made the economic boom possible. As Steve Kokotas, former executive director of Napa Economic Development Corporation observed:

> The most successful aspect of the project was the coming together of economic and environmental interests to satisfy the economic and environmental visions for the region. The successful marriage of the two visions can be seen strikingly on the riverfront itself. One side of the river is a beautiful, pristine natural scene. The other is an attractive, thriving urban landscape that both respects and protects the river that made it possible.

A Model Project

> This project is recognized around the world as a new way to think about flood protection.
>
> **Bill Dodd**
> *Chair, Napa County Flood Control and Water Conservation District*

Because of its groundbreaking approach in balancing urgent environmental, flood protection, and community needs, the Napa River project is already being used as a new model for environmental planning and disaster prevention. The project is unique in its effectiveness in building consensus among a diverse array of groups that were at complete loggerheads for decades. In bridging seemingly incommensurable philosophical differences among these groups, the project created a new, conciliatory model for planning that can be applied to any vital issue involving multiple stakeholders in conflict. The project is a model of how cities can use consensus-building to bring into balance the significant and often contradictory social, economic, environmental, and regulatory demands they face as they try to overcome persistent problems such as housing shortages, economic stagnation, environmental threats, and cultural decline.

The project also provides a model in its groundbreaking approach to balancing environmental concerns with flood management needs, proving that doing the right thing environmentally can both protect and profit the local community. Perhaps no one has benefited from seeing this approach in action more than the Army Corps of Engineers. The project resulted in nothing less than a paradigm shift in the way the Corps does business. According to Larry Dacus, a Corps member who has been closely involved in the process from the start, the project is a model for the Corps in the way it balances flood protection with environmental restoration. He also says that the agency is now using consensus-based decision making in several of its current projects (reported in the January 2005 issue of *Landscape Architecture*). Kathleen A. McGinty, Chair of the Council on Environmental Quality during the Clinton Administration, expressed a similar sentiment, calling the project a "new model of environmental decision-making."

The success of the Napa River project is being communicated to a larger audience through a variety of media. One of the best case studies of the project can be found in the book *The New Economy of Nature: The Quest to Make Conservation Profitable*, by Gretchen C. Daily and Katherine Ellison and published by Island Press. The case study includes this summary of the project:

> Napa, California, a city plagued for decades by floods, work has begun on an innovative effort to free the Napa River from its levees and dams and allow it to spill over onto its historical floodplain, proving natural flood protection. The US Army Corps of Engineers, famous for pouring concrete, began tearing it out, removing levees along a seven mile stretch. Napa residents, who have voted to raise their own taxes to pay for the plan, have seen immediate paybacks, with property values soaring in expectation of an enticing new waterfront district and a dry downtown.

Replicability

The concepts and lessons learned in the Napa project have already been transferred to consensus-based, environmental restoration projects across the United States and around the world in such places as China; Australia; Reno, Nevada; Santa Cruz, Marin, and Monterey Counties, California; and along California's American, Truckee, and San Gabriel Rivers.

Combining groundbreaking techniques in ecology, engineering, and public facilitation to successfully address the multifaceted environmental, economic, and public-safety issues of a region in crisis, the project is of wide interest to environmentalists, urban planners, public outreach facilitators, scientists, and engineers. The project's community-building and facilitation techniques serve as a model of how diverse interests ranging from public agencies, residents, business owners, and environmentalists can be brought into dialogue with each other to build a consensus and move forward on important projects benefiting everyone.

MIG, Inc., a consultant firm headquartered in Berkeley, California, was one of the key players in the process of consensus building for the Napa River project. Through projects like this, MIG has developed a toolbox of methods to make the process work, including participatory charettes, one-on-one meetings, hands-on, interactive workshops, feasibility studies, computer modeling, and well-designed, easy-to-understand visual materials that invite a critical discussion. These techniques each have a place and a use in the public process along with the hard science that is required to produce a viable outcome. Understanding how the social sciences, planning policy, and environmental sciences integrate with community values helps others use these concepts effectively.

Key Elements for a Successful Project

Even so, the Community Coalition model of a Living River planning and consensus development process does not work in every situation. At least seven key elements must be present to achieve the sort of success seen in Napa:

1. An emerging mission born from a crisis or mandate
2. Common knowledge resulting in shared meaning
3. A champion willing to take risks
4. A community of place
5. No better deals elsewhere
6. Primary parties participate in good faith
7. Multiple issues for trade-off resulting in multiple community benefits
8. Adequate resources

An Emerging Mission Born from a Crisis or Mandate

In order for the process to get started, there must be a deep and shared sense among the populace that *something must be done*. In the case of Napa, it was major floods in 1986 and 1995, combined with the unveiling of the third unacceptable Army Corps of Engineers design proposal. When natural disasters are the basis of a community crisis, you need to move quickly while the urgency is still in the minds of the locals. I suspect that when climate change begins to change weather patterns in local communities, there will be even more urgent calls for local action.

Common Knowledge Resulting in Shared Meaning

The consensus action planning process must invest in education and create a common understanding of the issues, science, and key dynamics together, so that everyone starts the process with common knowledge. This then evolves into shared meaning among the stakeholders. So often, government engineers and consultants do not adequately invest in educating the public and nonprofessionals about the underlying reasons behind design recommendations.

Visualization of complex principles helps in this area, as do professional facilitators. All ideas must be seriously considered, even if many of the stakeholders already know why. Everyone at the table needs to start with the same baseline information. Sometimes I refer to this success element as the need to love *every* idea—to death.

A Local Champion Willing to Take Risks

Generally, established bureaucracies and organizations are threatened by truly open, participatory democracy planning processes. Every successful community-consensus process I have participated in has included a key elected official who leads the charge to convince the government entity and community stakeholders to take a risk in how the design and decision process needs to occur.

In the Napa experience, the "normal process" had failed three times, and given that an acceptable plan would only be implemented if two-thirds of the voters agreed to a tax increase, the Flood Control District agreed to resource a community-based planning process. In doing so, it had to give up some power to the Friends of the Napa River, the Napa Valley Economic Development Corporation, and the Napa Chamber of Commerce and change the composition of its own governing body to add representation from the five cities in the county.

Additionally, there is usually at least one professional staff member who commits his or her full time and more to achieving the agreements necessary, meeting with all constituencies behind the scenes in order to identify deal-breaking positions before they come out in public planning sessions.

A planning process like this requires a few totally committed leaders who will live and breathe it for an intense and dedicated period of time.

Our experience with the flood-tax ballot for the Ross Valley Watershed in Marin County, California, shows the importance of having a local champion. County Supervisor Hall Brown (cousin of former Governor Jerry Brown) risked a lot politically in campaigning for the tax, along with a special engineer from the Flood Control District, Jack Curley, who is also a performing musician and readily accepted the job of communicating technical data in a compelling and dynamic way. The flood tax ultimately passed by 56 votes of 9000 cast. Curley spent nights and weekends for a year staying in touch with all of the stakeholder organizations, which, if opposed to the tax in any organized way, could have scuttled the Community Coalition. You need someone who will live the process.

A Community of Place

It is essential that the geographic scope of the consensus planning effort is appropriate. It is not realistic to conduct a participatory process on a state or national level. Stakeholders in Florida and California are too far apart to identify with a "place" on a scale that lends itself to consensus-based planning. A watershed is an ideal geographic scope for the purpose of agreeing on a flood-protection plan. Even though the upper watershed stakeholders have different interests than the downstream residents, the Napa Valley as a whole is a community of place.

Primary Parties Participate in Good Faith: No Better Deals Elsewhere

To be successful, a consensus-based community-planning process cannot allow key stakeholders to participate in an environment that allows for a better deal to be obtained elsewhere. Agreement needs to be obtained up front that participants will truly play by the rules of the process outlined by the sponsors.

A perfect example of how the lack of good faith can cripple a project may be found in the four-county Pajaro River Flood Protection Community Coalition process that has been under way since 2000. All players agreed to sit at the table to discuss how to fix their flood-control system, which is currently a system of deteriorating levees. Even with drastic river clearing, the levees can only contain 10-year flood flows. All key constituencies—the urban City of Watsonville, a strong environmental community, and strawberry and lettuce growers—sustain extreme damage in a major flood. The consensus plan is to set back the levees 100 feet on each side of the river in order to allow a reasonable low-flow channel and adequate vegetation for migrating steelhead into the upper reaches of the Pajaro River in San Benito and Santa Clara Counties.

Meanwhile, the agricultural interests believe they can get a better deal by forcing the boards of supervisors to set an unequivocal policy that forbids the Coalition from taking *any* land out of agricultural production. Some supervisors are willing to support that position for short-term political gain. Many growers believe that the Pajaro River is a Federal Flood Control Channel that can simply be channelized like the Los Angeles River, in spite of the Army Corps' clear statements that the federal government can only contribute funds to a project that balances flood control and environmental restoration. Both the local elected officials and the Corps are leading the major stakeholders to believe they can get a better deal outside of the Community Coalition process. As a result, there is no agreement about what to do, even after 8 years of effort.

Multiple Issues for Trade-Off Resulting in Multiple Community Benefits

Consensus-driven community-planning processes tend to succeed if there is more than one issue on the table. Multiobjective flood-protection projects meet this element of success because they usually involve a multitude of issues, including flood protection, environmental restoration, transportation system improvements, land-use planning, provision for river trails and passive recreation, community health and safety, and taxation. If success *on the ground* is dependent on all of the stakeholders getting something from the process, then the more community benefits included in a project, the easier it is to achieve compromise and consensus.

Adequate Resources

Having adequate resources to tap into is crucial for a project's success. Professional planning process management, design assistance, visualization, photo simulations, adequate budget for engineering and hydrologic and hydraulic modeling (at least at the feasibility level), and community polling are all critical in achieving effective community consensus planning for large-scale projects. No one will sustain his or her participation if the process just rehashes the same limited information meeting after meeting. A significant investment must be made by the sponsors to assure that the process will produce answers to hard questions and will not simply rely on the network of informed or uninformed opinion. It helps tremendously if there is a process to involve stakeholders in the selection of the professional support resources.

Managing large-scale consensus planning projects is never easy. If one or more of these seven elements is missing, it is difficult, if not impossible to achieve a plan that does justice to all of the issues involved or that will be supported by the majority of stakeholders. On the other hand, if, through either hard work or good fortune, all of these elements fall into place, you have a fighting chance to come out of the process with a community-supported, technically feasible, and fundable plan.

The Napa River project offers a great example of how a community of diverse and even contradictory interests can join forces and bring together all of these elements to achieve protection from nature while at the same time protecting nature. As we move into the reality of climate change and experience more and more its unpredictable, devastating effect on our communities, the need to develop successful consensus-based planning processes like the one in Napa will only become more urgent.

Update: Napa River Flood Control Project

Progress in implementing the Napa River Flood Control Project has continued since the publication of this case study in 2008. While work has yet to be completed, economic development benefits continue to accrue to the City of Napa and other communities along the Napa River.

A Case Study conducted by the Landscape Architecture Foundation noted, "The Napa River Flood Control Project was part of a monumental effort in the 1990s to reconceive the river's flood control strategy on a regional scale. Adhering to the 'living river' principles developed by an impressive community coalition, the city of Napa has implemented project phases along a 7-mile stretch of river, resulting in the restoration of 1011 acres of habitat that meet the primary project aim of protecting the functionality of the river's ecological systems while providing flood protection for many Napa residents. Though the project's goal of 100-year flood protection will not be met until the final phases of the project have been completed, many environmental, economic, and social benefits are already being realized by the community today" (Kondolf n.d.).

The Landscape Architecture Foundation Case Study examined the benefits to the community of the implementation of the Flood Control Project to date that, "Resulted in the creation of 1,373 temporary building construction jobs and 1248 permanent retail and administrative jobs related to properties developed in expectation of 100 year flood protection along the Napa River" (Kondolf n.d.).

A presentation developed by the City of Napa in May 2011 noted that construction investment and disclosed real estate purchases in Napa from 1997–2010 equaled $898 million with the city claiming that the catalyst was the Flood Protection Project. Several new restaurants and hotels have been constructed in the years since the Flood Control project was implemented (City of Napa 2011).

The Napa County Flood Control and Water Conservation District maintains a page on the Napa County website that provides "some frequently asked questions and answers about the Napa River/Napa Creek Flood Protection Project" (County of Napa.org 2015). Presented in the box "Frequently Asked Questions and Answers about the Napa River/Napa Creek Flood Protection Project" are excerpts from this web page describing progress in implementing the Flood Control Project.

The US Army Corps of Engineers in an August 7, 2012, post on www.army.mil by J.C. Delgadillo entitled "Residents say 'cheers' to flood protection in Napa," noted "The cost of civil works projects is often shared by local, state and federal agencies. Although Napa voters approved a sales tax increase to fund their portion of the Napa River Flood Protection Project in 1998, it took more than a decade to get the momentum on the project going due to a variety of factors including lack of federal funds. With the advent of the American Recovery and Reinvestment Act of 2009, the Napa River Flood Protection Project, deemed shovel-ready, was infused with the funds needed to make significant progress." "Rapid advancements have not come at the expense of the environment," explained (Napa County Flood Control & Water Conservation District's Julie Blue) Lucido. "There is a balancing act we are doing with our flood project here in Napa. Even though we are not an ecosystem restoration project by authorization, we are implementing strategies to restore the

FREQUENTLY ASKED QUESTIONS AND ANSWERS ABOUT THE NAPA RIVER/NAPA CREEK FLOOD PROTECTION PROJECT

I hear people say we've had three 100-year floods in the last 20 years. Is that true? Why is construction of the Project behind schedule?

Money drives construction forward, and there are many reasons why the Project has not received as much money from the federal government as needed to maintain the intended pace of work. The original plan was to have the work done as early as 2005 or 2006. While the local share of funding that comes from the half-cent sales tax has outperformed revenue projections, the Project has not received all the needed funding from the state and federal governments. When the money is delayed, the work is delayed. One of the primary reasons for lower than expected levels of funding is the terrorist attacks of 9/11/01 and the resulting increases in domestic security and military spending. In addition, the Napa Project competes each year with other Corps of Engineers projects, many of which are much larger in scope.

I hear people say that because we had flooding in 2005 that means the money we've spent so far has been wasted. Is that true?

No, it's not true. The money spent so far has gone to acquire property, build four bridges, relocate utilities, restore 600 acres of wetlands, construct about 2 miles of flood plain and marsh plain terrace, and build the Hatt to First Floodwall and Promenade. Unfortunately, the Project cannot function fully until it is all complete.

When will the Napa River-Napa Creek Flood Project be completed?

Current schedules show final completion in 2015 if all needed funding is received on schedule.

The Flood Project construction started in 2000. After five years of work, why didn't the Flood Project prevent flooding on 12-31-05?

The Napa River-Napa Creek Flood Protection Project was only about 40% complete at the time of the 12-31-05 event. Flood protection will be achieved using an integrated system that includes wetlands, terraces, bridge replacements, a bypass channel, floodwalls, and levees. All of the components must be constructed before the Project will work as designed.

Source: County of Napa.org. 2015. Here are some frequently asked questions and answers about the Napa River /Napa Creek Flood Protection Project. County Flood Control and Water Conservation District. Accessed January 31, 2015. http://www.countyofnapa.org/Pages/DepartmentContent.aspx?id=4294968678.

environment, to create those important ecosystem restoration pieces," she said. "Providing a safer town has allowed new businesses to come in, folks to restore their homes, especially the historic homes," said Lucido. "People are willing to invest money here, so (the project) is creating a safer, happier community; a more vital community," she said (Delgadillo 2012).

The final pieces of the funding puzzle have been coming together in the past 2 years. In June 2013, Napa secured $16.6 million in funding from the US Army Corps of Engineers to construct a bypass channel. A June 25, 2013, editorial posted on Napa Valley Register.com noted, "This bypass channel will provide much more than just flood protection. It will help create a more cohesive downtown hub. With the bypass, the Oxbow Public Market, the former Copia site, the Wine Train and numerous other businesses east of Soscol Avenue become more visible, more accessible and a more prominent part of the downtown experience" (*Napa Valley Register* Editorial Board 2013). The editorial concluded that, "New and established merchants throughout downtown will benefit from bypass construction, as will both residents and tourists. These next steps in Napa's flood work may well prove to be downtown's next big economic catalyst, the latest in a series of economic boons the flood project has wrought" (*Napa Valley Register* Editorial Board 2013).

REFERENCES

Association of Bay Area Governments. 1989. *The Liability of Local Governments for Earthquake Hazards and Losses: Report: A Guide to the Law and Its Impacts in the States of Alaska, California, Utah and Washington*, 52 pp, February.

Bullock, J.A., G.D. Haddow, and K.S. Haddow. 2008. *Global Warming, Natural Hazards, and Emergency Management*. Boca Raton, FL: CRC Press.

Chakos, A., P. Schulz, and L.T. Tobin. 2002. Making It Work in Berkeley: Investing in Community Sustainability. *Natural Hazards Review* 3(2).

City of Berkeley. Climate Action Plan, adopted 2009.

City of Berkeley. California Disaster Mitigation Plan, adopted April 28, 2004.

City of Berkeley. California General Plan: A Guide for Public Decision-Making, adopted April 2002, Safety Element Revised 2003.

City of Berkeley. Energy Assurance Plan, completed June 2013.

City of Berkeley's designation as a 100Rc community by the Rockefeller Foundation.

City of Berkeley's updated Local Hazard Mitigation Plan, adopted December 2014.

City of Napa. 2011. A Look at the Napa River—Napa Creek Flood Protection Project and its impact on Downtown Napa, May. Available at http://www.floods.org/ace-files/Conferences/Louisville/ppts/Tuesday%20Plenary/Jill_Napa_FloodPlainManagers_5-11.pdf.

City of Tulsa Flood Mitigation Voluntary Acquisition Plan, approved by the Public Works Department, June 21, 1994, and amended April 5, 1999.

City of Tulsa Public Works Department. 1990–1991. The City of Tulsa Flood and Stormwater Management Plan, 1990–2005.

City of Tulsa Public Works Department. 1990–1995. Annual Reports.

City of Tulsa. 2002. *Multi-Hazard Mitigation Plan*. Consultants Flanagan & Associates et al., November 25. Available at http://www.rdflanagan.com/tulsa/tulsa_NHM_book.pdf.

County of Napa.org. 2015. Here Are Some Frequently Asked Questions and Answers about the Napa River/Napa Creek Flood Protection Project. County Flood Control and Water Conservation District. Accessed January 31, 2015. Available at http://www.countyofnapa.org/Pages/DepartmentContent.aspx?id=4294968678.

Delgadillo, J.C. 2012. Residents Say "Cheers" to Flood Protection in Napa. Accessed August 7, 2012. Available at http://www.army.mil; http://www.army.mil/article/85109/Residents_say__cheers__to_flood_protection_in_Napa/.

Federal Emergency Management Agency. 1992. Background Information for Tulsa, Oklahoma CRS Presentation, February 24.

Federal Emergency Management Agency. 2009. Safe Rooms Take Tulsa by Storm. Mitigation Case Studies.

Federal Interagency Hazard Team. 1984. *90-Day Post-Flood Recovery Progress Report*, September 4.

Federal Interagency Regional Hazard Mitigation Team. 1981. *Flood Hazard Mitigation: Handbook of Common Procedures*, September.

Federal Interagency Regional Hazard Mitigation Team. 1984. *Interagency Flood Hazard Mitigation Report*, in response to the May 31, 1984, Disaster Declaration, State of Oklahoma FEMA-709-DR, June 15. FEMA, Washington, DC.

Flanagan, R. 1991. Mingo Creek, Tulsa, Oklahoma. *A Casebook for Managing Rivers for Multiple Uses*. US Department of Interior National Park Service, in conjunction with Association of State Floodplain Managers and Association of State Wetland Managers, October.

French & Associates. 1993. Floodproofing and Acquisition Program Design Paper, for City of Tulsa, OK, August 20.

Garland, G. 1997. Model Flood Control: Tulsa Moves from Flood Capital to Model of Control, Planning. *The Baton Rouge Advocate*, October 21.

Heifetz, R. 1994. *Leadership without Easy Answers*. Cambridge, MA: The Belknap Press of Harvard University Press, p. 128.

Hinshaw, R.E. 2006. Living with Nature's Extremes: The Life of Gilbert Fowler White. *Tulsa Case Study*. Boulder, CO: Johnson Books, a division of Big Earth Publishing, pp. 195–204.

Kondolf, G.M. n.d. Napa River Flood Protection Project: Methodology for Landscape Performance Benefits. Landscape Architecture Foundation. Landscape Performance Series. Available at http://landscapeperformance.org/case-study-briefs/napa-river-flood-protection.

Kusler, J. 1982. Tulsa, Oklahoma. *Innovation in Local Floodplain Management: A Summary of Community Experience*. Natural Hazards Research and Applications Information Center Special Publication 4. Boulder, CO: Prepared for US Water Resources Council.

McLaughlin Water Engineers. 1984a. Flood Hazard Mitigation: Tulsa, Oklahoma, May 26–27, 1984 Flood Disaster, for City of Tulsa, OK, June 12.

McLaughlin Water Engineers. 1984b. Revised Mitigation Plan, for City of Tulsa, OK, July 31.

Meo, M., B. Ziebro, and A. Patton. 2004. Tulsa Turnaround: From Disaster to Sustainability. *Natural Hazards Review* 5(1): 1–9. Available at http://ascelibrary.aip.org.

Mileti, D.S., ed. 1999. *Disasters by Design: A Reassessment of Natural Hazards in the United States*. Washington, DC: Joseph Henry Press.

Napa Valley Register Editorial Board. 2013. Bypass Money Will Offer Much More Than Flood Control. *Napa Valley Register.com*. Accessed June 25, 2013. Available at http://napavalleyregister.com/news/opinion/editorial/bypass-money-will-offer-much-more-than-flood-control/article_3f7e1b2a-ddf4-11e2-a3db-0019bb2963f4.html.

National Wildlife Federation. 1998. *Higher Ground: A Report on Voluntary Property Buyouts in the Nation's Floodplains*. Vienna, VA: National Wildlife Federation.

National Wildlife Federation and J. Warrick. 1999. Seeking an End to a Flood of Claims. *National Wildlife Magazine*: 30–33. Vienna, VA, June–July.

Oklahoma Water Resources Board. 1987. Innovative Tulsa Utility Sets Storm-Safe City as Its Goal. *Oklahoma Water News*, September.

Patton, A. 1975. *Flaws in the Laws and Gaps in the Maps. The Evolution of Floodplain Management in Tulsa, OK 1970–1975*. City of Tulsa, Tulsa, OK.

Patton, A. 1976. Will Someone Please Make Those Floods Go Away? *Tulsa Magazine*: 14–17, Metropolitan Tulsa Chamber of Commerce. Tulsa, OK, October 7.

Patton, A. 1977. Why Tulsa Floods. *Tulsa Home and Garden Magazine*: 8–11, 46, 48. Tulsa, OK.

Patton, A. 1984. Memo to Street Commissioner J.D. Metcalfe: Flood Hazard Mitigation Program. City of Tulsa, OK, May 30.

Patton, A. 1988. Flood-Hazard Mitigation in Tulsa, Oklahoma. *Proceedings of the Twelfth Annual Conference of the Association of State Floodplain Managers, Nashville, TN*. Natural Hazards Research and Applications Center Special Publication #19, May 16–19.

Patton, A. 1993. *From Harm's Way: Flood Hazard Mitigation in Tulsa, OK*. City of Tulsa, OK, December.

Patton, A., ed. 1994. *From Rooftop to River: Tulsa's Approach to Floodplain and Stormwater Management*. City of Tulsa, OK, May.

Patton, A. 2004. Together We Can! Public-Private Partnerships for Flood-Hazard Mitigation. Presentation to the Disasters Roundtable Workshop, Reducing Future Flood Losses: The Role of Human Actions. Washington, DC: The National Academies, March 2.

Patton, A., and R. Flanagan. 1980. Haikey Creek—Flash Floods and the Frontier Ethic. *Intergovernmental Management of Floodplains*, Rutherford H. Platt, ed. Program on Technology, Environment and Man, Monograph #30. Boulder, CO: Institute of Behavioral Science, University of Colorado.

Patton, A., G. Birt, D. Breed, and R. Kindy. 1976. In Harm's Way: Flooding in Tulsa: A Case Study in the Creation of Disaster. Presentation for University of Tulsa Symposium on Floodplain Management. Tulsa, OK.

Poertner, H.G. 1980, revised 1982. Stormwater Management in the United States: A Study of Institutional Problems, Solutions and Impacts. *Tulsa Case Study*. US Department of Interior, pp. 222–233, September. Washington, DC.

Rubin, C.B., and M. Nahavadian. 1987. *Details on Frequency of Disaster Incidents for Federally-Declared Disasters, 1965–1985*, March. George Washington University. Washington, DC.

Tulsa Metropolitan Area Planning Commission; Team One, R.D. Flanagan, and Environmental Sciences Corp. 1976. *Flood Information Study*. Tulsa, OK. Tulsa Metropolitan Area Planning Commission.

Tulsa Partners and City of Tulsa. 1998. Grant Application to FEMA for Tulsa Project Impact program. Tulsa, OK, November 4.

Tulsa Partners and City of Tulsa. 2002. Tulsa Project: Building a Disaster-Resistant Community. *Report to the Federal Emergency Management Agency on 1999–2002 Activities of Community-Based Partners Working to Reduce Disaster Risk*, March 31.

Tulsa Partners, Inc. 2007. *Business Plan and Case Statement: The Millennium Center for Green and Safe Living*. Tulsa, OK.

Tulsa Project Impact; Indian Nations Council of Governments. 2001. *Community Risk Assessment for City of Tulsa and Tulsa County, OK*. Tulsa, OK, November. Tulsa Project Impact.

US Army Corps of Engineers, Tulsa District. 1982. *Tulsa Urban Study Summary Report and 16 Reports on Study Findings*. Tulsa, OK, February. US Army Corps of Engineers.

US Army Corps of Engineers, Tulsa District. 1985. *Documentation, Flood of May 27, 1984, Tulsa Oklahoma Metropolitan Area*. Tulsa, OK, May. US Army Corps of Engineers.

University of Tulsa Urban Studies Department, and T. Filipovitch, ed. 1977. *Proceedings of the Floodplain Management Symposium*, October 14–16, 1976. Tulsa, OK. Proceedings of the Floodplain Management Symposium.

White, G.F. 1966. Task Force on Federal Flood Control Policy. *A Unified National Program for Managing Flood Losses*. US Government Printing Office, Washington, DC.

White, G.F. 1975. *Flood Hazard in the United States: A Research Assessment*. Boulder, CO: Institute of Behavioral Science, University of Colorado.

Wright Water Engineers et al. 1985. *Non-Structural Floodplain Management Hazard Mitigation Measures*, for City of Tulsa and FEMA, September.

Wright Water Engineers et al. 1986a. *Flood Hazard Mitigation Report, 1986 Floods*, for City of Tulsa, November 14.

Wright Water Engineers et al. 1986b. *Guidelines for Inventory and Removal of Structures from Flood Hazard Areas*, for City of Tulsa and FEMA, September.

Wright, J.M., and J.L. Monday. 1996. *Addressing Your Community's Flood Problems: A Guide for Elected Officials*. Madison, WI: The Association of State Floodplain Managers, Inc., and the Federal Interagency Floodplain Management Task Force.

Young, T. 1984a. City of Tulsa Stormwater and Floodplain Management. Position Paper Issued during Young's Campaign for Mayor. Tulsa, OK, February 23.

Young, T. 1984b. Transcript of KOTV Televised Speech on Flood Mitigation Program. Tulsa, OK, July 12.

7 Case Studies in Climate Change Adaptation

INTRODUCTION

This chapter includes three case studies that examine climate change adaptation programs in Flagstaff, Arizona, Grand Rapids, Michigan, and Norfolk, Virginia. According to the project brief included in this chapter, these case studies were prepared as part of a "Kresge Foundation initiated critical assessment of community-based adaptation in the United States. This assessment was structured to explore what communities are doing on adaptation and why, what they are and are not accomplishing, what barriers to adaptation they face, and whether they are overcoming these barriers. A team of researchers, led by Jason Vogel and Joel Smith of Stratus Consulting, has joined forces with the Kresge Foundation to conduct this assessment."

As noted in the project brief, "Norfolk, Virginia, developed a new flood and coastal zone ordinance that increased its freeboard requirement to 3 feet above flood stage. The Norfolk Planning Commission explicitly considered climate change projections and increased the freeboard requirement from 2 feet to 3 feet as a result. Neighboring communities have begun following Norfolk's lead.

"Grand Rapids, Michigan, developed Vital Streets and Sidewalk Spending Guidelines to reduce flooding from extreme rain and snowmelt events and to achieve other community objectives. This ordinance shifted the city's strategy for stormwater management from conventional engineering solutions to green-infrastructure solutions, such as parks development.

"Flagstaff, Arizona, implemented the Flagstaff Watershed Protection Project to mitigate fire hazard and associated flood risk in the ponderosa pine forests that surround Flagstaff. In 2012, Flagstaff citizens voted overwhelmingly to pass a $10 million bond measure to fund implementation of the project. The US Forest Service and other partners have contributed an additional $2 million to support the project."

Also included in this chapter is a case study of the adaptation measures designed and implemented in New Jersey and New York as part of their recovery efforts from 2012's Hurricane Sandy.

REDUCING VULNERABILITY TO CATASTROPHIC WILDFIRES AND RELATED FLOODING IN FLAGSTAFF, ARIZONA

Megan O'Grady, Karen Carney, and Jason M. Vogel
Stratus Consulting

Flagstaff, Arizona

Since the 1990s, northern Arizona has experienced several catastrophic wildfires. Damage from one of these fires, the 2010 Schultz fire, also contributed to severe flooding during heavy rains shortly after the fire. In November 2012, voters in Flagstaff passed a $10 million bond measure to

INTRODUCING THE KRESGE FOUNDATION FUNDED PROJECT, COMMUNITY-BASED ADAPTATION IN THE UNITED STATES

A Critical Assessment

Jason M. Vogel

Stratus Consulting and John Nordgren, The Kresge Foundation

Since the concept of climate change adaptation began to enter the mainstream around 2007, a rapidly growing field of adaptation practitioners has emerged, along with networks to connect them. However, this emergence has not kept pace with the explosion of interest in climate adaptation at the community level. Communities defined as either municipal or regional entities recently have engaged in resilience, preparedness, or climate adaptation efforts. Unfortunately, there are few concrete examples of specific actions implemented to adapt to climate change noted in the literature or in practice.

Yet communities have a long history of responding to extreme events such as coastal storms, droughts, floods, fires, and heat waves. In the vast majority of cases, communities undertook these activities to protect or restore the status quo against known climate risks. Such actions are not adaptation to climate change because they do not account for potential changes in future conditions and they lead to no net reduction in vulnerability. But in some of these cases the reduction in current risks has been so extraordinary that they can serve as examples of how communities can reduce risks from climate change. Furthermore, in select cases, communities have implemented new processes or policies that lead to a net reduction in vulnerability under current and potential future climate conditions. Whether these actions constitute climate adaptation is an open question, and one which this assessment will address.

In 2013, after investing significantly in community-based adaptation for 5 years, the Kresge Foundation initiated a critical assessment of community-based adaptation in the United States. This assessment was structured to explore what communities are doing on adaptation and why, what they are and are not accomplishing, what barriers to adaptation they face, and whether they are overcoming these barriers. A team of researchers, led by Jason Vogel and Joel Smith of Stratus Consulting, has joined forces with the Kresge Foundation to conduct this assessment. Technical advisors Ronald Brunner, Susanne Moser, and Missy Stults, together with a project advisory committee* of thought leaders, help ensure the assessment adds real value to the field of community-based adaptation.

We designed the assessment to be evidence-driven and empirically grounded. The team's research began with interviews of 50 thought leaders working at the nexus of climate adaptation and related fields, such as natural hazards management, natural resource management, law, community planning, sustainability, insurance, and environmental justice. Many of those interviewed were invited to be on the project advisory committee. Based on this foundational data, the team began exploring approximately 20 in-depth community-level case studies. The case studies focus on bottom-up efforts in communities that have decided that current or potential climate impacts put important community values at risk. The case studies offer

* Project advisory committee members include Steve Adams of the Institute for Sustainable Communities, Vicki Arroyo of the Georgetown Climate Center, Rosina Bierbaum of the University of Michigan, Garrett Fitzgerald of the Urban Sustainability Directors Network, Paul Flemming of Seattle Public Utilities, Hector Galbraith of the National Wildlife Federation, Nancy Gilliam of the Model Forest Policy Program, Tonya Graham of the Geos Institute, Lara Hansen of EcoAdapt, Kathy Jacobs of the University of Arizona, Jim Murley of the South Florida Regional Planning Council, Jacqui Patterson of the National Association for the Advancement of Colored People, Cara Pike of Climate Access, Julie Rochman of the Institute for Business and Home Safety, Jim Schwab of the American Planning Association, and Ed Thomas of the National Hazards Mitigation Association.

detailed accounts of how and why each community took a specific action intended to reduce vulnerability or increase resilience to climate impacts. We center each case study on a specific action taken by that community intended to address community vulnerability. The cases highlight the essential elements that tell the story of each community.

1. Norfolk, Virginia, developed a new flood and coastal zone ordinance that increased its freeboard requirement to 3 feet above flood stage. The Norfolk Planning Commission explicitly considered climate change projections and increased the freeboard requirement from 2 feet to 3 feet as a result. Neighboring communities have begun following Norfolk's lead.
2. Tulsa, Oklahoma, developed a program to acquire and remove properties repeatedly damaged in floods from a regulatory floodplain boundary enhanced beyond the FEMA floodplain boundary. Tulsa has acquired more than 1000 repetitive-loss properties, consequently reducing flood damage during extreme rain events.
3. Grand Rapids, Michigan, developed Vital Streets and Sidewalk Spending Guidelines to reduce flooding from extreme rain and snowmelt events and to achieve other community objectives. This ordinance shifted the city's strategy for stormwater management from conventional engineering solutions to green-infrastructure solutions, such as parks development.
4. Flagstaff, Arizona, implemented the Flagstaff Watershed Protection Project to mitigate fire hazard and associated flood risk in the ponderosa pine forests that surround Flagstaff. In 2012, Flagstaff citizens voted overwhelmingly to pass a $10 million bond measure to fund implementation of the project. The US Forest Service and other partners have contributed an additional $2 million to support the project.

Based on a detailed analysis of 20 ongoing case studies, including the four summarized above, this assessment will draw substantive conclusions about key issues facing each community. We will ask three central questions about each of the communities' experiences:

First, did the community's action effectively address the targeted problem? For example, if the goal is to increase resilience to coastal flooding, does a new freeboard requirement achieve that end for the community, or just for homeowners and businesses that can afford to make the change?

Second, is the community less vulnerable or more resilient to the impacts of climate change? This question seeks to put the outcome of a community's actions in the context of the increased variability and uncertainty that climate change will introduce.

Third, how did each community's action affect vulnerable or underserved community members and were these considerations part of the calculus for the action? For example, in considering the effects of heat on human health, did the community consider differential vulnerability to heat waves because of age or poverty?

This assessment will add value to the growing body of knowledge about approaches to climate adaptation that do and do not work, drawing lessons learned from empirical examples and communicating those examples for the benefit of the field and communities across the country. This assessment differs from other efforts to evaluate adaptation because it (1) allows the empirical evidence of 20 case studies to drive the analysis, (2) focuses on community actions intended to reduce vulnerability, and (3) explores community-driven efforts. We will examine the relevance of each community case by working through questions that we hope will uncover insights that help practitioners understand the role of climate change considerations in reducing vulnerability. For example, to reduce their vulnerability, must communities adapt to climate change or can they instead take actions based only on historical events? Can we find commonalities among communities that are addressing their vulnerability to climate, in a way that offers insight for engaging other communities on this issue? Do climate adaptation plans, resilience plans, or climate action plans play a significant role in action at the community level? The assessment team will address these and other questions when it issues its complete set of case studies and analysis in late spring or summer of 2015.

fund forest thinning activities to reduce the risk of catastrophic wildfire in the watersheds around Flagstaff, thus creating the Flagstaff Watershed Protection Project (FWPP). The FWPP involves forest thinning treatments in critical but hard-to-treat areas of the forests around Flagstaff, including in the city's two main watersheds, Rio de Flag and Walnut Creek-Upper Lake Mary. The majority of the project area is comprised of National Forest (approximately 10,544 acres), but Arizona State lands and city lands are also included. Currently, the Forest Service, the City of Flagstaff, and the Arizona State Department of Forestry have conducted some treatments in preapproved areas and the Forest Service is conducting analyses on the remaining acres through the National Environmental Policy Act (NEPA) process (see Figure 7.1).

Why and How Flagstaff Implemented the Flagstaff Watershed Protection Project

Flagstaff and the Forest

Flagstaff sits in the middle of the largest contiguous ponderosa pine forest in the world. In the 1990s, these forests faced a series of record-breaking wildfires that began to increase awareness in the community about how past management practices had increased the forests' susceptibility to large-scale, intense, and devastating fires. Other factors, such as the bark beetle outbreak, had also contributed to the declining health of the area's forests (see Figure 7.2).

Flagstaff's economy and culture are closely linked to the forest, so local experts realized they needed to act. In 1997, individuals from the Coconino National Forest of the US Forest Service (USFS), the City of Flagstaff, Northern Arizona University, and environmental and business groups came together and formed the Greater Flagstaff Forest Partnership (GFFP) (see http://www.gffp.org and Appendix 7A for a list of GFFP partners). In the same year, the City of Flagstaff created its Wildland Fire Management division. GFFP began to raise awareness about the importance of forest management and identify key areas in need of treatment.

Schultz Fire and Subsequent Flooding

Following a series of record-breaking fires throughout the 2000s, in 2010, an abandoned campfire started the Schultz fire in the Coconino National Forest just north of Flagstaff. The fire burned over 15,000 acres. A month after the fire was contained, Flagstaff and the surrounding area experienced the fourth wettest monsoon event on record. The Schultz fire had damaged vegetation and ground

Vulnerability	Severe wildfires
Outcome	Land use
US geographic location	Southwest
Population	65,660
Median household income	$48,676
Political environment	Moderate

FIGURE 7.1 Flagstaff community profile.

FIGURE 7.2 Dry Lake Hills from the northern edge of the City of Flagstaff. (Source: M. O'Grady, 2014.)

cover, diminishing the forests' capacity to absorb much of the rain that fell. Thirty million gallons of water ran off the Schultz fire hillsides, resulting in severe flooding as far away as 10 miles from the fire site (Klassen and Howard 2011; Youberg et al. 2011). The flooding impacted several residential areas, damaged a main water pipeline, cutting off approximately 20 percent of the city's water supply, damaged 320 Native American cultural sites, and killed a 12-year-old girl (Brehl et al., personal communication 2014; Klassen and Howard 2011; Youberg et al. 2011).

The Schultz fire and subsequent flooding raised awareness in the community not only about residents' vulnerability to wildfire, but also to fire's secondary effects, such as flooding, erosion, and impacts to the water supply, that could affect the whole community. Every individual interviewed for this case study believed that the Schultz fire and subsequent flooding were a major factor in the formation and support of the FWPP (Brehl et al., personal communication 2014; Elson and Phelps, personal communication 2014).

Peer Learning from Santa Fe at the Southwest Fire Science Consortium

In late February 2012, GFFP representatives attended a conference in Santa Fe, New Mexico, sponsored by the Southwest Fire Science Consortium. At this conference, they heard a presentation about an effort in Santa Fe to identify the community's willingness to pay to help protect their watershed resources from catastrophic fire damage. Santa Fe has secured a grant to conduct the Payment for Watershed Services (PWS) approach to forest management, which included a significant amount of community outreach and education (Brehl et al., personal communication 2014). Paul Summerfelt, GFFP's president and the city's fire management officer at the time, attended the conference and realized that the GFFP and others in Flagstaff could motivate their community to help fund critical forest management as well.

Developing a Bond Measure and Building Support among Stakeholders

When Mr. Summerfelt and his colleagues returned to Flagstaff, they organized a local workshop to pitch the idea of PWS to the sustainability commissioners, city staff, department heads, and the city manager, Kevin Burke. Mr. Burke was supportive of the idea but proposed that Flagstaff raise the funds for their forest management project in a different way. He knew of a city bond that was set to expire later that fall. If the city had allowed the bond to expire, it would have decreased Flagstaff residents' taxes. However, Mr. Burke proposed that the city renew this bond and redirect the funds to wildfire management, which would provide support for forest management without increasing taxes. The bond would raise $10 million for the management of approximately 15,000 acres of forest around Flagstaff that was not currently covered under other projects (NAUERI 2013); this land included the Coconino National Forest, Arizona State land, and city land.

While the Forest Service retains decision-making authority on what activities would be allowed on National Forest land, the city maintains decision-making authority on the bond dollar (e.g., what the money will fund). Transferring the bond funds would require both a city council and a citywide vote of approval that would create community ownership of the work. Using bond funding would avoid the need for the city to conduct a costly and lengthy willing-to-pay study. This idea was presented to the City Council for vote in the spring of 2012. The City Council voted in favor of the initiative and bond question 405 was created for the ballot of the next fall citywide election.

As soon as the City Council approved putting the bond on the ballot, a small team of individual Flagstaff citizens organized a political action committee (PAC) in support of the measure. They hired a campaign consultant and raised $8000–$9000 (Brehl et al., personal communication 2014). The PAC conducted a hands-on, on-the-ground campaign that included distributing yard signs, soliciting support door-to-door, and handing out flyers at local events and farmers' markets.

In November 2012, bond question 405 was presented to the voters. The bond would provide support for forest management "[t]o prevent flood damage to the City of Flagstaff and to protect the city water supply from damages which occur from large-scale and/or severe wildfire(s) in town watersheds in the city" (City of Flagstaff 2012). Money raised through the bond would support activity on city, state, and federal lands. Each agency would be in charge of treatment on their land, but treatment options would be explored collaboratively through a partnership. The bond funding would be held by the city and distributed as needed to other agencies. The initiative passed with 73 percent voter approval (NAUERI 2013).

Exit polling conducted by Northern Arizona University's Ecological Restoration Institute (ERI) found that the majority of voters supporting the bond issue were primarily concerned with decreasing the risk of postfire flooding and protecting city water resources. The majority of supporters also "strongly agreed" or "agreed" that the project "partnership is a model of how to accelerate needed forest restoration" (84%) and that the "Forest Service would be accountable to the City" of Flagstaff with this project design (70%) (NAUERI 2013).

The planning effort on the National Forest is led by the FWPP Interdisciplinary Team (IDT), which is comprised of approximately 16 core members and eight consulting members, including a representative of the City of Flagstaff. The city was granted Cooperating Agency status early in the planning process to enable the city to have a seat at the planning table. An Executive Committee is also in place to discuss overarching decisions related to FWPP, including funding and implementation. The Executive Committee is comprised of representatives from the US Forest Service, the city, Coconino County, and the Arizona State Department of Forestry. Although each agency is responsible for the treatment of their land, the committee meets once a month to help facilitate overall project coordination (see "Next Steps" below for more information on the FWPP).

Effect of the FWPP

Although large-scale implementation of the FWPP has not yet begun, interviewees reported on several project successes. The bond issue had remarkable success in the election, and that high level of public interest and involvement has carried through to the planning process. Those involved with the campaign and FWPP members credit its high voter approval rating in part with the strength of public engagement, in addition to the still recent memory of the Schultz fire (Brehl et al., personal communication 2014; Elson and Phelps, personal communication 2014). This widespread support is even more meaningful considering that ERI's exit polling found that the majority of voters incorrectly believed that the bond was going to increase their taxes (NAUERI 2013).

Interviewees reported that the high voter approval rating has continued to be important to the project because it demonstrated strong public support for the project and empowered the community to take ownership of it. This high level of public support has potentially influenced opposition of the FWPP and has sent a signal to national, state, and local leaders that the community is dedicated to the project's success. This has also encouraged the project team to continuously inform the public and encourage public input into the project as much as possible, well beyond what is required by law (Brehl et al., personal communication 2014; Elson and Phelps, personal communication 2014). For example, the FWPP has developed a monitoring plan to track whether and how key goals of the project are achieved over time. The public noted the importance of such a plan, and key themes to address in it, during the bond campaign. The FWPP has held 20 public workshops to refine the plan and ensure that key voter concerns have been included (see http://www.flagstaffwatershedprotection.org/wp-content/uploads/2013/03/FWPP-Monitoring-Plan.pdf).

In the year following the passage of the bond, the project has raised an additional $2 million in additional direct and in-kind funding. The formal partnership between Flagstaff and the USFS, which was established soon after the bond passed, has led to substantial USFS contributions to the project; as of fall 2014, the USFS has contributed approximately $1.6 million to cover planning costs, wildlife surveys, hand thinning, prescribed burning, archaeological surveys, noxious weed surveys and treatments, and road work (Elson and Phelps, personal communication 2014). An additional $400,000 has come from the State of Arizona, Coconino County, Northern Arizona University, volunteers, and local citizens (Flagstaff Watershed Protection Project 2014; Summerfelt, personal communication 2014).

Moving Forward and the Broader Context

Next Steps

The project momentum has not waned since the passage of the bond. Much has already been accomplished and work is still continuing daily. However, although the project idea and public support came together quickly, treatment work in the Rio de Flag and Lake Mary watersheds could take up to 8 to 10 years to fully complete based on the complexity of the project, contracting processes, and seasonal operating limitations.

As mentioned previously, Flagstaff has decision-making authority over the bond dollars and the USFS has decision-making authority over what occurs on the National Forest. However, a wide diversity of stakeholders is involved in the FWPP. For example, government participants alone include the City of Flagstaff, the Arizona State Forestry Division, the USFS (Coconino National Forest), the US Fish and Wildlife Service, and Arizona Game and Fish Department. The immediate focus of the FWPP IDT is to finalize a treatment plan through a Record of Decision (ROD) so that large-scale treatment work can begin on the National Forest as soon as possible. To that end, the NEPA planning process began in November 2012 and is expected to end with the release of the Final Record of Decision in September 2015; this is roughly half the time it typically takes for a project of this scale (Summerfelt, personal communication 2014).

In April 2013, a Notice of Intent (NOI) to prepare an Environmental Impact Statement (EIS) was published in the Federal Register, and the FWPP IDT released a Proposed Action to the citizens of Flagstaff outlining different treatment options. Based on the comments the FWPP team received, the USFS released a Draft Environmental Impact Statement (DEIS) in July 2014, which analyzed four treatment alternatives: no action, minimal treatment, proposed action with cable logging, and proposed action without cable logging (e.g., using helicopter logging and specialized steep-slope equipment) (see Appendix 7B).

The DEIS was open for public comment for 45 days, during which time the FWPP team also held public meetings and outreach events to raise awareness and solicit public feedback. Between these events, and email and hard-copy submissions, the team received comments from 107 individuals who raised 530 separate topics. Every comment was in support of the project as a whole, but raised various concerns about how the treatments would be conducted. It helped the team realize that there was a need to create an implementation plan, a transparent document to address public concerns regarding contractor oversight, mitigation measures, and area closures. As of fall 2014, the FWPP IDT was in the process of drafting this plan along with a decision that addressed public comments by developing a blend of all three action alternatives. The IDT is also finalizing the EIS and conducting formal consultation with the US Fish and Wildlife Service for activities related to the Mexican spotted owl, a threatened species.

In the meantime, the USFS has been able to conduct some work on portions of FWPP lands that were previously approved for treatment through the NEPA process. Treatment methods already used include controlled burns and hand thinning and piling of cut material (slash; see Figure 7.3).

The state conducted a mechanical treatment operation on a parcel of state land, and the city has been engaged doing hand-thinning and controlled burning on a number of city and private land parcels. This work has allowed the FWPP team the opportunity to provide visual examples of treatment options to help further inform the public.

Most recently, in October 2014, the Rural Policy Institute of Northern Arizona University released a Cost Avoidance Study that estimates the amount of financial impact avoided by implementing the FWPP project. The results of the study estimate a range of potential losses from catastrophic wildfire and postfire flooding to be between \$573 million and \$1.2 billion dollars if no treatment is done to the forests. This helps justify the estimated project treatment costs of \$10 million (Arizona Rural Policy Institute 2014).

Project Resistance

Even among supporters of the bond initiative, there is some uncertainty and ongoing discussion about how the work should be conducted. Concerns have been raised by some environmental organizations that the impacts of harvesting on steep slopes, via cable and helicopter logging, would affect the Mexican spotted owl, a threatened species. As required by NEPA, FWPP members have met with these groups throughout the process and made every effort to address their concerns, including ensuring that ample public outreach would be an integral component of the project. Despite active public outreach efforts, some FWPP members also acknowledge that they will likely face increased public concern as they begin the treatment work and the short-term impacts become more real to the community. For example, once people start seeing logging equipment in their favorite part of the forest, or that their favorite biking trail is temporarily rerouted, they may resist the implementation of portions of the plan. This is a typical pattern seen with other projects; however, such concerns are usually replaced with satisfaction once treatments are completed and forest visitors get accustomed to the new, more open conditions (Brehl et al., personal communication 2014; Elson and Phelps, personal communication 2014).

FIGURE 7.3 Two treatment demonstration areas in the Coconino National Forest: a controlled burn area (a) and a burn pile of cut debris (b). (Source: M. O'Grady, 2014.)

Broader Context

FWPP is not the only forest-thinning project underway around Flagstaff. Many acres of forest surrounding Flagstaff had previously been treated via the city, GFFP, the state, and the USFS efforts. Others, such as the Hart Prairie Forest Health Restoration Project and the Wing Mountain Forest Restoration Project, have been flagged for treatment under other projects. The FWPP is an essential compliment to these other projects because it targets critical areas not otherwise designated for treatment due to steep slopes, challenging terrain, unique vegetation types, and the presence of the Mexican spotted owl. The FWPP treatment areas, although significantly smaller in size to those treated under other projects, help ensure comprehensive treatment for the forests around Flagstaff.

In parallel to the FWPP effort, Flagstaff has also become increasingly aware of the need to better understand and prepare for climate change. In 2006, the City Council passed the US Mayors Climate Protection agreement, and in 2007 they hired a full-time sustainability director, Nicole Woodman (Smith 2012). In 2008, Ms. Woodman and Flagstaff's Utility and Economic Vitality directors went to Boston for an Institute for Sustainable Cities' Climate Leadership Academy. During the workshop, they played out different scenarios for Flagstaff, one of which was: What happens if there is a major wildfire in the Rio de Flag or Lake Mary watersheds? They considered the potential implications of a major wildfire for both the city's water supply and the economy (Brehl et al., personal communication 2014). It did not take long for this situation to be tested in reality with the Schultz fire.

During this same period, the city was also conducting a climate change vulnerability assessment. The effort was led by Ms. Woodman and a few individuals, such as Mr. Burke, were involved with both projects. The assessment effort looked at the ways in which climate might make Flagstaff vulnerable to changes in climate, or how climate might exacerbate issues the city was already facing. Their assessment identified catastrophic wildfire risk in the city's watersheds as Flagstaff's main vulnerability, which helped promote the cause for sustainable forest management. The city's climate change vulnerability assessment was completed in May 2012.

Lessons Learned

Has Flagstaff Reduced Current Risks of Catastrophic Wildfire and Related Flooding?

Wildfire is an integral part of the forest ecosystems surrounding Flagstaff, and no amount of forest management will entirely eliminate its occurrence. In fact, wildfire is critical to forest regeneration and allowing small-scale and low severity fires to burn in wild areas can help ensure long-term forest productivity. However, high-severity, large-scale fires are a threat to both forest health and local communities. Forest management techniques, such as thinning overgrown areas, reduce fuel loads and thereby help prevent small fires from becoming catastrophic and destructive.

Flagstaff's actions through FWPP, Four Forest Restoration Initiative (4FRI), and GFFP's work on private land are focused on doing exactly that—reducing fuel loads and thus the risk of catastrophic wildfire in the city's watersheds. However, because the FWPP work could take 8 to 10 years to complete, the reduction in wildfire risk will not occur for some time. Additionally, after the initial treatments are completed, maintenance burning will be necessary to sustain the reduced risk in the future. While the FWPP does not currently include the funding for ongoing maintenance, the USFS does conduct maintenance burning regularly as a part of other, similar projects. In addition, the need for this funding was identified early in the process to include an amount each year following initial forest treatments. This issue will be revisited with firm options and a recommended course of action in the near future.

In summary, once the program has completed its plans, it seems likely that it will contribute to reduced contemporary wildfire risks on the lands it helps manage. However, ongoing maintenance is crucial to ensuring reduced risks are sustained over the long term.

Is Flagstaff Less Vulnerable to Wildfires and Related Flooding under Climate Change?

Overall, the FWPP seems likely to help reduce vulnerability to future climate change. Climate change projections for the Southwestern United States indicate increased temperatures, reduced spring snowpack and soil moisture, and more severe droughts, all of which will likely contribute to more frequent and severe wildfires (Melillo et al. 2014). Forest health in the Southwest will also be further threatened by invasive species and pests, whose impact may be amplified under future climate (Environmental Protection Agency 2013). Reducing fuel loads, as is being done through the FWPP and other programs, will help reduce the risk of catastrophic wildfire. However, extreme drought and heat, which could be more frequently experienced in Flagstaff in the future, could result in catastrophic fires even in forests with reduced fuel loads. That said, the measures being undertaken through the FWPP could reasonably be classified as true no-regrets actions. That is, the group's formation and actions were initially prompted by nonclimate concerns and provide benefits regardless of climate change, but FWPP activities also will help reduce vulnerability under future climate conditions.

Because forest treatment projects cannot eliminate the risk of fire completely, especially under changing climate conditions, other approaches could also be undertaken to help reduce future vulnerability to climate change. For example, improving response mechanisms to extreme events could better equip communities to respond. The existence of the FWPP may have created conditions that would facilitate such rapid response actions in and around Flagstaff. For example, FWPP members have built a strong internal team and have a good rapport with the community. These strong ties and open communication lines could be critical when actions require coordination among the wide array of stakeholders that manage the lands surrounding Flagstaff. Also, by building partnerships now to manage forest health, Flagstaff has put itself in a better position to cooperatively manage climate-related impacts to forests as they emerge in the future.

Does the FWPP Affect Marginal Populations in the Region?

The FWPP area includes approximately 140 acres of land owned by the Navajo Nation. The City of Flagstaff has worked with the Navajo Nation and will conduct some treatment work on tribal lands as part of FWPP. There are also other tribes who will be affected by treatment work done on USFS lands. The FWPP members have consulted these tribes as part of the NEPA planning process (Summerfelt, personal communication 2014).

Acknowledgments

We would like to thank the following for participating in interviews as part of this case study:

- Mike Elson, Coconino National Forest
- Erin Phelps, Coconino National Forest
- Mark Brehl, Wildland Fire Management, City of Flagstaff
- Nicole Woodman, Sustainability Manager, City of Flagstaff
- Anne Mottek Lucas, Mottek Consulting
- Diane Vosick, Northern Arizona University
- Paul Summerfelt, Wildland Fire Management, City of Flagstaff

APPENDIX 7A: MEMBERS OF THE GREATER FLAGSTAFF FOREST PARTNERSHIP

Voting Members

Arizona Game & Fish
Arizona Forest Restoration Products
Arizona State Forestry Division
City of Flagstaff Fire Department
Coconino County–Community Development Department
Coconino Natural Resource Conservation District
Coconino Rural Environment Corps

Ecological Restoration Institute at Northern Arizona University
Mottek Consulting, LLC
Northern Arizona University School of Forestry
The Arboretum at Flagstaff
The Nature Conservancy
US Fish and Wildlife Service, Region 2
Wildwood Consulting, LLC

Associates
Arizona Public Service (APS)
National Park Service, Flagstaff Area National Monuments

Cooperators
Ponderosa Fire Advisory Council
USDA Coconino National Forest

APPENDIX 7B: FLAGSTAFF CASE STUDY

Actions	Alternative 1: No treatment	Alternative 2: Proposed action with cable logging	Alternative 3: Proposed action without cable logging	Alternative 4: Minimal treatment
Total treatment acres	Jack Smith Schultz/Eastside1	8937 acres 5963 acres DLH 2975 acres MM	8937 acres 5963 acres DLH 2975 acres MM	5802 acres 3459 acres DLH 2343 acres MM
Percentage of total project area to be treated	0%	85% 79% DLH 100% MM	85% 79% DLH 100% MM	55% 46% DLH 79% MM
Acres to be hand thinned	0 acres	846 acres 699 acres DLH 147 acres MM	832 acres 652 acres DLH 180 acres MM	438 acres 438 acres DLH 0 acres MM
Acres to be mechanically thinned	0 acres	7124 acres 4697 acres DLH2 2427 acres MM	7137 acres 4743 acres DLH 2394 acres MM	5264 acres 2953 acres DLH 2311 acres MM
Acres to be helicopter logged	0 acres	0 acres	973 acres 973 acres DLH 0 acres MM	0 acres
Acres to be cable logged	0 acres	1242 acres 1169 acres DLH 73 acres MM	0 acres	0 acres
Acres to be prescribed burned	0 acres	8937 acres 5963 acres DLH 2975 acres MM	8937 acres 5963 acres DLH 2975 acres MM	5802 acres 3459 acres DLH 2343 acres MM

REDUCING VULNERABILITY TO FLOODING IN GRAND RAPIDS, MICHIGAN

Missy Stults
University of Michigan

Jason M. Vogel
Stratus Consulting

For the last 30 years, the City of Grand Rapids has been devising solutions to manage stormwater. However, recent flooding events, combined with aging stormwater infrastructure and growing public discontent about the dilapidated state of roads, presented an opportunity to rethink how stormwater was managed in the city. To address these concerns, the City of Grand Rapids created the Vital Streets and Sidewalk Spending Guidelines (Guidelines), mandating the use of green infrastructure when upgrading road and stormwater infrastructure. Through extensive public engagement combined with the deteriorating state of roads, the Guidelines and funding to implement

The City of Grand Rapids is located in western Michigan on the Grand River near the shores of Lake Michigan. Originally incorporated as a municipality in 1838, Grand Rapids now has a population just over 192,000, with a median household income of $39,000. The community is 64.6 percent Caucasian with just over 28 percent of the population holding a bachelor's degree or higher and a poverty rate of 26.8 percent. In size, Grand Rapids is 44.4 square miles with 4235.6 persons per square mile.

FIGURE 7.4 Overview of Grand Rapids.

them passed with 66 percent of voter support in early 2014. As of mid-2014, all upgrades to existing road or stormwater infrastructure, as well as any new road or stormwater projects, are required to prioritize green infrastructure techniques or justify why these techniques are not technically feasible. According to stakeholders in Grand Rapids, the Guidelines and the associated move to more green infrastructure in the city are helping change the culture of stormwater management in the city while simultaneously enhancing local water quality, replenishing the city's aquifer, cleaning the local air, reducing the urban heat island effect, and enhancing the overall sense of community (see Figure 7.4).

Why and How Grand Rapids Implemented Its Vital Streets Guidelines

A Long History of Flooding and State Water Quality Mandates Push the City Toward Gray Stormwater Infrastructure

From as early as 1883, documents show that major flooding on the Grand River, which runs through Grand Rapids (from events such as log jams, spring snowmelt, ice jams, and heavy rains) damaged homes, businesses, and roads primarily along the city's west side (Olson 2014). Following a major flood in 1904, the city built its first floodwalls in 1911 with additional floodwalls built in 1927, 1934, and 1936 (Clark 2014). In 2003 the city completed its most recent floodwall project, which raised all existing floodwalls and embankments one foot above the 100-year flood mark established in the 1904 flood (Clark 2014). Despite these investments in physical infrastructure to reduce vulnerability to flooding, the Grand River has continued to overflow, with a historic flooding event occurring in April 2013 (see Figure 7.5).

The legacy of inundation led the city to investigate strategies in addition to floodwalls for minimizing the threat of both riverine and inland flooding. This work gained traction roughly 30 years ago when the city began efforts to remove all existing combined sewer systems and replace them with separate sanitary sewer and stormwater systems (Alibasic 2014). Motivation for this change

21.85 feet on April 21, 2013
19.64 feet on March 1, 1985
19.54 feet of May 27, 2004
19.50 feet on March 28, 1904
19.29 feet on March 8, 1976
19.25 feet on April 3, 1960
19.25 feet on September 4, 1986
18.83 feet on March 3, 1982
18.60 feet on June 9, 1905
18.50 feet on April 18, 2013

FIGURE 7.5 The highest Grand River crests on record for the City of Grand Rapids. (From National Weather Service. 2014. Available: http://water.weather.gov/ahps2/hydrograph.php?wfo=grr&gage=gdrm4. Accessed November 3, 2014.)

came from the State of Michigan, which was actively working with the city to reduce the amount of sewage being discharged into the Grand River.

When heavy rainfall events occur, combined sewer systems can get overwhelmed, making them unable to effectively manage and treat sewer and stormwater waste separately. In these cases, excess sewer combines with stormwater and flows directly into discharge basins. In Grand Rapids, this discharge basin is the Grand River. In the early 1980s, the State of Michigan began citing the City of Grand Rapids as being in violation of water quality requirements. In response, the city began working with the state to devise a strategy to reduce the amount of combined sewer and stormwater entering the Grand River. The result was the creation of a stormwater management plan in 1988 that called for the division of Grand Rapids' combined sewer system into separate stormwater and sewer systems. To date, over 99.7 percent of all the combined sewer pipelines have been replaced (55 out of 59 pipes), and plans are underway for replacing the four remaining combined sewer and stormwater lines by 2019 (Lunn 2014).

Continued Flooding Ignites Support for Green Infrastructure

To date, the city has invested hundreds of millions of dollars in the transition to a separate sewer and sanitary system and has successfully reduced the amount of sewage entering the Grand River. This work, however, has not eliminated the growing threat of localized flooding. In fact, a US Environmental Protection Agency 2008 report found that in the Great Lakes region, "projected long-term (2060–2099) changes in precipitation suggest that if combined sewer overflow (CSO) mitigation efforts are designed based on historical precipitation, many systems could experience increases in the frequency of CSO events beyond their design capacity resulting in increases in overflow volume discharged to receiving waters" (US EPA 2008). This realization, combined with the fact that localized flooding was occurring more frequently, led a number of citizens, nonprofits, and city staff to seek alternative strategies for managing stormwater.

According to Haris Alibasic, the director of the City of Grand Rapids Office of Energy and Sustainability, the movement to rethink how stormwater was managed in Grand Rapids crystallized during a series of community stormwater discussions held in partnership between the city and the West Michigan Environmental Action Council (WMEAC). Initiated in the late 2000s, these meetings were held with businesses, nonprofits, neighborhood associations, residents, and city commissioners to explore stormwater best practices from around the country and devise a plan for how Grand Rapids could integrate relevant practices into its operations. In addition to public meetings, a survey was administered to more than 600 residents to assess whether there was support for undertaking and financing improvements to the existing stormwater system. This effort, known as the Community Stormwater Master Planning Initiative, culminated in a 2012 report entitled *Sustainably Managing Stormwater in Grand Rapids.*

This report identified five priorities for the city, most notably the prioritization of green infrastructure over traditional hard physical infrastructure. Upon the report's release, organizations such as the Chamber of Commerce requested a more detailed look at what infrastructure improvements were needed (both gray and green), where they were needed, and how they would be funded (Occhipinti 2014). Based on this feedback, the City of Grand Rapids hired Tetra Tech for approximately $450,000 to help (1) create a Stormwater Asset Management Plan, (2) create a Stormwater Capital Improvement Plan, (3) update the city's Stormwater Technical Reference Manual, and (4) create an asset management software tool that allows the city to visually see where improvements to the stormwater system are needed. The results from these deliverables confirmed that significant investments were needed to upgrade the city's deteriorating stormwater infrastructure and that green infrastructure could, in many cases, provide an alternative to traditional gray infrastructure in meeting both stormwater needs and other community-wide goals. In addition to the work done by Tetra Tech, a series of other reports, as denoted in Figure 7.6, were instrumental in laying the foundation for community support of green infrastructure.

2010	Citizen Stormwater Initiative started
2010	Release of the Grand Rapids Sustainability Plan
2011	Release of the Green Grand Rapids Master Plan, a citywide green infrastructure master planning initiative
2012	Release of the Sustainably Managing Stormwater in Grand Rapids report
2013	Release of the Stormwater Asset Management Plan
2013	Release of the Stormwater Capital Improvement Plan
2013	Release of the Technical Stormwater Reference Manual
2013	Creation of the Vital Streets Task Force
2013	Release of the Sustaining Stormwater Investments in Grand Rapids report
2013	Release of the Grand Rapids Climate Resiliency report
2014	Passage of the Vital Streets and Sidewalk Spending Guidelines
2014	Creation of the Stormwater Oversight Commission

FIGURE 7.6 Key dates and actions related to green infrastructure transition in Grand Rapids.

Using Community Support for Road Upgrades to Finance Green Infrastructure and Create the Vital Streets and Sidewalk Spending Guidelines

Armed with this information, the next issue was how to begin implementing and financing key infrastructure improvements. In Grand Rapids and throughout Michigan, street infrastructure has been in dwindling disrepair for years. With the intense winter of 2013 leading to even more potholes and damaged roads, residents and businesses united in demanding the city invest in major repairs to city streets (Occhipinti 2014).

Recognizing that the "smartest time to do stormwater upgrades is when you're working on streets," Nick Occhipinti, director of policy and activism at WMEAC, notes that WMEAC worked with the city to leverage the political interest and will around street repairs to also update stormwater infrastructure. One idea for funding this work that quickly gained traction was the extension of an income tax that was about to expire (Occhipinti 2014). City staff, elected officials, nonprofits such as WMEAC, and many businesses agreed that pursuing an extension to the income tax was a politically viable solution for financing needed repairs. As such, the city set to work drafting specific details for how the income tax extension combined with other city funds could be used to simultaneously repair streets and upgrade key stormwater infrastructure.

The result of these efforts was the Grand Rapids Vital Streets and Sidewalks Spending Guidelines (Guidelines), which specifies how the income tax extension as well as other city funding would be spent, establishes stormwater performance standards, and creates a stormwater oversight commission to help guide the implementation of the Guidelines. Perhaps most importantly, the Guidelines require that, whenever feasible, the city must invest in green infrastructure as opposed to traditional hard infrastructure. In situations where green infrastructure is deemed technically unfeasible, the city must justify why this is the case and identify ways that green infrastructure can be used to supplement more traditional hard infrastructure approaches.

During a special May 2014 election, 66 percent of voters favored the Ordinance and the income tax extension required to finance it (LaFurgey 2014). While the clear and growing need for updated streets was a strong motivator for action, public education and outreach around the need for stormwater upgrades, including efforts such as public service announcements and community forums, helped to build the critical support needed to pass the Ordinance.

Building and Maintaining Community Support for Green Stormwater Management

Community support has been an essential element of the Guidelines and the associated income tax extension. While broad public support appeared to exist, a small but vocal number of citizens actively opposed new, large-scale stormwater investments and the movement to create the Guidelines. This group of opposition included constituents who supported the transition to green infrastructure but did not think the city was investing enough in making the requisite changes. As noted by Nick Occhipinti, "everyone is suspicious when you ask for new revenue."

To navigate citizen opposition, the city, WMEAC, and other partners "did our best to sit down with opponents and explain, in detail, why this is an issue, what the need is, and why we care about it. We also did public advocacy and education including an intentional citizen outreach effort in which we worked directly with community leaders to discuss the need." In addition, public meetings, listening sessions, and dozens of presentations were delivered to community stakeholders to help educate people on the need for stormwater upgrades and build support for the Guidelines. In the end, stakeholders came to a general consensus that (1) stormwater management is important, (2) investments in stormwater are going to be critical, and (3) to effectively manage stormwater, the city needed to have a solid understanding of what infrastructure currently exists, what state it is in, what needs to be done to properly manage it, and what upgrades are needed (Alibasic 2014; Lunn 2014). Moreover, by combining discussions about stormwater upgrades with road repair, the city was able to make a more holistic case about why public investment in critical infrastructure was essential.

To maintain support, the city uses hands-on events where city commissioners and the public get together to plant trees, clean bioretention islands, and help maintain rain gardens. As Mike Lunn notes, "we try to involve the community as much as we can and really highlight our projects that exist as stormwater amenities. We try to give people ownership over these projects so they feel invested and engaged."

While a number of individuals were supportive of stormwater management efforts, and more specifically, the move toward green infrastructure, a few elected and appointed officials and staff were particularly instrumental. These included Mayor George Heartwell, Commissioner Ruth Kelly, Deputy Mayor Eric Delong, Haris Alibasic, Mike Lunn, Suzanne Schulz, and the West Michigan Environmental Action Council.

Through these extensive engagement and outreach efforts, the City of Grand Rapids was able to successfully pass both the Vital Streets and Sidewalks Spending Guidelines and the income tax extension needed to help finance the implementation of the Guidelines. These two efforts passed with 66 percent of voter support during a special May 2014 election (LaFurgey 2014). While the clear and growing need for updated streets was a strong motivator for action, public education and outreach around the need for stormwater upgrades, including efforts such as public service announcements and community forums, helped to build the critical support needed to pass the Guidelines (Mike Lunn, personal communication).

Accomplishments of the Guidelines to Date

With wide and growing support from residents and the City Council, the City of Grand Rapids is now implementing the Guidelines. According to Nick Occhipinti, the city's Environmental Services Department is "already experiencing better funding for stormwater." "It makes sense to have these projects done using green infrastructure. And it also helps communities (residents, businesses) to better understand the value of green infrastructure," says Haris Alibasic. When you have a project like Joe Taylor Park, where you can incorporate green infrastructure elements, certainly the community can appreciate the value of onsite stormwater management" (see Figure 7.7). The cobenefits associated with the use of green infrastructure also make the strategy enticing for the city. For example, better water quality, replenishment of the city's aquifer, cleaner air, heightened walkability, heat island reduction, and energy savings are all cobenefits associated with green infrastructure (Alibasic 2014; Occhipinti 2014).

Has the transition to green infrastructure been a success? "Yes and no" says Nick Occhipinti, "we are not done yet." Haris Alibasic shares this sentiment, noting "there are always opportunities for progress in everything that you do. We have to continually invest and evaluate our progress as opposed to saying that we have succeeded. In this case, I couldn't call our work a failure; I wouldn't call it a complete success. I would call it a positive step in the right direction."

Mike Lunn, director of Environmental Services for the City of Grand Rapids, notes that "yes, I think we are very successful around stormwater and we are starting (especially with Joe Taylor Park, Mary Waters, and Tremont Parks) to have a lot of things to talk about. We just don't talk about them enough." Measuring success is hard, but for Mike Lunn and his team, one simple strategy is to see if "three years later, we still have vegetated solutions in place and they are still working, and if they are, then chances are they are going to continue working for much longer."

Even though the city does not currently have specific metrics to track the success of green infrastructure, it does use two ultimate screens to measure the success of any project: Did the strategy reduce greenhouse gas emissions, and was the strategy financially sound? While these metrics do not embody all of the possible ways of measuring success, they do provide a litmus test for ensuring projects are in alignment with broader city goals. In addition, Haris Alibasic notes that the city is currently looking for metrics of resilience that can be incorporated into the next round of updates to the Grand Rapids Sustainability Plan, the city's guiding document pertaining to all aspects of social, environmental, and economic sustainability, and these new metrics are likely to include more specific measurement related to stormwater management (Alibasic 2014). Perhaps more than any physical metric of success, the indicator that Grand Rapids is doing something right is the fact that, despite extremely limited staffing, stormwater management appears to "be a part of the city's culture" (Lunn 2014).

JOE TAYLOR PARK

Situated in the east side of Grand Rapids, Joe Taylor Park was created in 2011 and is one of the city's newest parks. Originally noted as being park-poor or underserved when it came to park space, the Baxter neighborhood and surrounding area had been flagged as in need of park space for years. In addition, localized flooding in the area had been raised as a concern necessitating action.

Initially, to address flooding-related concerns, the city looked at installing a 10-million gallon concrete holding basin. However, through a series of stakeholder meetings and reviews, the city decided, instead, to put in an infiltration basin composed of green infrastructure. A pervious parking lot, light-emitting diode (LED) security lighting, and a cistern to capture rainwater from the park to irrigate the park's trees and lawn were also installed.

In addition to the installation of stormwater features, "city officials asked neighbors to help design a kid-friendly park" (Harger 2011). Through this process, residents requested a picnic shelter, playground, seating areas, and a water park. The park caters to smaller children as well as families and single residents. More recently, residents installed a series of urban gardens.

Today, Joe Taylor Park is 2.2 acres of what was once blighted land that now creates an urban landscape more accessible and comfortable to the residents abutting the area. In addition, the site collects, treats, and drains stormwater from the entire neighborhood (approximately 40 acres) and is able to manage the first inch of water falling during a storm directly on-site.

FIGURE 7.7 Joe Taylor Park.

Moving Forward and Broader Context

While climate change has not been explicitly integrated into stormwater management efforts in Grand Rapids to date, Haris Alibasic and Mike Lunn both note that climate change was and remains an underlying theme that drives the city's investment decisions. As noted by Mike Lunn, the city has been incorporating future projections into stormwater planning for years by doing things such as "designing to a 25 or 100 year storm as opposed to a 10-year storm." "The idea is to try to get the rain to stay where it lands. And we know that the climate always changes so we prefer using green infrastructure because it will help us keep the water out of the river and lessen our impact on the environment," noted Mike Lunn. According to Haris Alibasic, "we did not use exact projections in our work to date. That's why we are moving to the next phase where we are looking at the exact science and future projections and incorporating these into our stormwater management planning initiatives."

Recognizing this need, the city has extended a contract with Tetra Tech to assess key issues such as how climate change could affect rainfall patterns and what areas of infrastructure are the most vulnerable to issues such as extreme weather and increased precipitation. The final product developed by Tetra Tech is meant to help guide the City of Grand Rapids in making decisions on which projects to undertake, how best to integrate green infrastructure into these projects, and when the projects should be undertaken to optimize the efficiency of the system.

"There is a recognition that climate change will inevitably have an impact on stormwater infrastructure," said Haris Alibasic. "And the way that infrastructure is planned in Grand Rapids is about taking into consideration the long-term as opposed to the short-term view and perspective. This goes along with climate change resilience preparedness, preparation, and planning because it affords us an opportunity to look at projections of climate change and integrate them into our infrastructure planning."

Even though climate change was not explicitly factored into the city's existing stormwater management plans, Grand Rapids does have a resilience report that has been helping lay the foundation for integrating climate change into other city planning efforts. The Grand Rapids Resilience Report was created by WMEAC in partnership with the city and was instrumental in helping the city acquire localized climate data (Alibasic 2014). The report discusses 22 sectors or topics within Grand Rapids likely to be affected by climate change, one of which is stormwater. To make the findings of the Resilience Report more applicable, Haris Alibasic and his team in the Office of Energy and Sustainability are working to integrate key findings from the Resilience Report into the updated Grand Rapids Sustainability Plan (due to be released in 2015).

Moving forward, the city has three priorities: (1) installing more green infrastructure while maintaining what already exists, (2) formally integrating climate change and resilience into citywide plans and operations, and (3) finding additional funding to complete the needed upgrades to the stormwater system. The city's newly created (summer 2014) stormwater management oversight commission will play a key role in helping to ensure these priorities are achieved. According to Mike Lunn, "The more successful you are, the more opportunities you get to continue doing these things." In addition, an opportunity exists for more regional work, but what this opportunity looks like and how it is seized remains to be seen (Lunn 2014).

Lessons Learned

Has Grand Rapids Reduced Current Risks of Flooding?

For decades, the City of Grand Rapids has been working to prevent flooding. In the mid-1900s, this work centered on addressing riverine flooding. In the late 1900s and early 2000s, the City focused on stormwater management. While important, these efforts missed the opportunity to holistically address flooding by also tackling inland flooding, which has been occurring more frequently due to urban development, increases in impervious surfaces, and an increase in extreme precipitation, among other things. The Vital Streets and Sidewalk Spending Guidelines helps fill this omission by ensuring that the city is preparing for flooding of all types (riverine, inland, and stormwater).

Given the nascence of the Guidelines, however, it still remains to be seen how effective they will be in spurring projects that ultimately reduce the community's vulnerability to flooding. As the city begins implementing projects, monitoring and evaluation will be essential to determine how effective the Guidelines are and to make revisions, as needed, to ensure that all infrastructure investments are contributing to actual reductions in vulnerability. Additionally, green infrastructure alone will be incapable of managing all future flooding in the city. As such, removing properties from the floodplain, investments in flood-proofing existing infrastructure, and strong flooding alert systems will all be important to helping reduce the flood risks to Grand Rapids' built, natural, and social systems.

Is Grand Rapids More Resilient to Climate Change?

While the Guidelines are new, their implementation will likely help the city build resilience to climate change. Specifically, the requirement to invest in green infrastructure will help the city be more resilient to increases in extreme precipitation events, which are projected to take place more frequently in the region (Melillo et al. 2014). Additionally, increases in green infrastructure can help to mitigate the effects of increasing temperatures and the associated urban heat island effect.

While green infrastructure might be considered a no-regrets adaptation strategy due to its many immediate and long-term benefits, it should be noted that the City of Grand Rapids did not explicitly integrate climate change into the design of the Guidelines. Instead, stakeholders involved in the process noted that climate change was always an underlying issue of concern that drove a significant amount of their decision-making. The effectiveness of an informal integration of climate change into decision-making remains to be seen. However, there could be value in subtly embedding climate change considerations into decision-making processes where certain stakeholders may be averse to openly discussing climate change.

Despite the fact that climate change was not formally integrated into the first draft of the Guidelines, the city has since commissioned consultants to help them more holistically understand how changes in climate could affect existing and future infrastructure investments citywide. In order for the city to be more resilient to climate change, the results from this study may need to be integrated into revisions to the Vital Streets and Sidewalk Spending Guidelines as well as into individual street and stormwater projects. If the current guidelines prove insufficient and the climate change information being pursued is not integrated into future revisions of the Guidelines, the city risks investing in infrastructure that is ill-prepared for extreme events as well as future changes in climate.

Do the Guidelines Affect Marginal Populations?

The Guidelines are meant to be applied citywide, meaning that all individuals within the city should receive equal access to updated streets and green infrastructure. The reality of this statement, however, has yet to be tested. The challenge for the City of Grand Rapids will be to ensure that implementation of the Guidelines is based on infrastructure need (i.e., the infrastructure in the greatest need of repair gets first priority), as opposed to the prioritization of the wealthy or those with the most influence. To date, green infrastructure projects in the city have emphasized working with underserved neighborhoods, but continuing this process going forward, especially in light of the extreme need and budgetary limitations, will present a challenge for the city.

Acknowledgments

We would like to thank the following for participating in interviews as part of this case study:

- Haris Alibasic, director of the Grand Rapids Office of Energy and Sustainability
- Mike Lunn, director of the Grand Rapids Department of Environmental Services
- Nick Occhipinti, director of policy and activism at the West Michigan Environmental Action Council

REDUCING VULNERABILITY TO COASTAL FLOODING IN NORFOLK, VIRGINIA

Alexis St. Juliana
Jason M. Vogel
Stratus Consulting

Norfolk experiences the impacts of coastal flooding regularly, but community members report that coastal flooding has become more frequent, more severe, and now occurs in places where flooding has not traditionally been a problem. Extreme weather events such as nor'easters and hurricanes generate large storm surges and heavy precipitation that exacerbate tidal flooding. As a result, the City of Norfolk passed changes to its flood and coastal zone ordinance to reduce the city's vulnerability to coastal flooding. Beginning in 2014, new structures in flood and coastal zones must be built with their lowest level 3 feet above the 100-year flood plain. Existing structures in flood and coastal zones that incur two flood damage events each worth 25 percent or more of their value must be elevated to meet this same 3-foot standard. Newly constructed properties are complying with the new standard; however, it will take more time to know the full impact of Norfolk's action. Community members feel that the standards will reduce the city's vulnerability for several decades into the future. The city is considering additional actions to create a more robust response to coastal flooding (see Figure 7.8).

How Norfolk Implemented Its Flood and Coastal Zone Ordinance

Increased Flooding Builds Motivation for Action in Norfolk

Norfolk is a coastal community that has been struggling with the impacts of both recurrent tidal flooding and inland flooding for decades. Due to subsidence and sea-level rise, episodes of tidal flooding are becoming more frequent, more severe, and are occurring in places where flooding has not traditionally been a problem. For example, certain streets flood regularly with high tides, including streets that did not flood in the past (see Figure 7.9). In 2014, there were several heavy rainstorms that shut down traffic across the city for hours at a time due to roadway flooding. These heavy rainfall events exacerbate flooding at high tide, when many drainage or outflow pipes are submerged or partially submerged and rainwater cannot properly drain.

In an effort to characterize the problem and identify solutions to Norfolk's flooding, the City of Norfolk funded a series of studies beginning in 2007. These included the *Preliminary Citywide Coastal Flooding Mitigation Concept Evaluation and Master Plan Development,* a *City-Wide Drainage Master Plan,* and long-term tidal and precipitation flooding analyses (City of Norfolk 2012; City of Norfolk et al. 2013; Fugro 2012; Timmons Group 2012). These studies recommend close to $1 billion in infrastructure improvements such as floodwalls, tide gates, pump stations, road elevation projects, drainage improvements, berm construction, and structural elevation projects (City of Norfolk 2014b).

Vulnerability	Tidal flooding
Outcome	Design standards
US geographic location	Southeast
Population (2013)[A]	246,392
Median household income[B]	$44,164
Political environment[C]	Liberal

A. Cooper Center, 2014
B. American Community Survey, 2012
C. Politico, 2012

FIGURE 7.8 Norfolk community profile.

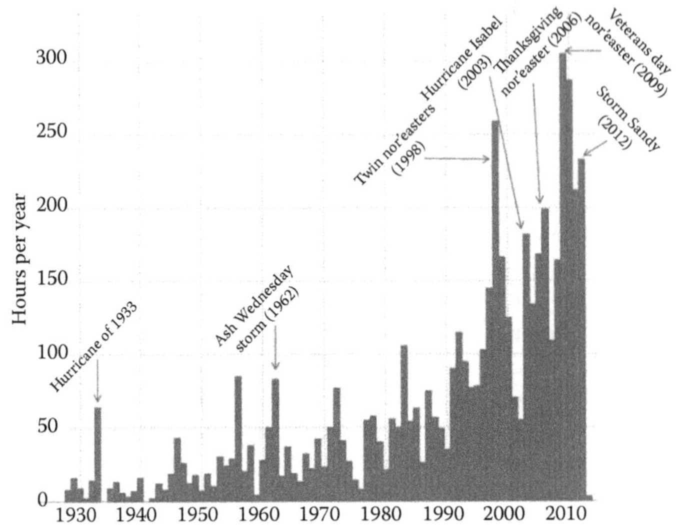

FIGURE 7.9 Street flooding in Norfolk caused by extreme high tides or storm surge. (From Atkinson et al. 2013. Sea Level Rise and Flooding Risk in Virginia. *Sea Grant Law and Policy Journal* 5(2):3–14. Available: http://www.ccpo.odu.edu/~tezer/PAPERS/2013_SGLPJ_Atkinson_SLR.pdf. Accessed November 3, 2014.)

Increased Flood Insurance Claims Motivate Norfolk's Department of City Planning to Take Action

Beginning in 2009, Norfolk's Department of City Planning saw an increase in insurance claims to the US Federal Emergency Management Agency's (FEMA's) National Flood Insurance Program (NFIP). Figure 7.10 shows the number of properties filing multiple flood damage claims to the NFIP. The Department of City Planning is the link between the City of Norfolk and the NFIP. The department is also the key player in Norfolk's overall flood mitigation strategy, floodplain and coastal zone mapping activities, and efforts to raise awareness on Norfolk's flooding issues, although several other city departments and regional entities play a role.

This increase in flood insurance claims was a very serious concern for the Department of City Planning for several reasons. First, they were concerned about Norfolk's citizens and the impact of flooding on their lives and properties. The department wanted to find a way to minimize impacts to properties so that owners would be protected from worsening flooding. Second, they were concerned about Norfolk's rating

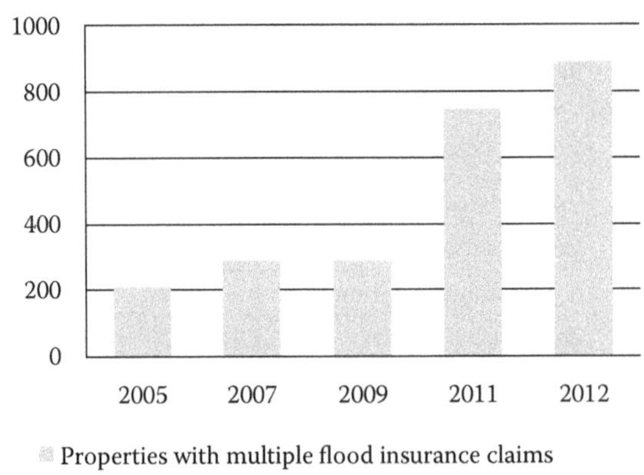

FIGURE 7.10 FEMA repetitive flood claim records. (Adapted from City of Norfolk, 2013b. Revisions to Floodplain Regulations. July 23. Available at http://www.norfolk.gov/documentcenter/view/9106. Accessed November 3, 2014.)

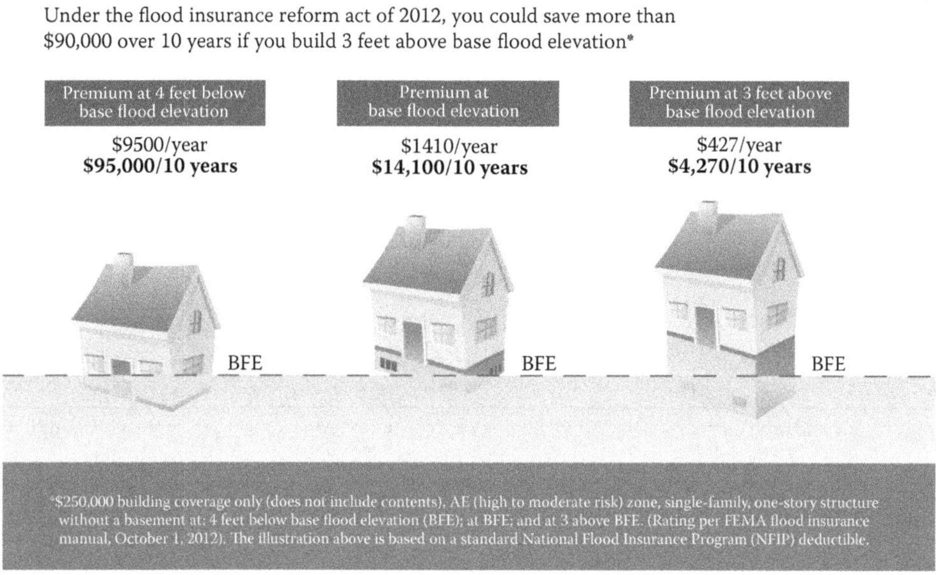

FIGURE 7.11 Example flood insurance premiums for a home below, at, and above base flood elevation. (From FEMA. 2013b. If Your Home or Business Has Been Flooded Build Back Safer and Stronger. Available: http://www.fema.gov/media-library-data/20130726-1858-25045-7797/build_back_stronger02_2013.pdf. Accessed December 22, 2014.)

within the NFIP Community Rating System (CRS) and the insurance premiums available to the Norfolk property owners under that program. The Department of City Planning wanted to ensure that flood insurance for its residents was affordable for all residents so they would be properly insured in the event of a serious flood. Third, the city was aware of citizens' concerns that flooding was getting worse and that action needed to be taken. Finally, in 2012 Congress passed the Biggert-Waters Flood Insurance Reform Act (the Act), which reformed the NFIP and included a schedule to reduce flood insurance premium subsidies in 2014 (FEMA 2013a).* Norfolk was already exploring options to improve its CRS rating and maintain or lower flood insurance premiums when the Act was passed, but the Act possibly sped up their actions. Figure 7.11 shows insurance premiums for a home below, at, and above base flood elevation.

Norfolk's Department of City Planning Drafts Proposed Ordinance Changes

In 2010 and 2011, the Department of City Planning reviewed options for changing the current ordinances in FEMA-defined flood and coastal zones. As part of this review, the department also consulted sea-level rise projections from the Virginia Institute of Marine Sciences. The department devised a series of proposed changes, which would protect citizens from flooding and reduce flood insurance premiums by improving Norfolk's status on the CRS. Although there were several proposed changes, the key changes are outlined in Figure 7.12.

To explain in further detail, increasing the city's standard from 1 foot above base flood elevation (the 100-year floodplain) to 2 feet above base flood elevation means that the lowest level of new construction would need to build two feet above FEMA's mapped base flood elevation; this structural elevation is also called freeboard. The second provision would change the regulations on existing buildings in flood and coastal zones. Previously, a 1-foot freeboard was required when a property incurred a single damage event or implemented improvements worth 50 percent of the building's insurance replacement value.† The new provision would look at cumulative damages

* Congress later delayed and rolled back some provisions of the Biggert-Waters Act with the Homeowners Flood Insurance Affordability Act of 2014.

† Insurance replacement values are typically less than the purchase price of properties.

Selected Proposed Changes to Norfolk's Zoning Ordinance by the Department of City Planning	
Existing regulation	**Proposed regulation**
Require the first floor level be elevated one (1) foot above the anticipated flood levels.	Require the first floor level be elevated two (2) feet above the anticipated flood level.
Require existing buildings come into compliance with current regulations if damage or improvements from a single event exceeds 50% of market value of the structure.	Require existing buildings come into compliance with current regulations once the sum of damage and/or improvements exceeds 50% of market value of structure over a 10-year period (Cumulative Loss/Improvement). Or If flood damage from two events, on average, each equals or exceeds 25% the market value of the structure (Repetitive Loss).

Source: Adapted from City of Norfolk. 2013b. Zoning Ordinance Text Amendment to Chapters 2, Definitions, and 11–3, Floodplain/Coastal Hazard Overlay District (FPCHO), to amend terms, simplify language, and to add requirements for properties located within the floodplain and to revise standards for calculating building height.

FIGURE 7.12 Selected proposed changes to Norfolk's zoning ordinance by the Department of City Planning.

and improvements over a 10-year period to determine if a building should come into compliance with a 2-foot freeboard. The Department of City Planning proposed two catalysts: (1) damages and improvements totaling 50 percent of the building's insurance replacement value, or (2) two flood damage events each totaling 25 percent or more of the building's insurance replacement value.

The Planning Commission Considers Climate Change and Revises the Proposed Standards

In 2011, the Department of City Planning brought its proposal to the appointed Planning Commission. The Department presented background information, including sea-level rise projections (see Figure 7.13). The Planning Commission reviewed the Department's recommendations and recommended one significant change: that the 2-foot freeboard standard be raised to 3 feet based on sea-level rise projections for Virginia. In particular, Commissioner Martin Thomas, Jr. suggested that 2 feet of freeboard

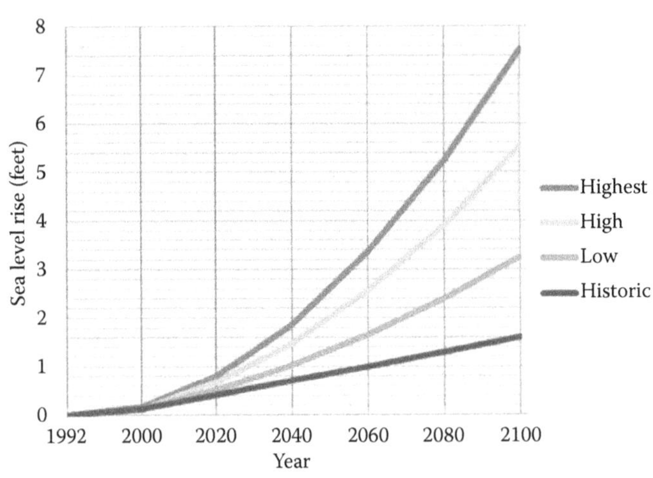

FIGURE 7.13 Southeast Virginia sea level rise scenarios. (From City of Norfolk, 2013b. Revisions to Floodplain Regulations. July 23. Available at http://www.norfolk.gov/documentcenter/view/9106. Accessed November 3, 2014; Virginia Institute of Marine Sciences. Undated. Sea-Level-Rise Scenarios. Available at http://www.vims.edu/newsandevents/topstories/slr_scenarios.php. Accessed December 9, 2014.)

was insufficient to protect structures in the flood and coastal zones based on the available sea-level rise projections. He felt that raising their current standard by just 1 foot would only benefit the city for the next 15–20 years, and that by increasing the freeboard standard to 3 feet, there was greater potential to protect properties for a longer time period. The Planning Commission unanimously agreed to increase the proposed standard to include 3 feet of freeboard for structures in flood or coastal zones.

Norfolk Seeks Stakeholder Feedback and Revises the Ordinance

With the Planning Commission's recommendations, the Department of Planning arranged a series of public meetings in May and June 2013 to discuss the proposed changes. Through this process, the city heard several concerns from a few attendees about the proposed changes. For instance, the Tidewater Builders Association (TBA), Willoughby Civic League, Hampton Roads Realtor Association, and others were concerned about the 10-year cumulative damage and improvement provision (City of Norfolk 2013b; Willoughby Civic League 2013). There were several arguments against this. First, with an estimated average cost of $100,000 to raise a structure, most property owners could not afford to comply. Second, property owners making improvements unrelated to flood damage (including damage from other natural events like high winds) would not be eligible for federal assistance. Third, damage or improvement values would be transferred to new property owners. For example, if a property owner made improvements worth 35 percent of a building's value, then when a new owner attempts to repair damage or make improvements worth 15 percent or greater, the building would need to come into compliance. Disclosure of such damages or improvements are not required at the time of sale. Fourth, TBA had concerns about the cost to developers and buyers associated with the 3-foot freeboard. The additional height contributes to higher building costs. There were also concerns about the historic character of buildings and neighborhoods with structural elevation (City of Norfolk 2013b). However, even with these concerns, all parties recognized that Norfolk should take action to reduce flood damage and improve its CRS rating. The concerned parties felt that the city could improve the rating through other means.

The Department of City Planning revised the proposed zoning ordinance based on public discourse and internal decision-making. In particular, they removed the 10-year cumulative loss provision in favor of a requirement that "existing buildings come into compliance with current regulations if flood damage from two events, on average, each equals or exceeds 25% the market value of the structure and the building is structurally damaged or altered" (City of Norfolk 2013a, p. 1). In other words, existing structures that incur two flood damage events worth 25 percent or more of their value must be elevated. In addition to the concerns expressed at public meetings, the Department of City Planning realized that they did not have an effective mechanism to track cumulative losses and improvements to properties. From here, the Planning Commission voted and passed the proposed zoning ordinance changes on June 27, 2013. The City Council formally approved the changes on November 26, 2013, which became effective on January 1, 2014 (Council 2013). The text box summarizes the final provisions of the zoning ordinance changes.

Accomplishments of Norfolk's Freeboard Standard

Those interviewed for this study (see the "Acknowledgment" section for a listing of interviewees) agree that the revised zoning ordinance is a success and that Norfolk is headed in the right direction. They feel that the new standards will help to protect structures and are evidence that the city is working to address a very real problem. Three feet of freeboard will take Norfolk an estimated 60 or 65 years in the future, giving them time to find big-picture solutions to recurrent tidal flooding.

Norfolk's revised flood and coastal zone ordinance went into effect on January 1, 2014. Over 50 new construction buildings have had to comply. The number of existing structures with two flood damage events worth 25 percent or more of their value is not available. These types of claims are reported to the city by FEMA but cannot be made public. Still, in 2014 there were no major storms in the region, and therefore there is not a strong likelihood that many existing structures would need to comply. Those interviewed for this study agree that it will take some time to gauge the new standards' effectiveness. One person mentioned that it may take 30 years to know if these actions were well-founded;

SUMMARY OF NEW FLOOD AND COASTAL ZONE REGULATIONS IN NORFOLK

- Three feet of freeboard
- Measure the height of structures from the design flood elevation or the ground level at the entrance of the structure, whichever is greater
- Require elevation certificates based on proposed construction with new development and additions in the Special Flood Hazard area
- Require all development within a 0.2% annual chance of flood to have a finished floor or be flood-proofed 18 inches above grade
- Require a 20-foot setback from the landward edge of mean high water for principal structures
- Prohibit the construction of subgrade crawl spaces within a Special Flood Hazard area
- Prohibit the use of breakaway walls in Coastal High Hazard and Coastal Floodplain districts
- Identify Coastal Floodplain Districts and provide regulations to match construction requirements for Coastal High Hazard Districts
- Require existing buildings to come into compliance with current regulations if flood damage from two events, on average, each equals or exceeds 25% the market value of the structure and the building is structurally damaged or altered

Source: City of Norfolk. 2013a. Flood Zone Regulation Changes, December 6. Available: http://www.norfolk.gov/DocumentCenter/View/12688. Accessed November 3, 2014.

the degree of sea-level rise will determine the standards' efficacy. Still, there is anecdotal evidence of a new construction home selling more quickly due to a lower flood insurance premium relative to other properties in the flood or coastal zones despite the additional construction cost.

Other cities in the region have looked to Norfolk as a leader on coastal flood mitigation issues. Several communities have raised their freeboard standards to 3 feet, including Hampton (Hampton 2014; Hampton Roads Regional CRS Workgroup 2014). Some examples of similar ordinances that are being considered or have passed include

- City of Portsmouth (considering 3-foot freeboard) (Hampton Roads Regional CRS Workgroup 2014)
- City of Newport News (considering 2-foot freeboard) (Hampton Roads Regional CRS Workgroup 2014)
- City of Poquoson (3-foot freeboard) (Hampton Roads Regional CRS Workgroup 2014)
- Gloucester County (2-foot freeboard) (Gloucester County 2011)
- City of Chesapeake (1.5-foot freeboard) (City of Chesapeake 2013)

Notably, Virginia Beach tried to raise its freeboard standard to 3 feet but could only achieve a 2-foot standard due to pressure from the building industry. Virginia Beach has more undeveloped land and the building industry sees an additional foot of freeboard as a more significant cost than in Norfolk. Also, Virginia Beach's more conservative planning commission was less willing to consider climate change.

Moving Forward and the Broader Context

Although there is agreement that Norfolk is moving in the right direction, there are concerns about the effectiveness of Norfolk's revised standards when considering Norfolk's overall vulnerability to flooding. Wetlands Watch, an environmental advocacy group in the region, felt that adaptation will occur very slowly. Since Norfolk's flood and coastal zones are very built up, there is a limited amount of new construction that will need to comply. Instead, adaptation will occur in a piecemeal way as existing structures are elevated. A second concern from Wetlands Watch is FEMA's ability to disperse funds in a timely manner; property owners could wait for years for funds to elevate their structures. In fact, Wetlands Watch estimates that if FEMA continues to disperse funding at its current rate, repetitive loss property owners already waiting for FEMA funds (for repairs or elevation) could wait 188 years, not including new property owners that are added to the list (Stiles et al. 2014). In the meantime, they could experience additional flooding events with limited means to repair their property. Wetlands Watch's final concern is that the revised standards only focus on structures. They will not address the greater flooding problems in Norfolk. Buildings may be safe, but the city will still flood, leaving some properties as virtual islands among flooded areas. As well, Wetlands Watch and TBA noted that zoning ordinances are not a topic of high public interest and there were relatively few people actively engaged or aware of the topic. They felt this might change in the future as existing structures need to come into compliance.

Those interviewed for this study agree that Norfolk needs a more robust response to address tidal flooding, inland flooding, and sea-level rise. This is also evidenced by *Coastal Resilience Strategy*, released by Norfolk (City of Norfolk 2014b). Several smaller-scale projects are being considered or implemented (see the text box). Even though Norfolk has taken an initial step to reduce its vulnerability, more substantial actions are needed. The city's *plaNorfolk2030* encourages the integration of sea-level rise into development activities (Schechtman and Brady 2013). As well, the city is involved in the Rockefeller Foundation's 100 Resilient Cities and RE.invest initiatives to identify adaptation actions and financing opportunities. Through these efforts Norfolk is reconsidering land-use planning priorities. The city has also considered transfers of development rights, which would equate to buyouts of properties in the most flood-prone neighborhoods in the city. However, this is a difficult action to take because despite flooding, properties adjacent to the water still have high property values and bring in significant tax dollars. As well, some property owners may be unwilling to relocate. The city has an array of possible infrastructure projects to consider, but it needs federal support to help fund these. Some of those interviewed are hopeful that federal funds will reach Norfolk due to the large naval presence in the area, and the Navy's reliance on sound infrastructure in and around Norfolk to maintain its operations. However, specific funding options have not been identified. In a similar vein, some of those interviewed felt that if the Navy were to take concrete action on adaptation or request the city to take some concrete action, that would help move the Norfolk adaptation process forward since the Navy base brings significant economic activity and influence to the region.

OTHER ADAPTATION ACTIVITIES IN NORFOLK

- Brambleton Avenue and Surrey Crescent road elevation projects
- Studies for an engineered shoreline at Ocean View Beach
- Living shoreline efforts along Haven Creek
- Replacing and elevating a bulkhead 1.5–2 feet above the existing bulkhead
- Installing a mobile pump to deal with tidal flooding at Lea View and 15th Street
- Multiple localized neighborhood level drainage improvements

Source: Applegate, A. 2014. For City of Norfolk, Park Becomes Wetlands Once Again. The Virginian Pilot, February 20. Available: http://hamptonroads.com/2014/02/city-norfolk-park-becomes-wetlands-once-again#. Accessed November 4, 2014; City of Norfolk. 2014a. Brambleton Avenue and Colley Avenue Improvements. Available: http://www.norfolk.gov/index.aspx?NID=203. Accessed November 3, 2014; City of Norfolk. 2014c. Ocean View Shoreline Project Update, January 7. Available: http://www.norfolk.gov/DocumentCenter/View/12969. Accessed November 3, 2014; Schechtman, J., and M. Brady. 2013. Cost-efficient Climate Change Adaptation in the North Atlantic. Sea Grant and North Atlantic Regional Team. National Oceanic and Atmospheric Administration. Available: http://www.regions.noaa.gov/north-atlantic/wp-content/uploads/2013/07/CEANA-Final-V11.pdf. Accessed November 4, 2014.

Lessons Learned

This section includes the author's evaluation of Norfolk's revised flood and coastal zone ordinance based on three questions: (1) Did Norfolk's action address its broader coastal flooding problem? (2) Did Norfolk reduce its vulnerability to climate change? (3) How did Norfolk's action affect marginal populations?

1. Did Norfolk's Action Address Its Broader Coastal Flooding Problem?

   Norfolk has taken an action that will help newly constructed structures and existing property owners reduce their vulnerability to flooding. While this is an important first step, it does not address the prevalence of flooding in Norfolk. Norfolk will continue to experience flood-induced road closures and consequent economic disruption or infrastructure damage along with their related costs. Notably, Norfolk's revised flood and coastal zone ordinance reduces long-term costs to homeowners. This ordinance will do very little to reduce costs or economic impacts to the city.

   However, Norfolk is actively looking at options to manage flood risks in the city through large-scale infrastructure projects and other alternatives. A suite of actions is needed in Norfolk to effectively reduce the community's risk to coastal flooding. The new 3-foot freeboard standard is just one piece of what will hopefully be a larger adaptation strategy.
2. Did Norfolk Reduce Its Vulnerability to Climate Change?

   Norfolk has taken one step to reduce its vulnerability to worsening coastal flooding due to sea-level rise, subsidence, and other factors. Norfolk's revised flood and coastal zone ordinance reduces the vulnerability of new construction to climate change by taking future sea-level rise into account. However, this ordinance helps existing property owners only if they experience repeated flood damage. Under this reactive approach to climate change it may take decades, and potentially a catastrophic flooding event, before this standard will significantly reduce the vulnerability of existing structures. As time passes, it will be important to revisit sea-level rise observations and projections to ensure that the 3-foot freeboard standard provides adequate protection.
3. How Did Norfolk's Action Affect Marginal Populations?

   Norfolk's flood and coastal zones tend to be comprised of high-value properties, which do not generally accommodate marginal populations. Still, some of Norfolk's actions might help to protect lower income residents. First, Norfolk's actions were motivated in part by its CRS rating. As noted, the city was concerned about the affordability of flood insurance and protection of its residents in the flood and coastal zones. The city recognized that some property owners may not be able to afford flood insurance with the city's current rating. The city is going through an update with the NFIP and hopes to lower its rating and make insurance more affordable. Second, the city considered but did not adopt a provision that could put the full cost of structural elevation on owners. Through stakeholder feedback the city recognized the inability of some property owners to bear this cost.

Acknowledgments

We would like to thank the following for participating in interviews as part of this case study:

- Joshua Clark, Tidewater Building Association
- Michelle Covi, Old Dominion University
- William (Skip) Stiles, Wetlands Watch
- Robert Tajan, City of Norfolk
- Martin Thomas, Jr., Norfolk Planning Commission

THE PLAN TO REBUILD AFTER HURRICANE SANDY: A ROADMAP FOR THE WORLD

Kim S. Haddow

On October 29, 2012, Hurricane Sandy slammed into New York City. Forty-four people were killed. The storm surge "smashed buildings and engulfed entire communities. It flooded roads, subway stations, and electrical facilities, paralyzing transportation networks and causing power outages that plunged hundreds of thousands into darkness. Fires raged. Wind felled trees. Heartache and hardship—and at least $19 billion in damage—[were] the storm's legacy" (PlaNYC 2013).

"By any measure, Sandy was an unprecedented event for New York City. Never in its recorded history had the city experienced a storm of this size. Never had a storm caused so much damage. Never had a storm affected so many lives" (PlaNYC 2013). Hurricane Sandy inundated 51 square miles of the city—17 percent of the city's total land mass (see Figure 7.14).

After the storm, New York was a changed city. New Yorkers realized their vulnerability to the consequences of climate change. "Sandy was in many ways a defining moment in the history of New York City. For the first time, the resiliency of New York (and America) was seriously questioned" (Mondkar 2014) (see Figure 7.15).

It would not be enough to restore New York to what it had been before the hurricane hit. Mayor Michael Bloomberg challenged his city to plan for more than the "next Sandy," but for the "defining challenge for our future"—climate change. The mayor explained: "We can do nothing and expose ourselves to an increasing frequency of Sandy-like storms that do more and more damage, or we can abandon the waterfront. Or, we can make the investments necessary to build a stronger, more resilient New York—investments that will pay for themselves many times over in the years to come…I strongly believe we have to prepare for what the scientists say is a likely scenario. Whether you believe climate change is real or not is beside the point—we can't run the risk" (NYC .gov 2013a).

FIGURE 7.14 Sea Bright, New Jersey, February 20, 2013: Homeowners stand and watch their home being elevated. This family has started the rebuilding process after being forced out of their home by Hurricane Sandy. (From Rosanna Arias/FEMA, https://www.fema.gov/media-library/assets/images/69546.)

FIGURE 7.15 New York City, January 15, 2014: Laura Phillips (with microphone), executive director of the Sandy Recovery Office, chaired a recent Regional Infrastructure Resilience Coordination Initiative group meeting that included 11 federal agencies. The States of New York, New Jersey, and Connecticut along with the City of New York and the Port Authority of New York & New Jersey also attended. Federal resources are supporting the states as they repair or rebuild infrastructure systems devastated by Hurricane Sandy. (From K.C. Wilsey/FEMA, https://www.fema.gov/media-library/assets/images/91015.)

The 440-page plan that emerged just 8 months after the storm, *A Stronger, More Resilient New York,* includes 257 initiatives to shore-up the coastline, upgrade buildings, protect city infrastructure and critical services, and addressed "not just current risk but also future risk from climate change" (PlaNYC 2014).

The 10-year plan to protect New York City from future superstorms and other climate-related threats has been called "the most ambitious and scientifically accurate plan of its kind in the world" (Bagley and Gallucci 2013). It has been praised for its foresight and thoroughness: "I think it would be hard to be more comprehensive than New York," said Heather McGray, codirector of the vulnerability and adaptation initiative at the World Resources Institute (Bagley and Gallucci 2013), and hailed for setting "a new global standard for long-term resilience planning" (C40Cities 2013). "Ultimately, *A Stronger, More Resilient New York* is not just a roadmap for long-term comprehensive resiliency planning for New York, but for cities worldwide" (National Geographic 2013).

Updated Climate Change Projections

In the weeks following Hurricane Sandy, Mayor Bloomberg asked New York City Panel on Climate Change (NPCC) "to produce updated climate projections that could become the foundation for a recovery plan that would fortify the city against future global warming impacts" (Bagley and Gallucci 2013). The panel, which is made up of an independent group of scientists and engineers, was established in 2008 as part of the city's sustainability and climate strategy.

Their finding—that the next few decades will be substantially worse for the city than was estimated only 4 years ago, the last time the panel modeled climate change—informed the development of recommended actions outlined in the report. The Panel's data "revealed that sea levels could rise at a faster rate than forecasted just four years ago—by more than 2.5 feet, potentially, by the 2050s. By that same decade, the city could have three times as many days at or above 90 degrees, leading to heat waves that threaten public health and the power system, among other infrastructure systems. The data also showed that the number of days with more than two inches of rainfall will grow from three in the last century to five in the 2050s" (Bagley and Gallucci 2013).

In addition, revised flood maps from FEMA place hundreds of thousands more New Yorkers in the 100-year floodplain. "We expect that by mid-century up to one-quarter of all of New York City's land area, where 800,000 residents live today, will be in the floodplain. If we do nothing, more than 40 miles of our waterfront could see flooding on a regular basis, just during normal high tide," Mayor Bloomberg noted (NYC.gov 2013b).

Public Engagement in the Plan's Development

The agency responsible for developing the plan, the Special Initiative for Rebuilding and Resiliency (SIRR) modeled the extensive stakeholder engagement process involved in the creation of PlaNYC. SIRR created a special task force to draft the plans, worked closely with an array of city, state, and national agencies, technical experts, and nongovernmental organizations, and "sought extensive input from dozens of elected officials, hundreds of community groups, and more than 1000 New Yorkers who participated in public workshops" (C40Cities 2013).

The stakeholders engaged in developing the Plan included

- 30+ city, state, and federal agencies
- 65+ elected offices and 19 community boards
- 320+ organizations including businesses, community, environmental, faith-based, and labor organizations
- 1000+ New Yorkers attending meetings or the 11 public workshops held in areas affected (C40Cities 2013)

The Plan

A Stronger, More Resilient New York calls "for everything from constructing 'adaptable floodwalls,' storm-surge barriers and dune systems, to revising building codes throughout the city." It is enormous in scope, "comprising the construction of sea walls and bulkheads, beach replenishment, the creation of parks as buffer zones, the retrofitting of apartment buildings, commercial structures and single-family homes, and the redesign of power stations, subway tunnels, sewage treatment plants, hospitals, utility poles and even ordinary streets" (Feuer 2014).

According to Daniel Zarrilli, the director of the City's Office of Recovery and Resiliency, "you'd have to look back to the rebuilding after the San Francisco earthquake for any real comparison" (Feuer 2014).

The plan is divided into two sections: a section on improving the city's infrastructure and systems and a section on rebuilding the five communities most impacted by the hurricane.

In section one, the plan outlines three layers of defenses: strengthening coastal defenses, rebuilding beaches, and elevating and constructing drainage systems, among other actions. The second layer of defense focuses on upgrading the city's building stock and calls for the retrofitting of the 68,200 buildings in the floodplain. The third layer spotlights the need to protect the city's critical infrastructure and services, most notably its energy, transportation, and heath care networks.

According to the plan, "these initiatives will further protect the coastline—our first defense against storms and rising sea levels—as well as strengthen the buildings in which New Yorkers live and work, and all the vital systems that support the life of the city, including our energy grid, transportation systems, parks, telecommunications networks, healthcare system, and water and food supplies" (PlaNYC 2013).

The first phase of the coastal protection plan includes 37 initiatives to protect vulnerable areas from waves and flooding and is based on four key strategies:

- Increase coastal edge elevations
- Minimize upland wave zones
- Protect against storm surge
- Improve coastal design and governance

The plan also includes detailed recommendations for protecting buildings, including

- Strengthen city building code to increase flood and wind resiliency for new buildings.
- For 68,200 existing buildings in floodplain, initiate a $1.2 billion incentive program to help building owners to complete flood resiliency projects, including elevating critical building equipment.
- Work with FEMA to reform the National Flood Insurance Program and provide for premium credits for risk reduction measures. The plan proposes to develop a system that would allow homeowners to undertake flood mitigation measures other than elevating their buildings—which is not feasible in much of New York—in order to qualify for reduced flood insurance premiums.

The city is working with and calling on utility providers to make infrastructure more resilient by

- Hardening existing infrastructure to increase flood and heat resiliency.
- Reconfiguring networks for greater redundancy.
- Reflecting climate risk in system designs and to include extreme weather when evaluating performance. During Hurricane Sandy, a swath of Manhattan lost power when a Con Edison substation was swamped with floodwaters. Today, 53 percent of the city's power plants are currently in the 100-year floodplain. By the 2050s, however, that would rise to 97 percent.
- Working to reduce energy demand and increase energy options for city consumers.

The plan also calls for investing in green elements of infrastructure that can reduce extreme weather impacts, including

- Improving and expanding drainage infrastructure (e.g., build sewers in areas with limited drainage, installing high-level sewers citywide, and expanding green infrastructure)
- Promoting redundancy in and flexibility of the water supply network
- Rethinking parks and open spaces to serve as buffers for adjacent neighborhoods
- Expanding tree networks across the entire city

In the report's second section, plans to make neighborhoods safer and more vibrant center on rebuilding actions to be taken in the five communities most impacted by Sandy: the 683,000 people and 42,000 businesses residing in the Brooklyn-Queens Waterfront, the East and South Shores of Staten Island, South Queens, Southern Brooklyn, and Southern Manhattan. "For the areas of New York that Sandy hit especially hard, this plan proposes local rebuilding initiatives that will help these communities emerge safer, stronger, and better than ever" (PlaNYC 2013).

"For example, in Southern Manhattan, a multilayered plan provides opportunities to retrofit the area's most vulnerable building stock and focuses investments in protecting areas with business and historic importance. The plan also seeks to protect Southern Manhattan's transmission and distribution substations, which are critical to the City's electric system" (C40Cities 2013).

Financing the Plan

Analysis from the Special Initiative for Rebuilding and Resiliency showed that the costs of storms will increase: in 2012 Sandy totaled $19 billion in damage and economic loss; in 2025 that cost grows to $35 billion in losses from an equivalent storm event, and to $90 billion by 2055 (NYC.gov 2013).

Mayor Bloomberg argued the city could not afford not to "make the investments necessary to build a stronger, more resilient New York—investments that will pay for themselves many times over in the years to come" (NYC.gov 2013b). He noted that FEMA reports that for every $1 spent on mitigation, $4 is saved in recovery and repair costs, and the "plan could save New York City untold billions in protection from future risk" (C40Cities 2013).

The price tag for implementing the more than 250 recommendations in the plan was estimated to be $19.5 billion. To pay for the initiative, Bloomberg said $15 billion has already been provided through previously allocated city capital spending and federal disaster relief measures. That leaves a gap of $4.5 billion, which the mayor said he intends to seek from the federal government. "We'll press the Federal government to cover as much of the remaining costs as possible," he said (Freedman 2013).

Progress on Implementation

The implementation of *A Stronger, More Resilient New York* is now the responsibility of Mayor Bill de Blasio, who created the Office of Recovery & Resiliency, "the first-ever City office dedicated to resiliency, which is aggressively moving forward the plan, integrating resiliency into how the City operates, securing critical federal funds and collaborating with government partners, and using the plan to expand economic opportunity for New Yorkers" (NYC.gov 2014).

On the 2-year anniversary of Sandy, in October 2014, the de Blasio administration declared the city has made "dramatic progress strengthening coastal defenses, upgrading buildings, protecting infrastructure and critical services, and making homes, businesses and neighborhoods safer and more vibrant" (NYC.gov 2014).

The city's news release includes progress on these fronts:

- Short-term measures were quickly put in place to immediately reduce risk. For example:
  - 4.15 million cubic yards of sand placed on city beaches
  - 26,000 linear feet of dunes on Staten Island alone, with more to come
  - 10,500 linear feet of bulkhead repairs around the city
  - Updated building codes, including 16 new local laws to improve residential and commercial resiliency
  - $1 billion in resiliency investments being made by ConEd to harden critical assets like substations and switches
  - Reforms to the national flood insurance program
- Longer-term measures are in development now across the entire city. For example:
  - Over $400 million to construct new armored levees and other infrastructure along Midland Beach and Staten Island's East Shore to substantially reduce risk in the future
  - Substantial investment in the next phase of coastal protection in the Rockaways and the communities surrounding Jamaica Bay
  - T-groins in Sea Gate
  - Dunes and other coastal protection in Breezy Point
  - Flood protection in Red Hook
  - Over $15 million in nature-based resiliency projects funded by the Department of Interior
  - Coastal protection projects funded by the federal Rebuild by Design program, including
    - Lower East Side flood protection, for which the city started surveying work this fall
    - Hunts Point food supply protection
    - Living Breakwaters off of Staten Island's South Shore, being implemented with the state
- Major investments in the Staten Island Bluebelt and other green infrastructure across the city to better accommodate stormwater
- Key resiliency upgrades at critical facilities, such as hospitals like Staten Island University Hospital (SIUH)
- New York City Housing Authority (NYCHA) recovery and resiliency funds to elevate boilers and build emergency generators and flood protection systems; recently, the mayor announced the model for this funding at Coney Island Houses
- Agency recovery and resiliency funds to restore and protect critical city agency services

- Major flood and coastal protection studies, including at Coney Island Creek, Gowanus Canal, Southern Manhattan, and Newtown Creek, and a raised shoreline study to protect the most vulnerable nonbeach parts of the coastline
- Department of City Planning Resilient Neighborhoods studies to ensure that the city tailors long-term solutions to support the vitality and resiliency of individual communities in the flood zones
- Support for new resiliency technologies to be applied to small businesses through the NYC: RISE competition

Daniel Zarrilli, director of the Mayor's Office of Recovery and Resiliency, explained, "The City is safer today than it was two years ago, and that's due to our aggressive implementation of a comprehensive climate resiliency plan. We have much work left to do, and the next phase of our $20 billion program includes a wide array of short-term measures and long-term investments that are strengthening our coastal communities and preparing our city for the future. Working collaboratively with many stakeholders across the five boroughs, we are strengthening the coast, upgrading buildings, protecting infrastructure, and making communities safer and more connected as we build a stronger, more resilient New York" (NYC.gov 2014).

REBUILD BY DESIGN

After Hurricane Sandy, President Obama appointed the Secretary of Housing and Urban Development Shaun Donovan to lead a task force "to radically rethink" the rebuilding of the region's infrastructure (Shorto 2014). Donovan asked Henk Ovink, a Dutch water-management expert, to oversee an international design competition intended to generate innovative plans to protect New York against the next big hurricane.

According to Ovink, "Every policy in the books right now is based on the science of yesterday. We have to try to find a way to escape the current boundaries of those policies. Rebuild by Design was that escape to rethink how we relate our self to these calamities. It's only then we can change the lives of communities that live in our cities" (Mondkar 2014).

One of the key principles Rebuild by Design encouraged architects and engineers to rethink was New York's relationship to the water. Instead of keeping the sea at bay as the city had done for 300 years, the time had come to invite it in. "We didn't want to just build barriers; we wanted to build an ecosystem. For that to happen, we have to live with the water, to understand it, while still understanding our vulnerabilities," Ovink explained (Feuer 2014).

The competition generated a pool of 148 proposals; six plans were chosen including "a network of levees, a waterfront greenway and a new power plant to protect the Hunts Point food market in the Bronx, and the planting of oyster beds and reefs off Staten Island's shore to mitigate the destructive force of oncoming waves" (Feuer 2014).

The most ambitious of the plans calls for the construction of the Big U—an 8-mile series of linked defenses wrapped like a chin strap around the coastline of Manhattan. "The project relies mainly on 10-foot-tall slurry-filled, earth-topped berms meant to guard the edges of the island, acting as a barrier to water while blending into a newly imagined string of waterfront parks" (Feuer 2014).

According to Big U designer Kai-Uwe Bergmann, "The idea was to create a public amenity that also had a protective element. We could have built walls, but walls are only used .01 percent of the time, during crises. We wanted something that was aesthetically pleasing, well designed and was useful all the time." In Battery Park, for instance, the plan calls for a series of "upland knolls" where people could sunbathe, garden, or even farm most of the time. During a storm, the built-up landscape would fend off the sea (Feuer 2014).

Conclusion

The plan to build "a stronger, more resilient New York" set a new global standard for long-term resilience planning "with its focus on the future risk from climate change and comprehensive approach" (C40Cities 2013). Praised for its foresight and thoroughness, the plan resulted from the combination of almost all the classic characteristics needed to create a resilient city—leadership, driven by a disaster, a climate mitigation plan, and a planning process were already in place to build on, a broad array of stakeholders—including the public—was involved in its development, designated funding was found and a mandate that the plan delivers current co-benefits—that the protective elements developed also include public amenities to improve the quality and security of life now and shield the city from the changes in the climate that are coming.

New York's plan also had assets that would not be possible to replicate everywhere—specifically, the fact that it has its own Panel on Climate Change made up of scientists and engineers who have been thinking about the city's sustainability and climate strategy, and the fact that Mayor Bloomberg had long been on the frontline of the fight to build the causes and consequences of climate change into his city's day-to-day management and planning for the future.

REFERENCES

Alibasic, H. 2014. Interview with Haris Alibasic, Director of the Office of Energy and Sustainability with the City of Grand Rapids, October 27.

Applegate, A. 2014. For City of Norfolk, Park Becomes Wetlands Once Again. *The Virginian Pilot*, February 20. Available: http://www.hamptonroads.com/2014/02/city-norfolk-park-becomes-wetlands-once-again#. Accessed November 4, 2014.

Arizona Daily Sun. 2010. Schultz Fire, Day 3, June 23. Available: http://azdailysun.com/news/local/schultz-fire-day/collection_be2f256c-7ecb-11df-9e5b-001cc4c03286.html#0.

Arizona Rural Policy Institute. 2014. Flagstaff Watershed Protection Project Cost Avoidance Study, October. Available: http://www.flagstaffwatershedprotection.org/wp-content/uploads/2014/10/Final-FWPP-Cost-Avoidance-October-27.pdf.

Bagley, K., and M. Gallucci. 2013. Is NYC's Climate Plan Enough to Win the Race against Rising Seas? *Insideclimatenews.org*, June 20. Available: http://insideclimatenews.org/news/20130620/nycs-climate-plan-enough-win-race-against-rising-seas.

Brehl, M., A. Mottek Lucas, P. Summerfelt, D. Vosick, and N. Woodman. 2014. Interview with Mark Brehl, Wildland Fire Management, City of Flagstaff, AZ; Anne Motek Lucas, Mottek Consulting; Paul Summerfelt, Wildland Fire Management, City of Flagstaff, AZ; Diane Vosick, Northern Arizona University; Nicole Woodman, Sustainability Manager, City of Flagstaff, AZ. Flagstaff, AZ, October 21.

C40Cities, Siemens. 2013. Climate Leadership Awards: New York Climate Close-Up. Available: http://www.cityclimateleadershipawards.com/2013-city-climate-leadership-awards-winners/.

City of Chesapeake. 2013. An Ordinance amending Chapter 26 of the Chesapeake City Code, entitled "Environment," Article IV, Floodplain Management, by amending Sections 26–86 through 26–105 to affirm floodplain districts, require the issuance of permits for residential and commercial development and redevelopment, imposing development restrictions on new and existing residential and commercial construction, establishing enforcement and penalties for violations, and standards for considering variances, June 3. Available: http://www.cityofchesapeake.net/Assets/supporting_docs/actions_planning/2013/2013-06-12/ta-z-11-10_staff_report.pdf. Accessed November 3, 2014.

City of Flagstaff. 2012. City of Flagstaff Bond Question, November. Available: http://www.flagstaff.az.gov/documentcenter/view/40301.

City of Norfolk. 2012. Norfolk Flooding Strategy Update, March 27. Available: http://va-norfolk.civicplus.com/documentcenter/view/1751. Accessed November 3, 2014.

City of Norfolk. 2013a. Flood Zone Regulation Changes, December 6. Available: http://www.norfolk.gov/DocumentCenter/View/12688. Accessed November 3, 2014.

City of Norfolk. 2013b. Zoning Ordinance Text Amendment to Chapters 2, Definitions, and 11–3, Floodplain/Coastal Hazard Overlay District (FPCHO), to amend terms, simplify language, and to add requirements for properties located within the floodplain and to revise standards for calculating building height.

City of Norfolk. 2014a. Brambleton Avenue and Colley Avenue Improvements. Available: http://www.norfolk.gov/index.aspx?NID=203. Accessed November 3, 2014.

City of Norfolk. 2014b. Coastal Resilience Strategy. Available: http://www.norfolk.gov/DocumentCenter/View/16292. Accessed November 3, 2014.

City of Norfolk. 2014c. Ocean View Shoreline Project Update, January 7. Available: http://www.norfolk.gov/DocumentCenter/View/12969. Accessed November 3, 2014.

City of Norfolk, Moffat & Michol, and Fugro. 2013. City of Norfolk Coastal Flood Mitigation Program Presentation, March 13. Available: http://www.norfolk.gov/DocumentCenter/View/3774. Accessed November 3, 2014.

Clark, C. 2014. 2013 Flood: Experts Describe How Close Grand Rapids Was to Crippling Floodwall Breach, January 5. Available: http://www.mlive.com/news/grand-rapids/index.ssf/2014/01/2013_flood_experts_describe_ho.html. Accessed November 3, 2014.

Council, J. 2013. Norfolk Raises Standards in Its Floodplain Law. *Inside Business*, November 27. Available: http://insidebiz.com/news/norfolk-raises-standards-its-floodplain-law. Accessed November 3, 2014.

Elson, M., and E. Phelps. 2014. Interview with Mike Elson and Erin Phelps, Coconino National Forest. Flagstaff, AZ, October 21.

Environmental Protection Agency. 2013. Climate Impacts in the Southwest. Available: http://www.epa.gov/climatechange/impacts-adaptation/southwest.html.

FEMA. 2013a. Biggert-Waters Flood Insurance Reform Act of 2012 (BW12) Timeline. US Department of Homeland Security, April 17. Available: http://www.fema.gov/media-library-data/20130726-1912-25045-8239/bw_timeline_table_04172013.pdf. Accessed November 3, 2014.

FEMA. 2013b. If Your Home or Business Has Been Flooded Build Back Safer and Stronger. Available: http://www.fema.gov/media-library-data/20130726-1858-25045-7797/build_back_stronger02_2013.pdf. Accessed December 22, 2014.

Feuer, A. 2014. Building for the Next Big Storm; After Sandy, New York Rebuilds. *New York Times*, October 25. Available: http://www.nytimes.com/2014/10/26/nyregion/after-hurricane-sandy-new-york-rebuilds-for-the-future.html.

Flagstaff Watershed Protection Project. 2014. Frequently Asked Questions. Available: http://www.flagstaffwatershedprotection.org/faqs/.

Freedman, A. 2013. New York Launches $19.5 Billion Climate Resiliency Plan. *Climate Central*, June 11. Available: http://www.climatecentral.org/news/new-york-launches-20-billion-climate-resiliency-plan-16106.

Fugro. 2012. Preliminary City-wide Coastal Flooding Mitigation Concept Evaluation and Master Plan Development, May. Available: http://www.norfolk.gov/DocumentCenter/View/7614. Accessed November 3, 2014.

Gloucester County. 2011. Disaster Preparedness Guide. Available: http://www.gloucesterva.info/Portals/0/es/documents/hurricane2011.pdf. Accessed November 3, 2014.

Hampton. 2014. Ordinance to Amend and Re-Enact Chapter 2.1 of the Zoning Ordinance of the City of Hampton, Virginia Entitled "Definitions" by Amending Section 2.1–2 Pertaining to the Definition of Height, Chapter 17.3, Article V of the Zoning Ordinance of the City of Hampton, Virginia Entitled "Flood Zone District" by amending sections 17.3–31.2, 17.3–32, 17.3–33.1, 17.3–33.2, 17.3–33.3, 17.3–34.1, 17.3–34.2, 17.3–34.3, 17.3–34.8 and Adding a New Section 17.3–34.9 Pertaining to Construction Requirements for New Construction and Substantial Improvements for Properties Located within Flood Zone Districts, September 10. Available: http://council.hampton.gov/Documents/Document.aspx?q=sEDpEXYoG8JgDF8Ziq6NZWTYsXpHhyWzt8KM0xHafKTXFXhtVRuifwl9DiobA%2B5F. Accessed November 4, 2014.

Hampton Roads Regional CRS Workgroup. 2014. Bi-Monthly Newsletter, December.

Harger, J. 2011. Grand Rapids Opens Joe Taylor Park; an Oasis for Kids in a Once-Blighted Neighborhood, June 28. Available: http://www.mlive.com/news/grand-rapids/index.ssf/2011/06/grand_rapids_opens_joe_taylor.html. Accessed November 4, 2014.

Klassen, K., and D. Howard. 2011. The Schultz Fire & Subsequent Flooding. *Firefighter Nation*, January 1. Available: http://www.firefighternation.com/article/features-0/schultz-fire-subsequent-flooding.

LaFurgey, J. 2014. GR Road Tax Extension Passes. Available: http://woodtv.com/2014/05/06/turnout-low-for-gr-road-tax-vote/. Accessed November 3, 2014.

Lunn, M. 2014. Interview with Mike Lunn, Director of the Environmental Services Department at the City of Grand Rapids, October 27.

Melillo, J.M., T.C. Richmond, and G.W. Yohe, eds. 2014. Climate Change Impacts in the United States: The Third National Climate Assessment. US Global Change Research Program, 841 pp. doi:10. 7930/J0Z31WJ2.

Mindock, C. 2013. Flagstaff Moves Forward on the Watershed Protection Project. *Arizona Daily Sun*, April 30. Available: http://azdailysun.com/news/local/flagstaff-moves-forward-on-the-watershed-protection-project/article_80695e24-b1c6-11e2-b3e9-001a4bcf887a.html.

Mondkar, B. 2014. Cities by the Water: What NYC Can Learn from Copenhagen's Urban Planning. *untappaedcities.com*, May 6. Available: http://untappedcities.com/2014/05/06/cities-by-water-what-nyc-can-learn-from-copenhagens-urban-planning/.

National Geographic. 2013. A Stronger, More Resilient New York. Posted by the C40 News Team in City Solutions, July 26. Available: http://voices.nationalgeographic.com/2013/07/26/a-stronger-more-resilient-new-york/.

NAUERI. 2013. Forest Health and Water Supply Protection Project Ballot Measure: Exit Poll Results, May. Available: http://www.flagstaffwatershedprotection.org/wp-content/uploads/2013/06/Prop405_ExitPoll_FactSheet_May20131.pdf.

NYC.gov. 2013a. Mayor Bloomberg Outlines Ambitious Plans to Protect City Against the Effects of Climate Change to Build a Stronger, More Resilient New York. *NYC*, the Official Website of the City of New York, June 11. Available: http://www1.nyc.gov/office-of-the-mayor/news/201-13/mayor-bloomberg-outlines-ambitious-proposal-protect-city-against-effects-climate-change.

NYC.gov. 2013b. Mayor Bloomberg Presents the City's Long-term Plan to Further Prepare for the Impacts of a Changing Climate. *NYC*, the Official Website of the City of New York, June 11. Available: http://www1.nyc.gov/office-of-the-mayor/news/200-13/mayor-bloomberg-presents-city-s-long-term-plan-further-prepare-the-impacts-a-changing.

NYC.gov. 2014. Mayor de Blasio Marks Second Anniversary of Sandy, Announces Major Progress on City's Recovery and Resilience Work. *NYC*, the Official Website of the City of New York, October 29. Available: http://www1.nyc.gov/office-of-the-mayor/news/498-14/mayor-de-blasio-marks-second-anniversary-sandy-major-progress-city-s-recovery-and#/0.

Occhipinti, N. 2014. Interview with Nick Occhipinti, Director of Policy and Activism at the West Michigan Environmental Action Council, October 27.

Olson, G. 2014. A Short History Grand Rapids, Michigan. Available: http://grcity.us/administrative-services/customer-service/Pages/A-Short-History-of-Grand-Rapids.aspx. Accessed November 3, 2014.

PlaNYC. 2013. *A Stronger, More Resilient New York*. NYC Special Initiative for Rebuilding and Resiliency, June 11. Available: http://www.nyc.gov/html/sirr/html/report/report.shtml.

PlaNYC. 2014. *Progress Report, 2014. A Greener, Greater New York. A Stronger, More Resilient New York*. The City of New York, Mayor Bill De Blasio. 140422_PlaNYVP-Reprot-Final-Web.pdf.

Schechtman, J., and M. Brady. 2013. Cost-efficient Climate Change Adaptation in the North Atlantic. Sea Grant and North Atlantic Regional Team. National Oceanic and Atmospheric Administration. Available: http://www.regions.noaa.gov/north-atlantic/wp-content/uploads/2013/07/CEANA-Final-V11.pdf. Accessed November 4, 2014.

Shorto, R. 2014. How to Think Like the Dutch in a Post Sandy World. *New York Times Magazine*, April 9. Available: http://www.nytimes.com/2014/04/13/magazine/how-to-think-like-the-dutch-in-a-post-sandy-world.html?_r=0.

Smith, D.M. 2012. Flagstaff Sustainability Chief Nicole Woodman Keeps Cool Head as Temperatures Rise. *Grist*, October 5. Available: http://grist.org/cities/flagstaff-sustainability-chief-nicole-woodman-keeps-a-cool-head-as-temperatures-rise/.

Stiles, W., S.H. Jarbeau, S. Hughes, and M.-C. Stiff. 2014. The Challenge of Mitigating Virginia's Flooding and Sea Level Rise Impacts. *Wetlands Watch*, November. Available: http://www.wetlandswatch.org/Portals/3/WW%20documents/sea-level-rise/ww_flood_white-paper1-1.pdf. Accessed December 1, 2014.

Summerfelt, P. 2014. Email to Megan O'Grady, Stratus Consulting re: Study follow-up questions. City of Flagstaff, Flagstaff, AZ, December 6.

Timmons Group. 2012. City-wide Drainage Master Plan. Final Submittal, November 8. Available: http://www.norfolk.gov/DocumentCenter/View/3641. Accessed November 3, 2014.

US EPA. 2008. A Screening Assessment of the Potential Impacts of Climate Change on Combined Sewer Overflow (CSO) Mitigation in the Great Lakes and New England Region. National Center for Environmental Assessment, Office of Research and Development, Washington, DC.

Willoughby Civic League. 2013. Minutes of the Regular Meeting, September 12. Available: http://www.willoughbycivicleague.com/september-2013.html. Accessed December 22, 2014.

Youberg, A., K. Koestner, and D. Neary. 2011. Wildfire, Rain and Floods: A Case Study of the June 2010 Schultz Wildfire, Flagstaff, Arizona. *Arizona Biology* 40(3). Available: http://www.azgs.az.gov/arizona_geology/winter10/article_feature.html.

8 International Case Studies in Climate Change Adaptation

George D. Haddow

INTRODUCTION

A March 2011 Climate Report prepared by CDC Climat and entitled *Drawing Up a National Climate Change Adaptation Policy: Feedback from Five European Case Studies* noted that "in terms of practical measures taken by governments, which have so far mainly focused on the mitigation side, adaptation was considered until recently as a secondary issue among climate policy priorities, and one that was more related to developing countries." The report also noted, "Since the turn of the century, and more specifically since 2005, adaptation has been attracting increasing interest from political decision-makers in developed countries, which are already seeing the first effects of climate change (severe droughts, flooding, etc.). Prompted by public opinion in their countries, and warned by scientists about the increasing magnitude of these events, decision-makers are beginning to draw up and implement adaptation policies and measures at all government levels, from local to international" (Dumollard and Leseur 2011).

An information paper developed jointly by the International Institute for Sustainable Development (IISD), the Stockholm Environment Institute (SEI), the World Conservation Union (IUCN), and Intercooperation and published in December 2003 entitled *Increasing Community Resilience to Climate-Related Disaster through Sustainable Livelihoods: A Conceptual Framework for Climate Change Adaptation* noted, "Climate change, therefore, threatens to exacerbate existing vulnerabilities and further entrench development disparities. Those with the least stand to suffer the most. Thus, with regional changes and impacts already being observed, the need for adaptive response measures is imperative. For the poor and other vulnerable people, the need is urgent." The paper also noted, "While it has been difficult to define what role adaptation should play in the international negotiations, it has become increasingly apparent that all countries will need to develop thorough and sustainable adaptation strategies if the impacts of climate change are to be effectively addressed. Regardless of successes in negotiating emission reductions, climate change is already occurring and will become an even more serious issue in the future" (IISD et al. 2003).

A report developed by the World Bank Group entitled *Economics of Adaptation to Climate Change (EACC)—Vietnam* and published in 2010 noted, "The major concern is the extent to which climate change will hit poor households, partly because of the decline in agricultural incomes and partly because of an increase in food prices relative to the general cost of living. The lowest 20 percent of households—either urban or rural—arranged by household expenditure per person will experience larger reductions in real standards of living due to climate change than the top 20 percent of households." The report also noted, "Thus, the primary focus of policies to adapt to climate change for the sectors covered under the EACC studies, should be to protect the poor, the vulnerable, and those least able to respond to changing climatic stresses. The goal should be to provide farmers and others with the tools and resources that will enable them to respond to climate change itself and to the new risks that will accompany climate change" (EACC 2010).

The Australian Department of Industry, Innovation, Climate Change, Science, Research and Tertiary Education developed a series of Case Studies entitled *Adaptation Good Practice* that examined climate change adaptation practices across Australia that were published in 2013. The Synthesis Report for this series, entitled *Key Lessons from Practitioners' Experiences*, noted, "Climate change adaptation is most likely to be implemented at a regional or local scale. Over the last decade a number of local governments, corporations and community groups have taken steps to manage the impacts of unavoidable climate change. As a result, there is an emerging body of practices that provide potential exemplars for practitioners and organisations considering ways to deal with the risks from climate change impacts" (Rissik and Reis 2013).

As these reports indicate, climate change is recognized internationally as a threat to those most vulnerable populations, local, regional, and national economies and the environment. Adaptation plans and strategies have been developed and in many cases implemented to secure livelihoods, protect critical infrastructure, and to reduce the impacts on natural resources. A number of case studies have been developed by government agencies and nongovernmental organizations in recent years that examine how local, regional, and national adaptation plans and strategies were developed and implemented that identify both the keys to success and the barriers that had to be overcome.

The purpose of this chapter is to examine these case studies in order to understand how climate change adaptation practices around the world occur and what guidance can be gleaned from these practices that can help chart the course of climate change adaptation in the United States and other parts of the world. In addition, a brief discussion concerning integration of climate change adaptation and disaster risk reduction is included at the end of this chapter in order to highlight what we believe is an opportunity to leverage the efforts in both practices to achieve more resilient communities across the globe.

LIVELIHOODS AND CLIMATE CHANGE

The joint IISD, SEI, IUCN, and Intercooperation publication entitled *Increasing Community Resilience to Climate-Related Disaster through Sustainable Livelihoods: A Conceptual Framework for Climate Change Adaptation* noted, "Today's climate-related disasters foreshadow the likely adverse future impacts of climate change, signaling an urgent need to minimize current vulnerabilities. For poor communities living on fragile and degraded urban and rural lands, such as steep hillsides, drylands and floodplains, actions must address the deteriorating environmental conditions that undermine their livelihoods and capacity to cope with disasters." The report also noted, "Protecting and enhancing natural services through activities such as watershed restoration, mangrove reforestation and rangeland rehabilitation, can help these communities secure their livelihoods and improve their capacity for adapting to the impacts of climate change" (IISD et al. 2003). The box "Responding to Climate Change: The Case for Adaptation" presents excerpts from this paper.

AUSTRALIA'S CLIMATE CHANGE ADAPTATION GOOD PRACTICE PROJECT

Australia's Climate Change Adaptation Good Practice Project developed 16 case studies of a "range of projects and programs that were selected to ensure: a geographic spread across Australia; a variety of scales of adaptation practice; a range of organisations (local and state government, regions, industry and infrastructure)" (Rissik and Reis 2013). The two boxes "Case Study: Mornington Peninsula Shire Council's Community Engagement Programs" and "Case Study: Green Cross Australia: Harden Up—Protecting Queensland" provide snapshots of the case studies developed by

RESPONDING TO CLIMATE CHANGE: THE CASE FOR ADAPTATION

From the beginning of climate negotiations, it has been accepted that adaptation has some role to play in reducing vulnerability to climate change. While it has been difficult to define what role adaptation should play in the international negotiations, it has become increasingly apparent that all countries will need to develop thorough and sustainable adaptation strategies if the impacts of climate change are to be effectively addressed. Regardless of successes in negotiating emission reductions, climate change is already occurring and will become an even more serious issue in the future.

Much of the early work on adaptation focused on future climate scenarios using general circulation models (GCMs), which helped to identify potential impacts. These models proved to be extremely limited in telling us about regional impacts of climate change. A new generation of research is now addressing the issue by looking at vulnerability and adaptation within the context of current climate change and variability as well as longer-term climate change, since communities who are vulnerable to today's climate stresses will only become more vulnerable as global temperatures rise.

It follows then, that adaptation must start with actions that target current vulnerabilities, allowing communities to build more resilient and secure livelihoods that can confront the impacts of climate change.

REDUCING VULNERABILITY THROUGH SUSTAINABLE LIVELIHOODS

If reducing current vulnerabilities is the starting point of adaptation, then poverty reduction is essential to the process, since poverty is both a condition and determinant of vulnerability. Yet poverty reduction requires an understanding of how local livelihoods are conducted and sustained, as the assets and capabilities that comprise peoples' livelihoods often shape poverty and the ability to reduce it. Moreover, by understanding the dynamics of poor people's livelihoods, we can begin to understand how they will be affected by climate change impacts, how they might respond with the resources they have, and how these conditions can be reflected and built on for successful adaptation strategies.

Given the reliance of the poor on environmental services for their livelihoods, a central element of this adaptation approach should be ecosystem management and restoration activities such as watershed rehabilitation, agroecology, and forest landscape restoration. By protecting and enhancing the natural services that support livelihoods, vulnerable communities can maintain local safety nets and expand the range of options for coping with disruptive shocks and trends. Thus, restoring mangroves along the cyclone-battered coasts of Vietnam, for example, not only restores degraded ecosystems and increases physical protection against storms, but also boosts aquaculture production, which generates much-needed income for local communities. This combination of a secured natural resource base, reduced exposure to natural hazards, and diversified livelihood activities has increased resilience to future threats, including climate change. In fact, this sustainable livelihoods approach to adaptation has the advantage of meeting immediate development needs while contributing to longer-term capacity development that will create a basis for reducing future vulnerabilities.

Source: IISD et al. 2003. *Increasing Community Resilience to Climate-Related Disaster through Sustainable Livelihoods: A Conceptual Framework for Climate Change Adaptation.* International Institute for Sustainable Development (IISD), the Stockholm Environment Institute (SEI), the World Conservation Union (IUCN), and Intercooperation. December 2003. http://www.iisd.org/pdf/2003/envsec_livelihoods_1.pdf.

CASE STUDY: MORNINGTON PENINSULA SHIRE COUNCIL'S COMMUNITY ENGAGEMENT PROGRAMS

SNAPSHOT

Mornington Peninsula Shire Council's *Sustainable Peninsula Initiative* in 2001 provided a framework and key drivers for community action on adaptation to reduce vulnerability to the potential effects of a changing climate in tandem with actions to mitigate greenhouse gas emissions. Early actions included addressing issues such as water and energy efficiency, renewable energy, land sustainability, and protection of the Green wedge.

In 2006 the Shire commenced a serious discussion with the community about climate change at a forum entitled "Your Community Your Future." This was followed in 2008 by a series of climate change Community Conversations supported by an information kit entitled "Climate Change: What Are We Doing about It," which presented scientific evidence from a study of the climate change impacts on the Western Port Region and gave credibility to the Conversations.

The Conversations motivated residents and local community groups to contribute to the Shire's climate change adaptation agenda. The latest stage of the community engagement journey includes behaviour change programs based on resolving barriers identified by residents, as well as other actions around managing key climate change risks. Some of the Shire's actions to respond to the risks and opportunities associated with climate change include committing $30 million to develop and implement an Integrated Drainage Strategy focused on managing increased rain intensity and sea level rise; doubling its Fire Management budget; and achieving a 60% reduction in potable water use.

RISKS AND IMPACTS ADDRESSED

- Average annual temperatures will increase by up to 1.3 degrees in 2030 and 3.5 degrees in 2070
- Increasing frequency and severity of intense rainfall events, heat waves, droughts and bushfires
- Sea level could rise up to 170 mm by 2030 and 490 mm in 2070, putting coastal communities and infrastructure at risk of inundation and flooding

DRIVERS FOR ADAPTATION ACTION

In 2006, following strong leadership from Chief Executive Officer (CEO) the Shire had its first serious discussion with the community about climate change.

OUTCOMES ACHIEVED

- Tailored communication strategies and engagement activities that are socially inclusive
- Community action around climate change mitigation and adaptation
- Climate change adaptation opportunities built into Council processes
- Adaptation actions on the ground

Source: National Climate Change Adaptation Research Facility (NCCARF), http://www.nccarf.edu.au/localgov/case-study/mornington-peninsula-shire-councils-community-engagement-programs.

CASE STUDY: GREEN CROSS AUSTRALIA: HARDEN UP—PROTECTING QUEENSLAND

SNAPSHOT

Queensland is prone to climate related natural disasters such as floods, droughts, bushfires, tropical cyclones, storms and storm surges. A string of recent extreme weather events over the past few years has highlighted the need for households, businesses and communities to be better prepared for extreme weather.

In response, Green Cross Australia, a national environmental organisation that helps Australians adapt to climate change in ways that embrace sustainability and community resilience, in partnership with business, research, community and government organisations, developed and implemented *Harden Up—Protecting Queensland.*

This statewide, web-based approach encourages Queenslanders to assess their vulnerability to key natural disaster hazards, and take practical action to become more self-reliant and resilient in the face of extreme weather events. Through the use of online technology, social and digital media, *Harden Up* encourages Queenslanders to be aware (identify their personal risk exposure to cyclone, bushfire, severe storms and storm surge), prepare (take practical actions to reduce hazard exposure) and help others (build community resilience).

The *Harden Up* program provides information on historical weather at suburb/town scale, future projections, and detailed information for each specific weather event, to help people understand the history of severe weather in their community and consider any likely future risks. Armed with this information, *Harden Up* users can then tailor a plan to suit the climate hazards specific to their location, and circumstances.

RISKS AND IMPACTS ADDRESSED

Extreme weather events in Queensland.

DRIVERS FOR ADAPTATION ACTION

In 2010, GCA was approached by a natural disaster resilience program officer from the QLD Government Department of Community Safety to explore how they might work together to build climate resilience in QLD communities.

OUTCOMES ACHIEVED

Since launching in late 2011, the *Harden Up—Protecting Queensland* program has seen over:

- 35,500 people access the website
- 155,000 pages of content viewed with visits lasting around 5 minutes on average
- 18,000 discrete actions have been taken to prepare for extreme weather in QLD.

Source: National Climate Change Adaptation Research Facility (NCCARF), http://www.nccarf.edu.au/localgov/case-study/green-cross-australia-harden-up—protecting-queensland.

Australia's Climate Change Adaptation Good Practice Project. Access to the Synthesis Report and the full 16 Case Study reports is at http://www.nccarf.edu.au/localgov/map.

The Synthesis Report identified a generic framework developed by the Adaptation Good Practice (AGP) project that included four phases: Research and Assessment, Decision-Making and Planning, Implementation, and Reflect. The box "Adaptation Framework" presents brief explanations of each of the four phases included in the Synthesis Report.

It should be noted that the Synthesis Report included a table that tracked the progress of each of the 16 case studies through the four phases of the adaptation process and that only two of the 16 case studies had entered the implementation phase and none of the projects studied had completed implementation (Rissik and Reis 2013).

The Synthesis Report identified five critical success factors that included "leadership, good engagement with stakeholders, strong connectivity with other relevant activities, ongoing sustainability and consideration of short and long-term costs and cost-effectiveness" (Rissik and Reis 2013). The following boxes present excerpts from the report describing each of the five critical success factors and examples from the 16 case studies supporting each factor.

In addition to identifying critical success factors from the 16 case studies, the Synthesis Report also identified several lessons learned from the experiences in the 16 case studies, including

Time: Project developers need to be realistic about the complexity of the projects and allocate sufficient time for each part of the project whether it is data provision or engagement and consultation.

Funding/cost: Seed funding has been shown to play an important role in stimulating effective adaptation planning and most case studies accessed government funding programs.

Data/technology: Discussions with stakeholders in various NCCARF-related forums have indicated a preference for sector-specific or locally specific information and real-life examples. The case studies clearly articulate this need.

Project development/management: A transparent, open process is necessary to ensure ongoing commitment of project development and management after process and ownership at all levels within a council, organisation or community. The process also needs to be flexible to allow changes to the project scope and direction whilst still focusing on the intended outcome.

Collaboration and information sharing: Many of the case studies made the point that a holistic, integrated approach that brings together scientific and technical experts, strategic planners, risk managers, engineers, asset managers and other relevant interests is necessary to ensure that all aspects of climate risk and adaptation are addressed.

Engagement: Stakeholder engagement is vital to ensure that the issues and the outcomes are fully understood and agreed to by the community or organisation. The AGP case studies provide examples of the increasing use of social media via Facebook, Twitter and YouTube by councils and organisations to provide many new and vibrant opportunities to engage with all levels of the community.

Communication/dissemination: Whilst dissemination of the outcomes of these AGP projects is being actively undertaken by NCCARF, each organisation has responsibility to widely broadcast the outcomes and benefits of their work.

Leveraging: The value of the projects has been exemplified by leveraging key information or processes into other "spin-off" projects. For example, the Hunter region Decision Support Handbook process is being followed by the councils in developing a Heatwave Plan, and the Climate Futures for Tasmania project provided downsized climatic data for use by local organisations, businesses and farmers to assess climate change impacts such as the risk of fruit fly establishment.

ADAPTATION FRAMEWORK

RESEARCH AND ASSESSMENT

Research and data acquisition is fundamental to understand the extent of climate change risks and to determine realistic, cost effective options. Technical, climate, financial and social information are all necessary to support decision-making.

DECISION-MAKING AND PLANNING

Long-term planning and investment is fundamental to adaptation. Planning establishes the framework for decisions and actions that will be implemented to address climate risk. Planning establishes the vision, aims and objectives against which the success of the adaptation actions will later be monitored and evaluated. Plans must be tailored to reflect the location and levels of risk, the institutional environment in which they will be applied, and the values, capacity and resources of the organisation.

IMPLEMENTATION

Implementation includes the execution of adaptation actions and plans. There are two general approaches to mainstreaming climate risk and adaptation assessment into planning processes.

- Conduct an initial climate risk assessment separate from current planning processes and integrate the outcomes at a later stage.
- Identify climate risks through existing mechanisms, recognising that new data or longer timeframes may need to be considered.

There are four broad approaches to adapting to climate change particularly in relation to sea level rise in the coastal zone:

1. Do nothing, live with the impact and accept the risk
2. Hold the line—build community resilience and implement contingency plans
3. Limited intervention or accommodation—implement adaptation mechanisms such as beach nourishment, re-vegetation, rock wall, raising buildings, buffer zone
4. Managed retreat

REFLECT

This stage provides for monitoring and evaluation of the project, based on the targets and performance indicators established in the implementation phase. Monitoring can provide an early warning of trends or a new problem which needs to be addressed. This stage should also be used to obtain updated information on climate changes to factor into the planning process.

Evaluation should also consider whether the outcomes were met, what aspects of the project have been successful and what did not work and why. These learnings can then be built into future projects.

Source: Rissik, D. and N. Reis. 2013. *Key Lessons from Practitioners' Experiences.* National Climate Change Adaptation Research Facility. Australian Department of Industry, Innovation, Climate Change, Science, Research and Tertiary Education. 2013. http://bit.ly/1yjWqxZ.

CRITICAL SUCCESS FACTOR: LEADERSHIP

Leadership is considered in terms of the contribution of the executive leadership (strategists and project managers) and stakeholders. All projects exhibited strong leadership to ensure that the outcomes met user needs. The South Australia government has been bolder than other jurisdictions and has enshrined climate change adaptation in legislation through the Climate Change and Greenhouse Emissions Reduction Act 2007, following that with the South Australia Greenhouse Strategy.

A number of case studies outlined circumstances where individual councils, regional or industry organizations had taken the lead and developed frameworks and tools which can be used by other groups, so avoiding "reinventing the wheel." Examples include:

- The Sydney Coastal Councils Group developed information and tools that are readily transferable for use by local governments, land managers and other stakeholders. These include the outputs of the physical inundation and modelling methodology, and guidance on planning principles. A generic communication strategy, templates and tools have been specifically designed to be applicable to communicating information on other extreme weather hazards such as flooding or bushfires.
- The Townsville City Council took the lead in developing a systemic approach to identifying and assessing coastal hazards and provided a compendium of coastal hazards and relevant adaptation options which can be used by other councils.
- In the absence of adequate state or national level frameworks and guidance to manage existing coastal hazards and the potential impacts of climate change, seven councils in the New South Wales Hunter region developed a decision support handbook and guidance for councils.
- The Local Government Association of South Australia's tool quantifies the likely costs for maintenance and repair of three major road types from expected changes in the climate temperature and rainfall to 2100. The modelling and software has the advantage that it can be extended to include additional asset classes or climate variables.
- The AdaptWaterTM online climate change analysis tool, whilst developed specifically for water utilities to manage the impacts of extreme weather events, can be readily adapted to apply to other utilities and infrastructure.

Source: Rissik, D. and N. Reis. 2013. *Key Lessons from Practitioners' Experiences.* National Climate Change Adaptation Research Facility. Australian Department of Industry, Innovation, Climate Change, Science, Research and Tertiary Education. 2013. http://bit.ly/1yjWqxZ.

Responsibility: "Think Globally and Act Locally" is a catchcry which encourages all of us to take responsibility for actions to improve our environment whilst also contributing to the wider global good. Many of the case studies underscore the value of local activities and local responsibility such as the use of "climate champions" within organisations to raise awareness and help implement activities.

Ongoing action: Adapting to climate change is a journey involving many steps. It cannot be undertaken quickly but needs to be developed and built on as information is acquired, awareness is raised and stakeholder groups are empowered.

Performance monitoring: Ongoing monitoring is needed to ensure that the processes or activities established to adapt to climate change impacts are effective and do not result in maladaptation. Whilst many of the case studies were effective in achieving their intended outcomes, very few monitored the outcomes to determine their ongoing effectiveness (Rissik and Reis 2013).

CRITICAL SUCCESS FACTOR: ENGAGEMENT

Engagement addresses the extent and mechanisms of communication, engagement and collaboration with stakeholders and the partnerships established to develop good practices. Examples include:

- Green Cross Australia engaged with state and local governments, industry associations (insurance, emergency management, property), research and media organisations to ensure that online information was correct, comprehensive and widely available.
- The design process for the Brisbane Airport New Parallel Runway was supported by a comprehensive, 22-month engagement program comprising information kits, factsheets, a website, a freecall information line, media advertising, public information sessions, displays in shopping centres and libraries, and briefings for key groups.
- Mornington Peninsula Shire Council has a long-running community engagement program which encourages sustainable activities through "Climate Change Community Conversations," social media and newsletters, an Eco Living Display Centre, and a Green Business Network and partnerships to "encourage sustainable activities."

Source: Rissik, D. and N. Reis. 2013. *Key Lessons from Practitioners' Experiences.* National Climate Change Adaptation Research Facility. Australian Department of Industry, Innovation, Climate Change, Science, Research and Tertiary Education. 2013. http://www.nccarf.edu.au/localgov/sites/nccarf.edu.au.localgov/files/Climate%20Change%20Adaptation%20Good%20Practice_Synthesis%20Report_Key%20lessons%20from%20practitioners%20experiences.pdf.

CRITICAL SUCCESS FACTOR: CONNECTIVITY

Connectivity is about relationships and is considered in both internal and external contexts. Internal connectivity refers to the extent of integration of the project and outcomes with other processes (planning, risk management, implementation strategies) across the organisation whereas external connectivity refers to opportunities for collaboration with local, regional, state-wide or national organisations, the ability of the project outcomes to be transferred to or used by other organisations, and the relationship of the project.

Projects that displayed excellent connectivity were:

- Hunter Region Decision Support Handbook—the process is now being used to develop other plans and an e-workbook will allow the process to be easily accessed by other organisations and councils.
- Great Barrier Reef Climate Change Action Plan 2007–2012—the project established strong, ongoing relationships with partners from the public, private, industry, community, and research sectors. The Action Plan complemented government and industry climate change response strategies, connected with local councils, and enabled knowledge sharing relationships with over 35 world-leading research institutions.

- South Australian Integrated Climate Change Adaptation is based on partnerships and collaboration between government, industry and community organisations. Increasing recognition of the impacts of climate change and extensive collaborations among South Australian adaptation practitioners has resulted in legislation, frameworks, strategies, policies and clarification of roles and responsibilities which have established a consistent, state-wide approach to climate change risk assessment and adaptation.

Source: Rissik, D. and N. Reis. 2013. *Key Lessons from Practitioners' Experiences*. National Climate Change Adaptation Research Facility. Australian Department of Industry, Innovation, Climate Change, Science, Research and Tertiary Education. 2013. http://www.nccarf.edu.au/localgov/sites/nccarf.edu.au.localgov/files/Climate%20Change%20Adaptation%20Good%20Practice_Synthesis%20Report_Key%20lessons%20from%20practitioners%20experiences.pdf.

CRITICAL SUCCESS FACTOR: SUSTAINABILITY

SUSTAINABILITY

Sustainability refers to the provision of long-term benefits. The AGP project considers social, environmental, and economic sustainability in terms of the principles of intergenerational equity and subsidiarity (good governance at local levels) and ability to avoid maladaptation (adverse consequences). Projects that provide good examples of the consideration of sustainability principles include:

- Climate Futures for Tasmania determines local level climate change trends for key sectors and supports long-term sustainable management of resources and assets.
- Green Cross Australia: Harden Up—Protecting Queensland provides user-friendly, location-specific climate projections and information on extreme weather events to support long-term planning by individuals, households and businesses.
- Incorporation of the climate change impact of sea level rise into the design of the Brisbane Airport New Parallel Runway ensures that the facility will continue to be operational for generations.
- Great Barrier Reef Climate Change Action Plan 2007–2012 was underpinned by sound environmental, social, economic, and governance principles in determining actions to ensure the long-term sustainability of the reef and dependent communities, industries and economy.

Source: Rissik, D. and N. Reis. 2013. *Key Lessons from Practitioners' Experiences*. National Climate Change Adaptation Research Facility. Australian Department of Industry, Innovation, Climate Change, Science, Research and Tertiary Education. 2013. http://www.nccarf.edu.au/localgov/sites/nccarf.edu.au.localgov/files/Climate%20Change%20Adaptation%20Good%20Practice_Synthesis%20Report_Key%20lessons%20from%20practitioners%20experiences.pdf.

CRITICAL SUCCESS FACTOR: COST

Cost is considered in terms of the total cost of the project and the balance between expenditure and benefits of good adaptation implementation to meet short-term requirements and anticipate long-term needs. Some case studies include cost-benefit analysis (CBA) as a means of determining the cost-effectiveness of the project activities.

The value of the project and the availability of additional funding for further work is also recognised. Case studies mainly sourced funding from Australian Government grant programs such as the Coastal Adaptation Pathways Program (CAP), Local Adaptation Pathways Program (LAPP), Natural Disaster Mitigation Program (NDMP) and Natural Disaster Resilience Program (NDRP).

The case studies varied in cost and funding sources, for example:

- The total budget for the Quantifying the Cost of Climate Change Impacts project was $740,000 comprising $320,000 from NCCARF, $230,000 from collaborators and $190,000 in-kind support.
- The total cost of the Climate Futures for Tasmania project was about $3.2 million in direct funding over four years plus about $4.8 million of in-kind contributions from partners.
- Green Cross Australia received funding from the Australian Government through NDRP and the Queensland Government. As it is a web-based resource, it requires only minimal funds to maintain. A report calculated the social return on investment and concluded that "for every $22 invested in Green Cross, somewhere in Australia someone is inspired to take a practical step towards a secure and sustainable future."
- The cost to integrate sea level rise into the design of the Brisbane Airport New Parallel Runway was negligible in terms of the overall cost of development of a new runway.
- The Australian Government committed over $9 million to the Great Barrier Reef Climate Change Action Plan 2007–2012 with additional investment and in-kind contributions from key partners.
- The City of Melbourne Climate Change Strategy and Action Plan was partly funded by the (then) Department of Climate Change and Energy Efficiency under LAPP. The City of Melbourne committed about $30 million in 2010–2011 and $11.2 million in 2011–2012 to implement its climate change adaptation actions.

Source: Rissik, D. and N. Reis. 2013. *Key Lessons from Practitioners' Experiences.* National Climate Change Adaptation Research Facility. Australian Department of Industry, Innovation, Climate Change, Science, Research and Tertiary Education. 2013. http://www.nccarf.edu.au/localgov/sites/nccarf.edu.au.localgov/files/Climate%20Change%20Adaptation%20Good%20Practice_Synthesis%20Report_Key%20lessons%20from%20practitioners%20experiences.pdf.

ADDRESSING THE HUMANITARIAN CHALLENGES OF CLIMATE CHANGE: REGIONAL AND NATIONAL PERSPECTIVES—CASE STUDIES

In May–June 2009, the Inter-Agency Standing Committee (IASC) Task Force on Climate Change published *Addressing the Humanitarian Challenges of Climate Change: Regional and National Perspectives—Case Studies*. This volume included a total of 19 case studies of climate change adaptation activities sponsored by IASC member organizations in communities across the globe. A list of the IASC organizations that contributed to this report are listed in the box "IASC Members Contributing to *Addressing the Humanitarian Challenges of Climate Change: Regional and National Perspectives—Case Studies*" (Selby and Venton 2009).

The following box entitled "IASC Case Study: Community Based Adaptation in Bangladesh" was included in the IASC report and provides an example of the type of case studies included in the IASC report. The full report can be accessed at http://sheltercentre.org/library/addressing-humanitarian-challenges-climate-change-regional-and-national-perspectives.

IASC MEMBERS CONTRIBUTING TO *ADDRESSING THE HUMANITARIAN CHALLENGES OF CLIMATE CHANGE: REGIONAL AND NATIONAL PERSPECTIVES—CASE STUDIES*

The report and case study volume document a collective consultation process that was made possible by contributions from the following IASC organisations and the International Strategy for Disaster Reduction (UNISDR):

International Federation of Red Cross and Red Crescent Societies (IFRC)
Office for the Coordination of Humanitarian Affairs (OCHA)
World Food Programme (WFP)
Action by Churches Together (ACT) International
CARE International
Caritas Internationalis
Food and Agriculture Organization of the United Nations (FAO)
International Organization for Migration (IOM)
Norwegian Refugee Council (NRC)
Office of the United Nations High Commissioner for Refugees (UNHCR)
Oxfam International
Plan International
Save the Children
United Nations Children's Fund (UNICEF)
United Nations Development Fund for Women (UNIFEM)
United Nations International Strategy for Disaster Reduction (UNISDR)
United Nations Development Programme (UNDP)
United Nations Environment Programme (UNEP)
United Nations Human Settlements Programme (UN Habitat)
United Nations Joint Programme on HIV and AIDS (UNAIDS)
United Nations Population Fund (UNFPA)
World Health Organisation (WHO)
World Vision International (WVI)

Source: Selby, Sarah and Courtenay Cabot Venton. 2009. *Addressing the Humanitarian Challenges of Climate Change: Regional and National Perspectives—Case Studies.* The Inter-Agency Standing Committee (IASC) Task Force on Climate Change. May–June 2009. http://sheltercentre.org/library/addressing-humanitarian-challenges-climate-change-regional-and-national-perspectives.

IASC CASE STUDY: COMMUNITY BASED ADAPTATION IN BANGLADESH

Implementing Agency: Food and Agriculture Organization of the United Nations (FAO)

BACKGROUND

Due to its geophysical position and socio-economic context, Bangladesh is highly prone to regular natural hazards and the impacts of climate change. In particular, regions of Bangladesh are susceptible to drought, with adverse impacts for rice crops (which account for more than 80 percent of the total cultivated land) and jute (the country's main cash crop).

In 2005, the Food and Agriculture Organization of the United Nations (FAO) initiated a project at the request of the Bangladesh government that was designed to improve the adaptive capacities of rural populations and their resilience to drought and other climate change impacts. It was also aimed at sharing project findings with service providers and policy-makers, in order to improve support to future adaptation processes. Objectives of the pilot project were as follows.

Firstly, to develop a methodology to increase understanding of how:

- Results of climate change impact assessments, based on General Circulation Models and different climate change scenarios, can be translated into location specific agricultural impact outlooks and livelihood adaptation practices;
- Adaptation options can be tested and implemented in a participatory way with farmers; and
- Project results can be shared with agricultural researchers, policy-makers and the climate change community, in order to facilitate replication of success cases and avoid mal-adaptations.

Secondly, to initiate and facilitate field testing with farmers of:

- Livelihood adaptation strategies to better respond to disaster and climate risks; and
- Improved long-lead climate forecasting and responses to future climate change projections in agriculture.

Project activities included:

- In-depth local situation assessment;
- Identifying suitable entry point activities;
- Developing a participatory project implementation strategy, suitable for replication elsewhere;
- Putting in place or strengthening institutional mechanisms;
- Technical capacity building and training;
- Developing a set of viable location-specific adaptation options relevant to the agro-ecological and social settings of vulnerable groups; and
- Identifying lessons learnt from the community-based adaptation processes.

Local awareness raising strategies included orientation meetings, field days, folk songs, drama, and demonstration rallies. In the absence of farmers' field schools in the project region, the FAO initiated participatory field demonstrations on the farmer's own fields. More than 225 demonstrations of 15 viable adaptation practices were conducted for five seasons in four sub-districts of undulating upland areas in Northwest Bangladesh.

SELECTING ADAPTATION OPTIONS

Categories of good practice recommended for demonstration included agronomic management, water harvesting, water use efficiency, crop intensification, alternate enterprises, alternative energy sources, household level energy efficiency and post harvest practices.

Viable adaptation practices for each of the categories were selected through a sequence of evaluation processes involving a range of stakeholders from the community level through to local research institutions and development organisations.

Selection criteria included: drought mitigation potential, suitability for future climate scenarios, economic viability, sustainability, increased productivity, social acceptability, relevance to vulnerable communities, gender integration, household income, applicability to multiple sectors, seasonal relevance, immediate need, institutional support, and expert acceptance.

A wide range of livelihood adaptation options were then prioritised for demonstration as part of the project, including: (1) homestead gardening (providing additional income and nutritional security); (2) drought-tolerant fruit tree gardening (allowing crop diversification); (3) mini-nurseries for fruit trees (a community initiative providing income); (4) improved stoves for household use (fuel and time saving); (5) rain water harvesting (providing economic benefits and ensuring a stable income); (6) use of alternative cereals such as maize (drought tolerant and income stabilising); (7) use of compost and organic manures for water conservation (needs training and time for decomposition); (8) cultivation of non-conventional oilseeds (pest and disease infestation reduced yields); (9) dry seed bed to manage early dry spell risk (requires technical capacity); (10) water saving irrigation for rice (requires electricity).

Of the several adaptation practices demonstrated, farmers groups identified mini-ponds, cultivating jujube fruit trees, dry seedbeds for rice, and homestead gardens as their preferred adaptation options.

OUTCOME

The project successfully demonstrated, developed and is constantly updating a menu of diversified good practice adaptation options, which have been tested in the field to promote community based adaptation and build resilience to climate change through applying a livelihoods perspective. These include not only physical adaptive measures (such as irrigation) but introducing alternative enterprises/systems; adjusting existing practices; making socio-economic changes such as livelihood diversification; strengthening community resilience and formal institutional structures; formulating policy to catalyse enhancement of adaptive livelihood opportunities; and creating awareness and advocacy on disaster risk reduction and links with climate change and adaptation.

LESSONS LEARNED

- *Launch adaptation with a focus on current variability* and factor in climate change.
- *Introduce climate adaptation as a social learning process* to create the capacity to cope with climate change-related impacts.
- *Introduce multiple and integrated adaptation measures* across sectors.
- *Ensure that adaptation to climate change is location-specific.*
- *Develop clear definitions of roles and responsibilities* for institutional capacity building and organisational networking.
- *Apply a livelihoods perspective* to understand and promote local-level adaptation to climate change.

- *Promote better sustainable natural resource management* practices in the context of future risks.
- *Revitalise and strengthen research* and development links.
- *Monitor ongoing adaptation practices* and potential risks of mal-adaptation, and establish links with policy making.
- *Assess the value of indigenous knowledge* in the context of managing future risks.
- *Focus on adaptation practices related to crop diversification and income generation*, which are preferred at community level.
- *Promote public-private partnerships* in climate change adaptation.

Source: Selby, Sarah and Courtenay Cabot Venton. 2009. *Addressing the Humanitarian Challenges of Climate Change: Regional and National Perspectives—Case Studies.* The Inter-Agency Standing Committee (IASC) Task Force on Climate Change. May–June 2009. http://sheltercentre.org/library/addressing-humanitarian-challenges-climate-change-regional-and-national-perspectives.

It should be noted that of the 19 case studies included in the IASC report, the climate change adaptation actions identified in all but one of the case studies had been implemented (Selby and Venton 2009).

DRAWING UP A NATIONAL ADAPTATION POLICY: FEEDBACK ON FIVE EUROPEAN CASE STUDIES

The March 2011 Climate Report prepared by CDC Climat and entitled *Drawing Up a National Climate Change Adaptation Policy: Feedback from Five European Case Studies* included the following conclusion:

"Since the turn of the century, and more particularly since 2005, adaptation has been attracting increasing interest from political decision-makers in developed countries, and the number of public measures is growing. As they notice that their country has already been affected by the initial impacts of climate change (severe droughts, flooding, etc.), prompted by public opinion, and warned by scientists about the increasing gravity of developments, decision-makers are beginning to draw up and implement adaptation policies and measures at government levels, from the local to the international ones.

The five European countries we chose to include in this report are among the most advanced in terms defining adaptation policies. A comparative analysis has enabled us to highlight not only the differences between them, but also their common features, thus providing us with an indication of the key points that apparently need to be included in all adaptation policies.

Among those points, one can note the following: (i) high-level research into both local climate change impacts and socio-techno-economic research; (ii) an appropriate institutional framework, that includes the founding of institutes that are specifically dedicated to working on adaptation in order to coordinate research and/or define policy, and the involvement of stakeholders, which is institutionalised to a greater or lesser degree according to the country's economic and political environment, and (iii) the identification of key issues and of potential measures that can be implemented, which are often linked to existing sectoral or local policies aimed at ensuring the effectiveness of such measures.

Indeed, the research work performed on climate change represents the stage that comes before drawing up and implementing practical adaptation measures, and the studied countries have developed their own issue-based research. The understanding of climate change adaptation has been considerably increased by the major research efforts made regarding the issue, but must be further expanded. Despite inherent uncertainty on this issue, adaptation policies have adopted various decision-making principles that enable decisions to be made in this context (the selection of no-regrets or flexible measures, the use of probability calculations, recourse to multi-disciplinary and multi-player approaches, etc.).

The various institutional organisations (federal government, centralised or non-centralised government, with or without powerful local authorities) also determine the institutional process followed in order to draw up an adaptation policy. All the countries have emphasised the local government level and have devolved the work on drawing up and/or implementing adaptation policies to sub-national governments to a greater or lesser degree. They have also underlined the need of coordination with private players, and the United Kingdom has even been developing specific tools intended for those players.

In most cases, the countries included in the report are only just beginning to draw up measures, except in the case of the Netherlands, which is at the forefront where country planning and flood risk management is concerned. Except in the case of flooding, where policies are usually quite advanced in most countries, the other issues are often still at the impacts analysis stage, and have not really reached the stage where operating measures can be defined. However, the countries are ready to consider the full range of possible measures ("hard," regulatory and organisational measures, etc.). The economic cost-benefit analysis of the measures and their funding has still not been significantly addressed. Likewise, the setting up of a committee to monitor measures in the medium-term and developing appropriate monitoring indicators are solutions that have not yet been widely envisaged, although they are an essential step for correcting policies on an ongoing basis.

These countries still need to improve the way in which they include adaptation considerations into existing policies, to continue their research on priority issues and improve the dissemination of tools that can be used by public and private players, to seek sources of funding and to reassess policies and measures on a regular basis. Operating measures are expected to be drawn up shortly, since most countries are setting themselves a 2012 target. Likewise, an EU Community adaptation strategy is also expected to be drawn up by 2013. The next three years will therefore be crucial for climate change adaptation in Europe. Faced with such a wide variety of issues and approaches in the Member States, the European Union organisations may not succeed in drawing up joint measures. However, the Community remains a key player in terms of ensuring trans-border coordination and at the very least, of disseminating information on best practices between the Member States, and of encouraging the remaining Members to draw up their own national adaptation policies. These points ought to be central factors in the new Community adaptation strategy" (Dumollard and Leseur 2011).

CARE REPORTS

In August 2014, CARE published a report entitled *Using Analysis of Climate Change Vulnerability and Adaptive Capacity for Program Decision-Making: Lessons from CARE Ethiopia's Experience* that "aims to share what CARE Ethiopia has learned so far from its efforts to integrate analysis of climate change vulnerability and adaptive capacity into program decision-making. It draws on the analysis conducted in different livelihood zones across the country, illustrated by detailed profiles of specific communities in each zone" (CARE Ethiopia 2014). The report listed the following findings:

- Climate-related hazards can no longer be treated as an anomaly;
- Climate change is not a "component" of programming;
- Climate change adaptation is inextricably linked with sustainable natural resource management;
- Adaptation to climate change requires options, not "solutions";
- Informed decision-making is key to adaptive capacity;
- Efforts to increase social equity must underpin adaptation efforts; and
- A better understanding of cultural and social barriers to adaptive action is needed (CARE Ethiopia 2014).

Access to a copy of this report is at http://www.careclimatechange.org/publications/adaptation.

In November 2011, CARE published a report entitled *Understanding Vulnerability to Climate Change Insights from Application of CARE's Climate Vulnerability and Capacity Analysis (CVCA)*

Methodology in order to "provoke thought, dialogue and debate by practitioners and policy makers on the issue of vulnerability to climate change and on approaches to vulnerability assessment in policy and in practice. Three case studies provide snapshots of the realities of vulnerable communities in Vietnam, Peru and Ghana that are already feeling the effects of climate change. The case studies provide context for a series of reflections on vulnerability, which draw on the case studies as well as on CARE's broader experience in analyzing vulnerability to climate change with communities around the world, using our CVCA methodology" (CARE 2011). The report's Executive Summary noted the following "key insights" resulting from the report:

- Access to information is a major factor determining ability to act on adaptation.
- For adaptation policies and programs to reach the most vulnerable, they must have a voice in the process.
- Insecure rights to resources exacerbate vulnerability.
- Existing coping strategies are ineffective or unsustainable, and may exacerbate vulnerability to climate change over time.
- The most vulnerable people lack access to services that would facilitate adaptation.
- Gender matters when it comes to vulnerability to climate change (CARE 2011).

The report concluded that "In order for adaptation efforts to be successful, vulnerability assessment must go beyond identification of vulnerable countries. Processes for vulnerability analysis and adaptation planning must involve all relevant stakeholders, including local communities and in particular the most vulnerable members of those communities. These processes must examine the social, economic and political drivers of vulnerability in order to identify the most vulnerable people within countries and communities and ensure that their needs, priorities and aspirations are reflected. This enables policy makers and adaptation practitioners to target resources and interventions where they are needed most" (CARE 2011).

A full copy of the report can be accessed at http://www.careclimatechange.org/files/adaptation/CARE_Understanding_Vulnerability.pdf.

MALDIVES: NATIONAL ADAPTATION TO CLIMATE CHANGE

The Maldives is a Pacific Island nation that, like similar island nations in the Pacific, is severely impacted by climate change. A background paper prepared by the Ministry of Housing, Transport and Environment entitled *National Adaptation to Climate Change* noted that "Climate change threatens the very existence of the low-lying small coral islands which we call home, and poses significant threats to the survival of our society, cultures and traditions which have, since time immemorial, co-existed harmoniously with the fragile island ecosystem. Furthermore, climate change threatens to remove the very lifeblood of our economy; our treasured coral reefs and oceans on which our most important sectors—the tourism and fisheries industries are based on" (Government of Maldives 2009).

The paper provides background information concerning the National Adaptation Programme of Action (NAPA) and identifies the ways the Maldives are vulnerable to climate change including rise of sea-level rise, decreasing the amount and increasing the intensity of rainfall, increasing the minimum and maximum temperature, and increasing the intensity and frequency of the extreme events (Government of Maldives 2009).

The paper identified vulnerable areas, including the following:

- Land, beach and human settlements
- Critical infrastructure
- Tourism
- Fisheries

- Human health
- Water resources
- Agriculture and food security
- Coral reef biodiversity (Government of Maldives 2009)

The following box, "Maldives: National Adaptation to Climate Change" presents the list of Priority Adaptation Actions identified in the NAPA and a list of urgent and immediate adaptation projects identified in NAPA. These lists have been copied directly from the *National Adaptation to Climate Change* paper.

INTEGRATION OF DISASTER RISK REDUCTION AND CLIMATE CHANGE ADAPTATION

A paper published in January 2011 in the journal Natural Hazards and Earth Sciences entitled *Integrating Community Based Disaster Risk Reduction and Climate Change Adaptation: Examples from the Pacific* noted "It is acknowledged by academics and development practitioners alike that many common strategies addressing community based disaster risk reduction and climate change adaptation duplicate each other. Thus, there is a strong push to integrate the two fields to enhance aid effectiveness and reduce confusion for communities" (Gero et al. 2011).

At a Joint Meeting of the Pacific Platform for Disaster Risk Management and Pacific Climate Change Roundtable in 2013, Michael Petterson, director of the Secretariat of the Pacific Community (SPC) Applied Geoscience and Technology Division, noted "joint national action plans (JNAPs) integrating climate change and disaster risk management provide countries with a powerful planning tool to help them consider disaster and risk across a range of ministries" (Secretariat of the Pacific Community 2013).

At this same meeting, Charles Carlson, director of Emergency Management in Cook Islands, said disaster risk management and climate change are now managed under the Office of the Prime Minister, reflecting the priority now placed on disaster risk management and climate change in Cook Islands. Mr. Carlson says the country is now looking to develop joint legislation to cover all future climate change and disaster risk management activities (Secretariat of the Pacific Community 2013).

The United Nations International Strategy for Disaster Reduction (UNISDR), the United Nations Development Programme (UNDP), and the World Bank Global Facility for Disaster Risk Reduction published a report in 2012 entitled *Disaster Risk Reduction and Climate Adaptation Change in the Pacific: An Institutional and Policy Analysis* that noted "As experience with both disaster risk reduction (DRR) and climate change adaptation (CCA) grows, there is increasing recognition that these two fields share a common focus: reducing the vulnerability of communities and contributing to sustainable development" (UNISDR 2012).

Excerpts from the report's Executive Summary noting practical lessons learned are presented in the following box "Integrating Disaster Risk Reduction and Climate Change Adaptation."

Another significant lesson learned noted in the report concerned Community Level Integration of DRR and CCA and noted "The greatest potential for harmonizing DRR and CCA in Pacific Island Countries (PICs) is at the community level. Community-based adaptation (CBA), ecosystem-based adaptation and community-based disaster risk management (DRM) (CBDRM) are powerful approaches for transcending the unproductive distinction between CCA and DRR that still pervades policies and planning at national and regional levels in parts of the Pacific. A key feature of community initiatives is that they often occur in a policy vacuum, with little national budget support. Due to weak linkages at the policy level, governments are missing out on opportunities to ensure that the national-level enabling environment is supportive at community level" (UNISDR 2012).

MALDIVES: NATIONAL ADAPTATION TO CLIMATE CHANGE

PRIORITY ADAPTATION ACTIONS

The following priority adaptation needs have been identified in the NAPA:

1. Integration of future climate change scenarios in the development of selected population centres or islands to adapt sea level rise and extreme weather risks associated with climate change
2. Protect human settlements: innovative coastal protection for development of selected population centres or islands
3. Improve and promote eco-friendly sustainable housing technology
4. Build resilience of fisheries
5. Acquire technologies and appropriate tools to manage water resources
6. Strengthen capacity for health services
7. Strengthen agricultural production and increase food security

URGENT AND IMMEDIATE ADAPTATION PROJECT IDENTIFIED IN NAPA

1. Integration of future climate change scenarios in the safer island strategy to adapt sea level rise and extreme weather risks associated with climate change
2. Coastal protection of safer islands to reduce the risk from sea induced flooding and predicted sea level rise
3. Coastal protection of Malé International Airport to reduce the risk from sea induced flooding and predicted sea level rise
4. Improve resilience of Island communities to climate change and variability through sustainable building designs
5. Protection of human settlements by coastal protection measures on safer islands
6. Enhance capacities at the community level to conserve and protect ground water sources through recharge, safe disposal of excreta and other wastes and other appropriate measures
7. Enhance adaptive capacity to manage climate change related risks to fresh water availability by appropriate wastewater treatment technologies
8. Flood control measures for vulnerable Islands
9. Enhance adaptive capacity to manage climate change related risks to fresh water availability by appropriate wastewater treatment technologies
10. Increase the resilience of local food production through enhancing the capacity of farmers, local communities to address food security issues caused by climate change and climate variability
11. Improve the health status of the population by the prevention and management of vector-borne diseases caused by changes in temperature and flooding due to extreme rainfall
12. Develop capacity to assess and address the implications of climate change for health and health systems, including tools and methodologies
13. Develop and implement strategies and measures relating to climate change and their cost-effectiveness
14. Assess health impacts of potential adaptation and mitigation measures undertaken by other sectors
15. Improve resilience of Island communities to climate change and variability through sustainable building designs

16. Investigating alternative live bait management, catch, culture and holding techniques in the Maldives to reduce vulnerability of the tuna fishery sector to the predicted climate change and variability
17. Protection of human settlements by coastal protection measures on safer islands
18. Increase resilience of coral reefs to reduce the vulnerability of islands, communities and reef dependent economic activities to predicted climate change
19. Create awareness among the population on the impact of climate change on water resources, the need to use water efficient devices and promote behaviours that are water conscious

Source: Government of Maldives. 2009. *National Adaptation to Climate Change.* Ministry of Housing, Transport and Environment. March 2009. http://www.ifrc.org/docs/IDRL/National%20Adaptation%20Programme%20(Climate%20change))/MALDIVES%20Adaptation%20to%20Climate%20Change.pdf.

INTEGRATING DISASTER RISK REDUCTION AND CLIMATE CHANGE ADAPTATION

PRACTICAL LESSONS LEARNED

Rationale for integration of DRR and CCA:

- Easing the burden of programming assistance.
- Minimizing duplication of effort and redundancies.
- Reducing potential conflicts in policy development.
- Making efficient use of scarce resources, and
- The increasing recognition, especially at the community level, that there is little practical difference between the two.

Barriers to integration of DRR and CCA:

- Capacity constraints of Pacific Island Countries (PICs) (related to lack of coordination, communication, political will, insufficient funds and absence of expertise).
- Separate global and regional frameworks for CCA and DRR.
- Perceptions of development practitioners that DRR and CCA are not valuable.
- Difficulty quantifying the benefits of DRR and CCA.

Approaches to address barriers and facilitate integration:

- Improved access to practical weather and climate change information.
- Strong enabling environment and enhanced communications to practitioners in other fields and to the broader public.
- More emphasis on bottom-up approaches; and
- Information support for decision-making (both scientific and economic).

Source: UNISDR. 2012. *Disaster Risk Reduction and Climate Adaptation Change in the Pacific: An Institutional and Policy Analysis.* United Nations International Strategy for Disaster Reduction (UNISDR), United Nations Development Programme (UNDP) and the World Bank Global Facility for Disaster Risk Reduction. 2012.

EXAMPLES OF EFFORTS TO INTEGRATE DRR AND CCA FROM PACIFIC ISLAND COUNTRIES

THE KINGDOM OF TONGA

Tonga was the first Pacific Island Country that successfully developed and approved an integrated national policy on CCA, DRR, and Sustainable Development (SD). The integration process began in October 2009 with a statement by Prime Minister Dr. Feleti Sevele to his Cabinet wherein he explained that he felt it was inefficient for a country of such a small size to address DRR and CCA separately (Sikivou 2013). With the support of the two regional organizations SPC/SOPAC and SPREP, a Joint Climate Change and Disaster Risk Management National Action Plan, called a JNAP for short, was completed in early 2010.

The Tonga Cabinet approved the plan in July of 2010, and it was subsequently put into practice. Since that time, the government has implemented a number of the activities that are prescribed under the joint plan, and in less than four years they had already reached 60% of the five-year plan's identified goals (SPREP 2013a). These goals include:

1. Improved good governance for CCA and DRR
2. Enhanced technical knowledgebase information, education and understanding of CCA and effective DRR
3. Analysis and assessment of vulnerability to CC impacts and disaster risks
4. Enhanced community preparedness and resilience to impacts of all disasters
5. Technically reliable, economically affordable, and environmentally sound energy to support the sustainable development of the Kingdom
6. Strong partnerships, cooperation, and collaboration within government agencies and with civil societies, nongovernmental organizations, and the private sector (UNISDR 2013)

Tonga established a JNAP Secretariat in order to provide multi-stakeholder input into the development of the plan, to ensure that the perspectives and needs of all sectors were considered, to develop effective strategies to address CCA and DRR needs, and to serve as the core group to make informed decisions based on interdisciplinary data and risk reduction assessments during the actual mainstreaming process (SPREP 2013b).

The country has taken considerable steps to ensure that climate change issues are integrated into the national Strategic Development Framework. These actions are supported by a commitment of sustainable funding to establish a Climate Change Division that includes four staff within the Ministry of Lands, Environment, Climate Change and Natural Resources. This division has taken the lead on promoting CCA goals in cooperation with the National Emergency Management Office (which promotes the national DRR agenda). Since this initiative began, the two offices have continued to collaborate to better coordinate donor-funded projects that would otherwise address DRR and CCA in isolation or otherwise be unnecessarily duplicative (Sikivou 2013).

Integration of DRR and CCA goals into the policy and planning of some, though not all, of the different sectors in Tonga is currently happening as a result of the creation of the Climate Change Division. One example is the development of a national forestry policy of the Ministry of Education and Training that introduces CC and DRM topics into the national primary and secondary school curricula (SPREP 2013c). A working committee on Education on Climate Change and Disaster Risk Management was established to develop and support the program which is designed to link cultural awareness with CCA, DRM, and environmental sustainability. The Climate Change Division is also meeting on a bi-monthly schedule with a range of development stakeholders, including donors, nongovernmental organizations, and the private sector, to allow an exchange of information on different CCA and DRM projects currently underway. This has provided a forum for lessons learned and best practice information sharing, and has limited the duplication of efforts.

Among several integration "next steps" described during a speech before the May 2013 Global Platform for DRR, the CEO of Tonga's Ministry of Infrastructure described an effort to consolidate the National Disaster Management Office, the government meteorology function, and other climate change related activities and functions in order to form a single, standalone agency.

THE COOK ISLANDS

The Cook Islands was among the first three Pacific countries to have completed and implemented a Joint National Action Plan for DRM and CCA. The Cook Islands JNAP replaced the existing Disaster Risk Management National Action Plan (DRMNAP) while simultaneously incorporating climate change considerations. This process, initiated in 2010, took two years to complete. Integration of DRM and CCA is also prioritized in the nation's National Sustainable Development Plan, which seeks "[a] Cook Islands where our people are resilient to disasters and climate change to achieve sustainable livelihoods" (Cook Islands 2013). What is notable about the Cook Islands experience is its efforts to structurally integrate DRR and CCA in addition to the functional integration through mainstreaming.

To integrate CCA and DRR, and to provide high-level direction to the combined effort, a dedicated Climate Change Division entitled "Climate Change Cook Islands" (CCCI) was established in the Office of the Prime Minister. This office is funded through both regular government budgets and donor resources. This move consolidated DRR and CCA together as the Emergency Management Cook Islands (EMCI) also falls within the Office of the Prime Minister (Institute for Sustainable Futures 2013). The move has proven very effective at enabling joint programming of CCA and DRR projects.

To better support the integration effort, a National Platform for CC and DRM was established. This platform is working on the development of more coordinated CCA and DRR funding mechanisms, and was successful in supporting the establishment of a DRM and CCA Trust Fund with an initial US$165,000.

The Cook Islands has taken great effort to mainstream DRR and CCA throughout the various sectors and at all government levels. These include locally-focused capacity building initiatives that have resulted in the creation of DRM Committees at the village-level in Rarotonga, and the placement of disaster coordinators on the Outer Islands. At the national level, each ministry and agency has appointed a Disaster Management Focal Point to address DRR and CCA activities in ministry operations.

The Cook Islands is one of the only countries to have released a national-level information platform for DRR and CCA. The DRR/CC Project Portal for the Cook Islands launched in April of 2013. This resource provides access to a wide range of data on DRM and CC projects, which in turn supports the planning and implementation of new activities.

Source: Coppola, Damon. 2013. *Integration of Climate Change Adaptation, Disaster Risk Reduction, and Sustainable Development in the Pacific: Informing the HFA2 Process*. UNISDR-Pacific. Geneva.

A report entitled *Integration of Climate Change Adaptation, Disaster Risk Reduction, and Sustainable Development in the Pacific* developed for UNISDR in December 2013 noted the following concerning the experience among Pacific Island nations in integrating DRR and CCA, "Regional organizations, namely Secretariat of the Pacific Community/Applied Geoscience and Technology Division of the Secretariat of the Pacific Community (SPC/SOPAC) and Secretariat of the Pacific Regional Environmental Programme (SPREP), have played an integral role in promoting and enabling integration through their extensive advocacy, financial support, technical assistance, staff backfilling, facilitation, capacity building, research, and other assistance. They have joined their efforts to form an institutional platform from which effective guidance, lessons, best practices, strategies, and tools have flowed, and it is doubtful the high degree of progress that exists today would have been possible without their involvement. That being said, it is the progressive thinking, innovation, and persistence of national government integration champions in each of the PICTs that ultimately deserve credit for the successes described in this report. The merging of CCA, DRR, and Sustainable Development (SD) action and goals in the Pacific has required these thought leaders to challenge the siloed arrangements that have become conventional practice among donors and governments alike. There is still a long way to go before integration of these related policy goals is achieved throughout all sectors and at all government levels, but the actions that have been taken to date have set a course and a place that promises to maintain the Pacific's position as forerunner in this movement" (UNISDR 2014).

The box "Examples of Efforts to Integrate DRR and CCA from Pacific Island Countries (PICs)" is taken from this report.

CONCLUSION

The climate change problem may be a global one, but there exist great disparities between nations with regard to the implications of climate-related hazards and the available options to manage risk (including—but not limited to—adaptation). Especially for the small-island developing states (SIDS), and the low-lying nations, the consequences of climate change have already begun in earnest and threaten the very survival of the nations even in the short-term. Many of these nations, despite their relatively small size, have been recognized for innovative adaptation policies and programs—many of which have been credited for achieving sizeable reductions in population and economic risk. The body of knowledge from which climate change adaptation policies and solutions may be drawn may be broad in its scope, but it is shallow in terms of its age of experience. All nations, rich and poor alike, have much to learn from this great equalizer of problems.

REFERENCES

CARE. 2011. *Understanding Vulnerability to Climate Change Insights from Application of CARE's Climate Vulnerability and Capacity Analysis (CVCA) Methodology.* CARE International Poverty, Environment and Climate Change Network (PECCN), November. Available at http://www.careclimatechange.org/files/adaptation/CARE_Understanding_Vulnerability.pdf.

CARE Ethiopia. 2014. *Using Analysis of Climate Change Vulnerability and Adaptive Capacity for Program Decision-Making: Lessons from CARE Ethiopia's Experience.* CARE Climate Change Information Centre, August. Available at http://www.careclimatechange.org/publications/adaptation.

Cook Islands. 2013. National Progress Report on the Implementation of the HFA (2011–2013)—Interim.

Dumollard, G. and A. Leseur. 2011. *Drawing up a National Climate Change Adaptation Policy: Feedback from Five European Case Studies.* Climate Study No. 27. CDC Climat Research, March. Available at http://www.cdcclimat.com/IMG/pdf/11-03-31_climate_report_no27_-_drawing_up_a_national_adaptation_policy.pdf.

EACC. 2010. *Economics of Adaptation to Climate Change (EAAC)—Vietnam.* The World Bank Group. Available at http://peacewindsamerica.org/wp-content/uploads/2014/02/EACC_Vietnam.pdf.

Gero, A., K. Meheux and D. Dominey-Howes. 2011. *Integrating Community Based Disaster Risk Reduction and Climate Change Adaptation: Examples from the Pacific.* Natural Hazards and Earth Sciences, January 10. Available at http://www.nat-hazards-earth-syst-sci.net/11/101/2011/nhess-11-101-2011.pdf.

Government of Maldives. 2009. *National Adaptation to Climate Change.* Ministry of Housing, Transport and Environment, March. Available at http://www.ifrc.org/docs/IDRL/National%20Adaptation%20 Programme%20(Climate%20change))/MALDIVES%20Adaptation%20to%20Climate%20Change.pdf.

IISD et al. 2003. *Increasing Community Resilience to Climate-Related Disaster through Sustainable Livelihoods: A Conceptual Framework for Climate Change Adaptation.* International Institute for Sustainable Development (IISD), the Stockholm Environment Institute (SEI), the World Conservation Union (IUCN) and Intercooperation, December. Available at http://www.iisd.org/pdf/2003/envsec_live lihoods_1.pdf.

Institute for Sustainable Futures. 2013. Understanding the Pacific's Adaptive Capacity for Emergencies in the Context of Climate Change: Cook Islands.

Rissik, D. and N. Reis. 2013. *Key Lessons from Practitioners' Experiences.* National Climate Change Adaptation Research Facility. Australian Department of Industry, Innovation, Climate Change, Science, Research and Tertiary Education. Available at http://www.nccarf.edu.au/localgov/sites/nccarf .edu.au.localgov/files/Climate%20Change%20Adaptation%20Good%20Practice_Synthesis%20Report _Key%20lessons%20from%20practitioners%20experiences.pdf.

Secretariat of the Pacific Community. 2013. Pacific Island Countries Lead the World on Integration of Disaster Risk Management and Climate Change Adaptation, May 8. Available at http://www.spc.int/fr/technol ogies-de-linformation/actualites/1328-pacific-island-countries-lead-the-world-on-integration-of-disaster -risk-management-and-climate-change-adaptation-.html.

Selby, S. and C.C. Venton. 2009. *Addressing the Humanitarian Challenges of Climate Change: Regional and National Perspectives—Case Studies.* The Inter-Agency Standing Committee (IASC) Task Force on Climate Change, May–June. Available at http://sheltercentre.org/library/addressing-humanitarian -challenges-climate-change-regional-and-national-perspectives.

Sikivou, M. 2013. Personal Interview. SPC/SOPAC Deputy Director. DRP, February 13.

SPREP. 2013a. Tonga: Gaining Ground on the Climate Change Front at Home. UNFCCC, Warsaw, Poland, November 21.

SPREP. 2013b. JNAP Development and Implementation in the Pacific: Experiences, Lessons, and the Way Forward. Apia, Samoa.

SPREP. 2013c. Tonga: Gaining Ground on the Climate Change Front at Home. Available at http://www.sprep .org/climate-change/tonga-gaining-ground-on-the-climate-change-front-at-home.

UNISDR. 2012. *Disaster Risk Reduction and Climate Adaptation Change in the Pacific: An Institutional and Policy Analysis.* United Nations International Strategy for Disaster Reduction (UNISDR), United Nations Development Programme (UNDP) and the World Bank Global Facility for Disaster Risk Reduction.

UNISDR. 2013. The Pacific Experience in Developing Policy and Legislation on DRR and CCA. Available at http://www.unisdr.org/we/inform/publications/34003.

UNISDR. 2014. Integration of CCA, DRR, and Sustainable Development in the Pacific. *6th Asian Ministerial Conference for Disaster Risk Reduction*, Bangkok, Thailand.

9 Conclusions and Recommendations

George D. Haddow

INTRODUCTION

The evidence is overwhelming that climate change will have a dramatic impact on future disasters. In coming years, the frequency and severity of disasters will continue to rise and it is imperative that individuals and communities take action to reduce the impacts of these disasters. Dramatically reducing the greenhouse gas emissions that cause climate change is one form of mitigation that must start immediately and continue unabated if we are to effectively eliminate climate change. However, it is also important that actions be taken to reduce the impacts of future disasters caused or aggravated by climate change.

This book has attempted to address how we as a society can successfully reduce the impacts of future disasters caused or aggravated by climate change. We have presented a series of essays and case studies written by individuals with real-world experience in creating and implementing hazard mitigation and climate change adaptation measures designed to reduce the impacts of future disasters. What we hoped to accomplish was to assure decision makers at all levels of society that reducing the impacts of climate change is not as daunting a task as it appears and that there are clear examples of how this can be successfully accomplished. We have provided information on the need for planners, environmentalists, and an entirely new set of professionals to join this effort, the various federal government programs that have played a major role in hazard mitigation in the past and new federal programs designed to support climate change adaptation activities, and case studies from communities across the nation and internationally that have successfully designed and implemented hazard mitigation and climate change adaptation programs that can serve as examples for the future.

The purpose of this chapter is to summarize the information presented in the previous chapters and to highlight those factors that will be critical to dealing with climate change in the future. Also presented are a series of recommendations designed to guide the efforts of a full range of community stakeholders (i.e., local, state and federal government, the private sector, nongovernmental organizations) in designing and implementing effective hazard mitigation and climate change adaptation programs to reduce the impacts of future disasters caused or aggravated by climate change.

CONCLUSIONS: WHAT WORKS

The most important thing we have learned is that reducing disaster risks and losses is best done at the community level. Outside resources and technical assistance are critical but effective and sustainable hazard mitigation and climate change adaptation programs and actions must be designed and implemented at the local level where disasters strike. Without the full support and participation of all stakeholders of a community in this effort it cannot be successful.

Based on the information presented in the essays and case studies in this book, we conclude that the following factors are critical to building successful community-based hazard mitigation and climate change adaptation programs that we believe will allow communities to survive and thrive in a changing climate:

- *Involve all community stakeholders.* Building a successful hazard mitigation and climate change adaptation program cannot be done by a select few government officials. This work requires the involvement and efforts of all community stakeholders, including government officials (elected and appointed), emergency managers, first responders (fire, police, emergency medical technicians [EMTs]), city and county managers, planners, community and economic development officials, the local media, nonprofit groups, large and small business owners and associations, the Chamber of Commerce, environmental groups, developers, utilities, churches, community organizations, voluntary and nongovernmental organizations, community-based organizations, community-based foundations, and individual citizens.
- *Leadership and local champions.* Leadership in this effort is all-important and a local champion can make the difference between success and failure. Such a champion can come from the public sector, the private sector, the nongovernmental sector, the community leadership, or from everyday citizens. Carol Williams in Tulsa—an everyday citizen—was the first of many champions in Tulsa's drive to reduce the impacts of flooding. The University of California at Berkeley is a champion of earthquake hazard mitigation not only on its campus but in the community at large. The vineyard owners in Napa championed the community's 20-year flood reduction plan and lead the fight to secure a critical local funding source. As noted in the case studies, mayors across the country are leading the way in taking climate change adaptation actions to make their communities more resilient. Local champions come in all shapes and sizes and they are all critical.
- *Private sector involvement.* The sustained health of any community is tied closely to the health of its economy and its private sector. In the aftermath of a disaster, the pace of recovery in the community can often be pegged to how quickly businesses and employers come back on line. Local business leaders and the local Chamber of Commerce must be involved in a community-based hazard mitigation effort. Their needs must be considered and the risks they face addressed. At the national level, former New York City mayor Michael Bloomberg has continued to lead in our nation's business community through the Risky Business Project. The private sector is an excellent source for local champions and one potential source for resources.
- *Involving new stakeholders.* Hazard mitigation has long been the purview of emergency managers and operations. The time has come to expand the pool of government and nongovernment stakeholders beyond emergency management. Planners, community development specialists, economic development professionals, and public administrators all need to be involved in hazard mitigation and climate change adaptation issues. Public interest groups, community-based organizations, and environmental groups all add to the expanse and effectiveness of efforts to reduce disaster risks and the impacts of climate change. Climate change adaptation practitioners also bring new skills and relationships to the table that will expand the reach and enhance the design and implementation of community-based hazard mitigation and climate change adaptation programs.
- *Resources.* As with any hazard mitigation and climate change adaptation program, resources will dictate much of the success of the program. Multiple sources of funding must be pursued, including
    - Local funding source. A local funding source can come in many forms such as the city bond issues used in Berkeley, the half-cent sales tax increase in Napa, and the stormwater drainage fee in Tulsa. Officials in Grand Forks, Flagstaff, and Norfolk found ways to leverage existing resources and to raise new resources. A consistent and sustained local funding source provides reliable funding for a community-based effort and can be leveraged very effectively with funds from other government and nongovernment sources. Providing a local funding source sends a clear signal to the community-at-large and potential donors that a community is serious about hazard mitigation and climate change adaptation.

  - Federal and state government funds. In the past, the federal government has been a significant source for hazard mitigation funding. FEMA funds a variety of hazard mitigation programs, including its Hazard Mitigation Grant Program (HMGP) and the Disaster Hazard Mitigation Assistance (DMA) program. Other potential sources of federal funding can be found in the various federal departments and agencies that have direct or indirect involvement in disasters such as the Corps of Engineers, the Department of Transportation (DOT), Housing and Urban Development (HUD), Environmental Protection Administration (EPA), Health and Human Services (HHS), Department of Energy (DOE) and others. Nearly all government funding programs at the state and federal levels can be linked to hazard mitigation and climate change adaptation and with a little creativity accessed for use in a community-based hazard mitigation effort.
  - Leverage resources. The City of Seattle's Project Impact program leveraged its $1 million in seed money from FEMA with over $4 million in funding from the public and private sectors. Some of the resources were money and some were in-kind donations but they all added up to Seattle being able to expand its funding base fourfold. New York City leveraged recovery funding from Hurricane Sandy to build back better. As noted earlier, a community can use a local funding source to leverage additional resources from other sources as well. Napa successfully used the funds from a half-cent sales tax increase to match funds made available from state government and foundation programs. The key is to identify what sources of money and in-kind services are available among all potential sources, including government, foundations, the private sector, and the nonprofit sector.
- *Leadership from state and federal government.* Local champions are important but so are champions at the state and federal government levels. The leadership and support for Project Impact by President Clinton and his FEMA director James Lee Witt served to validate local hazard mitigation efforts and to promote these programs with state and local decision makers, the public, and the media. FEMA's endorsement of hazard mitigation efforts in Tulsa and Berkeley through its Project Impact Community of the Year award allowed both communities to further expand and enhance their efforts. The leadership President Obama has shown in advocating the mitigation of climate change and the creation of the $1 Billion National Disaster Resilience Competition has helped promote climate change adaptation. As Brian Cowan noted in his essay in Chapter 4, "The federal government's role [in Project Impact] was acknowledged as merely one of the partners to a local jurisdiction's efforts, though perhaps one with a unique ability to supply knowledge, help create incentives, and bring a measure of accountability and visibility to the effort."
- *Consensus building.* Coming to community consensus on risks and what can be done to mitigate those risks played a large role in the success of the programs presented in this book. Consensus building remains at the heart of the 20-year flood protection plan developed and implemented in Napa. This example highlights how the consensus building process ensures that all stakeholders are involved and all views and ideas are heard. The resulting plans reflect the needs of all parties and include measures that can be supported and promoted by all involved. In addition, the all-inclusive planning process built support for passing the sales tax increase that served as the basis for funding the Flood Control Project.
- *Environmental protection and enhancement.* The critical role that a healthy and vibrant environment plays in reducing the impacts of future disasters was clearly illustrated in the case studies involving flood issues. Restoring 900 acres of wetlands in Napa and creating additional open spaces and retention ponds in Tulsa are excellent examples of how the natural environment can be an effective ally in reducing flood impacts in a community. A recurring theme throughout the case studies presented in this book is how communities stopped fighting their natural resources and learned to live and adapt to their physical environment as a direct means for reducing their hazard risks.

- *Fold resilience into existing plans and planning processes.* The American Planning Association's James Schwab concludes in his essay in Chapter 3, "Although young and evolving, the system and methodology for integrating climate adaptation into both local hazard mitigation plans, under the federal Disaster Mitigation Act, and local comprehensive plans through hazard-related and other elements, is clearly taking hold in those jurisdictions with the largest stake in surviving the impacts that may have some connection to global climate change." Examples of these types of activities can be seen in the New York City, Flagstaff, Norfolk, and Grand Rapids case studies. Ultimately the goal is to institutionalize hazard mitigation and climate change adaptation into the way decision makers at all levels (elected officials, city managers, business owners, home owners, heads of families) determine how their community should develop or redevelop, where to site a new business, decide to move a business, or to retrofit an old business, and where to live or buy a home.
- *Take advantage of a crisis.* In each of the case studies presented in Chapter 6, each community profiled took advantage of a recent crisis to rally community stakeholders to do something to reduce future disaster impacts. The heightened awareness of the community risk and the efforts to respond and to recover from a recent event seem to provide community officials with an opportunity to advance hazard mitigation and climate change adaptation measures that may be viewed with more skepticism in a noncrisis period. The cumulative effect of a series of crisis events over time (think the recent severe drought conditions in over 90 percent of California's counties) or an event that strikes another community or region but many in the community believe could strike them are also opportunities for community officials to advance measures to reduce future impacts.
- *Cobenefits.* Risk reduction efforts almost always provide cobenefits beyond reducing the impacts of future events and reducing recovery costs. In Napa, the implementation of the 20-year Flood Control project has served as the catalyst for the revival of the City of Napa, resulting in new jobs, nearly $1 billion in new investments, and a new, prosperous future for its residents. Grand Rapids fixed its sidewalks and streets while at the same time addressing their flood risk. In developing hazard mitigation and climate change adaptation programs, officials must identify the full range of cobenefits that these programs are likely to deliver in addition to hazard risk reduction.
- *Regional approaches.* Hazard and climate change risks often cross jurisdictional and community boundaries. The Southeast Florida Climate Change Compact is a shining example of what can be done when multiple jurisdictions band together to address a common hazard risk. Interest is growing in California and the Northeast in creating similar multijurisdictional regional compacts.
- *Know your risk from climate change.* Damon Coppola states in his essay in Chapter 2 that "It is no longer feasible to calculate static likelihood and consequence factors when assessing risk, even for those hazards that have no obvious association with climate change such as landslides. The effects of climate change are varied and complex, and without due consideration emergency plans and risk reduction policies will both become insufficient as these trends persist." Risk assessment must be an ongoing exercise and new tools and technologies must be developed to help community leaders, business owners, and individual citizens to understand how their risks are changing and what to do to reduce these risks. As Brian Cowan stated concerning the National Flood Insurance Program, "By coupling the flood risk identification with financial incentive, it has fostered a cultural recognition of flood risk in the business of home ownership and community housing development." Knowing your risk provides the basis for creating risk reduction measures and the incentives that may help reduce these risks.
- *Don't call it climate change.* A New York Times/Stanford University/Resources for the Future nationwide poll released on January 31, 2015, found that the overwhelming majority

of Americans believe that government should be addressing the causes and impacts of climate change. However, there are still segments of American society that deny climate change, including a number of key members of Congress who can make it difficult for federal agencies to fund climate change adaptation measures. Mayors in several cities who have implemented programs designed to reduce climate change impacts have chosen not to talk about climate change to describe these efforts. Instead they emphasize the positive effects of their actions in reducing disaster impacts and protecting their community's economy, well-being, and development.

RECOMMENDATIONS: HOW TO MAKE IT WORK IN YOUR COMMUNITY

The following sections present a series of recommendations on steps communities can take to design and implement a hazard mitigation and/or climate change adaptation program designed to reduce the impacts of future disasters as well as recommendations on how the state and federal governments, the private sector, and the nonprofit sector can support communities in these efforts.

LOCAL GOVERNMENT

Hazard mitigation and climate change adaptation is best applied at the community level. If all politics are local so too is hazard mitigation. In order to effectively reduce the impacts of future disasters caused or exacerbated by climate change, communities must make a commitment to understanding their risks and taking action. A community-based approach to hazard mitigation was successfully implemented in FEMA's Project Impact initiative in the late 1990s and has been recommended for communities hoping to adapt to the impacts of climate change by the Rockefeller Foundation and the ICLEI–Local Governments for Sustainability (see box entitled "Examples of Community-Based Approaches"). As noted in Chapter 7, similar planning processes have been and are currently being used in nations across the globe.

The community-based approach is the spine of a hazard mitigation program designed to reduce the impacts of climate change that we propose to be supported by the actions of the state and federal governments, nongovernmental organizations, and the business sector.

This approach requires that communities take the following steps:

- Create a community partnership to lead the project that includes participation from all community members, including individual residents, government officials, the local Chamber of Commerce, large and small businesses, the local media, state and local emergency management officials, police, fire and emergency medical officials, community groups, churches, unions, nonprofits, environmental groups, social services, and educational organizations.
- Identify community and neighborhood hazard risks. What are the risks that your community faces (hurricanes, severe storms, tornadoes, flooding, drought, wildfire, etc.)? Which areas are vulnerable to these risks? Where are vulnerable and special needs populations located? Identify specific issues such as vulnerability of government facilities and the siting of emergency operations centers. Identify the potential impacts on local businesses, schools, child care centers, homes, and so on. How do you incorporate ongoing changes to the climate in your risk assessment? Federal and state emergency management officials should provide technical assistance in conducting the risk assessment. One vehicle that can serve this function is the Community Hazard Mitigation Plan that FEMA requires all communities wishing to receive federal disaster aid to develop and update every 5 years. FEMA now requires that climate change considerations be included in this ongoing hazard mitigation planning process.

EXAMPLES OF COMMUNITY-BASED APPROACHES

FEMA'S PROJECT IMPACT

The goal of Project Impact is to bring communities together to take actions that prepare for—and protect themselves against—natural and manmade disasters in a collaborative effort. To accomplish this goal, we have organized pre-disaster activities into four phases:

- Build a Community Partnership comprised of all community stakeholders.
- Conduct a Hazard Identification and Hazard Vulnerability effort to examine the community's risks from natural and manmade hazards and to identify vulnerabilities to those risks.
- Identify and Prioritize Risk Reduction Actions designed to mitigate identified risks and vulnerabilities and to reduce the impacts of future disasters.
- Communicate your Plan to your Community in order to generate the public, political and resource support needed to implement the Plan.

Source: Project Impact: Building a Disaster Resistant Community. Guidebook. Federal Emergency Management Agency 1998.

ICLEI–LOCAL GOVERNMENTS FOR SUSTAINABILITY

The purpose of *Preparing for Climate Change: A Guidebook for Local, Regional, and State Governments* is to help you as a decision-maker in a local, regional, or state government prepare for climate change by recommending a detailed, easy-to-understand process for climate change preparedness based on familiar resources and tools:

- Scope the Climate Change Impacts to Your Major Sectors
- Build and Maintain Support to Prepare for Climate Change
- Build Your Climate Change Preparedness Team
- Identify Your Planning Areas Relevant to Climate Change
- Conduct a Climate Change Vulnerability Assessment
- Conduct a Climate Change Risk Assessment
- Set Preparedness Goals and Develop Your Preparedness Plan
- Implement Your Preparedness Plan
- Measure Your Progress and Update Your Plan

Source: Preparing for Climate Change: A Guidebook for Local, Regional, and State Governments. Center for Science in the Earth System (The Climate Impacts Group), Joint Institute for the Study of the Atmosphere and Ocean University of Washington, King County, Washington and ICLEI–Local Governments for Sustainability 2007.

JUDITH RODIN: PRESIDENT OF THE ROCKEFELLER FOUNDATION

The Rockefeller Foundation recently announced a major climate change initiative that concentrates on building resilience to a changing, challenging natural environment. As we see it, resilience incorporates five dimensions:

- Information—effective adaptation will always be locally driven…communities need sophisticated measurement and assessment tools, integrated information about risks those tools reveal, and the best substantial approaches to minimize them.
- Infrastructure—more than roads and sewers. It includes all the institutions and processes put in place to manage society.

- Insurance—underserved populations need access to the social and economic security that comes from sharing risk. And the more people who share the risk, the lower the cost of coverage.
- Institutional Capacity—resilience require that individuals and communities be empowered for and respond to crises from the ground-up. But there's also a critical role for governments and institutions to play in supporting resilience from the top-down.
- Integrated Systems—successful adaptation strategies...integrate urban planning, land-use regulation, water management, infrastructure investment, especially in energy and transportation, early-warning systems, and emergency and disaster preparedness, among many other elements.

Source: Climate Change Adaptation: The Next Great Challenge for the Developing World. Remarks as Delivered by Judith Rodin, president of the Rockefeller Foundation, to the American Association for the Advancement of Science 2008 Annual Meeting. 2008.

- Identify and prioritize those actions, both structural and nonstructural, that can be taken by individuals, organizations, businesses, and the government in the community to lessen these risks and reduce the impacts of future disasters (see the box "Potential Hazard Mitigation Actions").
- Communicate the plan to the community/neighborhood and generate the political, financial, and public support needed to implement the plan. Plan and conduct events and activities to promote and publicize the community's actions within the community and with national media, Congress, federal agencies, and other potential partners. Organize support for sustaining community project efforts. Local traditional media outlets (TV, radio, and print) and social media outlets (Twitter, Facebook, YouTube, etc.) can be very helpful in promoting the community partnership and the hazard mitigation efforts.

In support of these actions, we recommend communities undertake the following activities:

- Establish community emergency networks (CENs) designed to communicate hazard mitigation and preparedness messages to residents and to collect and provide information from residents to government, business, and nonprofit sector decision makers. Incorporate social media outlets (Twitter, Facebook, YouTube, etc.) into the CEN. Conduct a community integrated demographic mapping project to identify hazard mitigation needs within the community, establish the CEN points of contact, and to ensure participation among functional needs groups such as seniors, disabled, non-English-speaking, and so forth.
- Identify a local funding source to provide the match for federal, state, and private funding for hazard mitigation projects. Napa voters approved a half-cent sales tax increase to support their 20-year flood protection plan and for the past 20 years Tulsa has used a stormwater drainage fee approved by the voters to help fund critical flood hazard mitigation actions. Flagstaff identified an existing local funding source to finance their Flagstaff Watershed Protection Project. In these and other communities across the country, locally generated funds have been used to match funding from federal and state governments, the business community, and the nonprofit and foundation communities to fund hazard mitigation and climate change adaptation actions.
- Establish an ongoing monitoring and evaluation process that truly measures the benefits of the hazard mitigation actions to the community. Metrics should be established in each community to measure the reduction in disaster relief costs achieved by the hazard mitigation

action, the economic benefits to the community of becoming more disaster-resilient, and the multiple benefits realized from a healthy natural environment. This process will also evaluate the effectiveness of the community partnership and provide insights into how to improve all of its functions.

Creating community partnerships provides a direct benefit to the community and also provides collateral benefits to both federal and state governments. Benefits such as reduced disaster costs, stability of the tax base, continued economic growth and development, and an overall increase in the health and safety of citizens lead to a more socially and economically healthy community.

POTENTIAL HAZARD MITIGATION ACTIONS

Potential Hurricane and Flood Hazard Mitigation Actions

- Conduct audits of homes, child care centers, schools and neighborhood businesses to identify low cost actions that can be taken to reduce damage from future hurricanes and floods;
- Restore and protect the natural environment to provide defense from storm surge and flooding including clearing streams of debris, restoring and protecting wetlands and creating open spaces in the community to soak up rain and flood waters;
- Buyouts of properties in the floodplain;
- Apply protective film to windows in schools, child care centers and low income and senior residences;
- Evaluate all existing structural hazard mitigation entities such as levees, drainage and water diversion channels and flood protection gates.

Source: Project Impact: Building Disaster Resistant Communities Guidebook. FEMA 1998.

Potential Drought Reduction Actions

- Connect regional water systems
- Develop new groundwater sources
- Implement new technologies such as reverse osmosis for desalinization
- Provide financial incentives (e.g., tax breaks, rebates) for switching to more efficient manufacturing processes, irrigation practices, and appliances
- Renegotiate transboundary water agreements where applicable
- Update drought management plans to recognize changing conditions
- Increase authority to implement water restrictions and other emergency measures as needed
- Expand use of climate information (e.g. seasonal forecasts) in water resources planning and management
- Conduct additional research on how climate change may impact your community's water supply
- Include information on climate change impacts to water supplies and how residents can reduce water use in utility inserts, newsletters, web sites, and local newspapers

Source: Preparing for Climate Change: A Guidebook for Local, Regional, and State Governments. Center for Science in the Earth System (The Climate Impacts Group), Joint Institute for the Study of the Atmosphere and Ocean University of Washington, King County, Washington, and ICLEI–Local Governments for Sustainability 2007.

FEDERAL SUPPORT FOR COMMUNITY-BASED HAZARD MITIGATION AND CLIMATE CHANGE ADAPTATION

The federal government's most important role in hazard mitigation and climate change adaptation is to support local and regional programs by providing technical assistance and funding. FEMA and other federal agencies continue to take steps to incorporate climate change impacts into their activities and their planning processes. President Obama's Climate Action Plan has set a course for ongoing support for community-based hazard mitigation and climate change adaptation.

The establishment in 2015 of FEMA's new standard for the use of federal recovery funding in helping communities to build back better after flooding events is a major step in adding new resources that communities can tap to fund hazard mitigation and climate change adaptation. We recommend that similar standards be implemented to incorporate disaster resistant measures in rebuilding community buildings, facilities, and infrastructure in order to build back better in the aftermath of a disaster.

In order to create new federal funding sources, we recommend that a Hazard Mitigation Trust Fund be created. A fee could be added to every contract for building or upgrading any facility or project supported by federal funding (the amount could be as low as $10 or as high as $1000; a sliding scale could be created based on the cost of the project or the use of the facility—schools could be less). This may be reasonable but the costs would end up coming back to the government unless it was written into the contract as a contractor contribution. Another option under this category would be to add $1–$1000 to every building permit for development, the size of the addition based on the size of the development, but since permitting is a local issue there may be some resistance from local governments.

We also recommend that a Hazard Mitigation Investment Bank be created. This would be an entity that would be supported by those private sector industries that benefit most from the investment in reducing disaster impacts (i.e., mortgage bankers, building and construction industries, architects, utility companies). In exchange for investment, clients would get some level of tax incentive or other incentive. It could be designed so they could get some level of return on client's investment.

STATE GOVERNMENT SUPPORT FOR COMMUNITY-BASED HAZARD RISK MITIGATION

Climate change has been acknowledged and recognized as having a significant impact on the economies and resiliency of state operations. We have seen states and governors take very proactive positions and provide leadership to address the impacts of climate change. Actions they have advanced have addressed the larger issues of climate change. Our proposals to address the immediate impacts of the increased severity and frequency of disasters as a result of climate change seek to address a more immediate problem faced by those states across the nation.

- Incorporate hazard mitigation elements in all state-supported building and infrastructure construction and require a percentage (somewhere between 10%–20%) of this construction to be green construction). This action can result in safer buildings and reduced state disaster costs in the aftermath of a hurricane. This is an important consideration since many states are self-insured and a large disaster can represent significant budget problems for a state even if eligible for federal assistance. Simple measures such as incorporating protective film on windows in schools, courthouses, administrative buildings, and so on dramatically reduces damages and has the collateral benefit of improving energy efficiency. Other ideas include consideration of risk in siting of facilities and design of facilities to deal with hurricane wind levels. The costs of incorporating hazard reduction into new construction are minimal and are offset by potential losses.

- Develop and support state university teams to provide technical assistance to communities in reducing the impact of climate change through hazard mitigation. Public/community service has become a very positive requirement in many university curricula. Within schools of engineering, architecture, planning, public administration, and other related disciplines, part of the public/community service opportunity could be in assisting communities to design and implement hazard mitigation and climate change adaptation strategies for the community, for particular elements of the community, or even for individual buildings and structures. Universities could support problem-focused research to identify new and improved hazard mitigation techniques to reduce disaster risks and impacts.
- Provide state tax incentives to the private sector to incorporate hazard mitigation in their facilities and capital improvement programs. Revise economic incentives for business development to provide a premium for incorporation of hazard mitigation and climate change adaptation in siting and construction of new facilities or retrofit of existing facilities. Two of the most effective tools that states have at their disposal are their ability to create tax/economic incentives for supporting/promoting improved construction and development practices. Introducing the concepts of hazard mitigation and climate change adaptation as cost-effective and energy and environmentally friendly at the state level makes sense and complements local processes where the incentive process is widely accepted and understood. Green spaces and public areas have long been part of the incentive package and hazard mitigation alternatives that complement these approaches are just logical. States can easily support the need for these incentives with public safety and cost-savings arguments, as there is data to support offering these types of incentives. Statistics indicate between 40 percent and 60 percent of small businesses never recover from a disaster. State can support business adoption of hazard mitigation incentives by providing technical assistance and education to businesses on the postdisaster economic benefits of reducing risks to cut losses.
- Revise statewide building codes to require cost-effective efforts to reduce disaster impacts in all new construction and in any level or remodeling or reconstruction that impacts 25 percent or more of the building. In the United States, most building codes are established at the state level with many states allowing for adoption of local codes. Injecting disaster impact reduction into new construction is the most cost-effective way to implement disaster resilience actions and in many cases can be used as an effective marketing tool. The states with support from the federal government should identify a menu of the most effective and cost-effective hurricane hazard mitigation measures for a variety of construction types. These need to be incorporated into state building codes. Second, most states require upgrades to current code based on a 50 percent or greater impact on the building. Consequently, many buildings come in assessed as only 49 percent impacted. If this threshold was lowered to 25 percent, almost all major renovations would be impacted and a much higher level of disaster resilience would be achieved.
- Review jobs programs administered at the state level with support from the Department of Commerce (DOC) and the Economic Development Administration (EDA) that focus on ensuring disaster-resistant jobs so people don't lose their jobs in the event of disaster.

BUSINESS COMMUNITY SUPPORT FOR COMMUNITY-BASED HAZARD RISK MITIGATION

Actions that the business community could conduct in support of community hazard risk mitigation efforts include

- Provide leadership at the national, state, and local levels for hazard mitigation and climate change adaptation efforts. Community champions are a critical element in the success of any community-based effort and often the most effective champions come from the business community. National organizations such as the US Chamber of Commerce, the Business Roundtable, the National Association of Home Builders, the Realtors, the International Economic Development Council, and others should work to promote these community hazard mitigation and climate change adaptation efforts and to encourage their members to take a leadership role at the community level.

- Become full members of the community partnership established in each community. Hundreds of corporations became Project Impact national partners. In Tulsa alone 345 local businesses joined Tulsa's Project Impact program.
- Provide financial, material (i.e., products and services) and technical support to community efforts. Across the country in the late 1990s, major corporations and local businesses provided funding and materials to support community hazard mitigation efforts.
- Allow skilled employees to take paid leave to assist community partnerships and to help implement hazard mitigation actions. Major employers in a community could allow their risk managers and business continuity planners to help small businesses identify low-cost hazard mitigation actions they can take to protect their business, local businesses could provide computer specialists to help with community mapping projects, and local businesses could contribute the services of their construction supervisors and employees to retrofit homes, child care centers, and small businesses.
- Take steps to reduce the impacts of climate change on their operations, facilities, and employees. Complete business impact analysis for all facilities and operations and assist vendors and suppliers in doing the same. Provide grants and low-interest loans to employees to finance low-cost hazard mitigation actions to protect their homes.
- Include hazard mitigation, climate change adaptation, energy conservation, and environmentally friendly techniques in siting, designing, and constructing future facilities and retrofitting existing facilities. Make business facilities and operations the model for all community members to follow in ensuring that the community's economy is well protected from future hurricanes.

VOLUNTARY AND NONPROFIT AGENCIES SUPPORT FOR COMMUNITY-BASED HAZARD MITIGATION AND CLIMATE CHANGE ADAPTATION

The voluntary agencies active in disasters (VOADs) have long played a major role in responding to hurricanes. The groups provide immediate food, shelter, and clothing to individuals and families impacted by hurricanes. They also help communities to rebuild after these events.

In recent years, VOADs have been joined by an increasing number of nongovernmental organizations (NGOs) that provide funding, staff support, and partnerships to help individuals and communities to recover. Corporate, family, and community foundations have also become more involved in response and recovery efforts to major disasters.

It is time that voluntary agencies support hazard risk mitigation and climate change adaptation efforts in communities across the country. Actions they might take include

- Provide trained and experienced staff to help organize community partnerships, to design community hazard mitigation and climate change adaptation plans, and to implement these plans. The nonprofit sector has volunteers and paid staff located in communities across the country who could become involved in community hazard mitigation and climate change adaptation efforts by adding this function to existing community activities and/or creating new programming to support hazard mitigation and climate change adaptation efforts. The relationships already established by these groups in communities could be leveraged very effectively to support a new community disaster-resistant initiative.
- Provide financial and technical support for the development of training and mentoring materials to provide guidance to new communities. Foundations could provide the funding

and community-based organizations and programs could provide the expertise needed to develop training and mentoring programs and to make these programs available to community leaders around the country.

- Provide financial support for staffing for the community partnerships. The Council on Foundations has published a report entitled *Reducing the Impacts of Disasters on Children: Opportunities for Foundations* that provides a list of 35 activities that a foundation could invest in that would help reduce the impacts of future disasters on children. The Council should produce a similar report for community-based hazard mitigation and climate change adaptation efforts and distribute and promote it to its members. Such a report could serve as a guide for how individual foundations could craft new funding programs to support community hazard mitigation and climate change adaptation efforts.
- Provide financial and technical support for the establishment of a monitoring and evaluation (M&E) program; nonprofit groups have extensive experience in designing and implementing M&E programs. This expertise should be used to design and implement the types of M&E programs needed to measure progress in community hazard mitigation and climate change adaptation efforts, measuring economic benefits realized through risk and impact reduction and to determine the savings realized by reduced losses from future disasters.
- Take steps to reduce the impacts of climate change on their programs, operations, and facilities. Just as the business community should take steps to reduce future impacts, so should the nonprofit sector. As this sector becomes more and more involved in disaster response and becomes more critical to the successful recovery of communities from disasters, it is important that their operations, services, and facilities remain functional after a disaster, and conducting an audit of these operations and facilities and taking action to reduce risk will ensure that they are better able to serve when needed.
- Incorporate hazard mitigation and climate change adaptation planning and actions into their existing community development programming. The nonprofit sector supports a myriad of development programs in low income and disadvantaged neighborhoods and communities. These are the areas and populations that are often the hardest hit by disasters. It is critical that the nonprofit sector understand the risks from disasters that their development programs face and take action to reduce these risks so that damage to these programs is limited when the next hurricane strikes.
- Ensure that all functional needs populations are represented in the community partnership and the needs of these populations considered in all planning and design functions. The nonprofit sector has a long history of working with special needs populations and should take an active role in ensuring that their needs are recognized and considered in all hazard mitigation efforts. The nonprofit sector must leverage its ongoing work with special needs populations to ensure their involvement in the community hazard mitigation efforts.

CONCLUSION

There are solutions to reducing the impacts of the changing climate; many are based on proven, effective, and cost-efficient activities that communities across the country have already taken to reduce the impacts of future disasters. These lessons and these processes must be applied now and urgently given what we know.

Index

G

H

I

N

O

P

T

U

V

W

Y

For Product Safety Concerns and Information please contact our EU
representative GPSR@taylorandfrancis.com
Taylor & Francis Verlag GmbH, Kaufingerstraße 24, 80331 München, Germany

www.ingramcontent.com/pod-product-compliance
Lightning Source LLC
LaVergne TN
LVHW080645011026
849160LV00014B/1323
* 9 7 8 1 4 9 8 7 2 5 3 6 1 *